The Maintenance
of Brick and Stone
Masonry Structures

I found that the Arch thereof looked shaky and insecure; moreover, that a Great and Irregular-shaped Cleft or Crack ran, after the fashion of a Lightning-flash in a Painted Seascape, athwart the structure thereof from Keystone to Coping. As I was regarding this unpleasing Portent, the Genius told me that this Bridge was at first of sound and scientific construction, but that the flight of Years, Wear and Tear, vehement Molecular Vibration, and, above all, Negligent Supervision, had resulted in its present Ruinous Condition.

From *On the Bridge,* a modernized version
of Addison's allegory *The Vision of Mirzah.*
Published in the issue of *Punch* for August 1, 1891

The Maintenance of Brick and Stone Masonry Structures

Edited by

A.M. SOWDEN

E. & F.N. SPON

An imprint of Chapman and Hall

LONDON · NEW YORK · TOKYO · MELBOURNE · MADRAS

UK   Chapman and Hall, 11 New Fetter Lane, London EC4P 4EE
USA   Van Nostrand Reinhold, 115 5th Avenue, New York
                NY10003
JAPAN   Chapman and Hall Japan, Thomson Publishing Japan,
                Hirakawacho Nemoto Building, 7F, 1-7-11 Hirakawa-cho,
                Chiyoda-ku, Tokyo 102
AUSTRALIA Chapman and Hall Australia, Thomas Nelson Australia, 480
                La Trobe Street, PO Box 4725, Melbourne 3000
INDIA   Chapman and Hall India, R. Seshadri, 32 Second Main
                Road, CIT East, Madras 600 035

First edition 1990

© 1990 E. & F.N. Spon

Typeset in 10½/12pt Sabon by
Witwell Ltd, Southport
Printed in Great Britain at the
University Printing House, Cambridge

ISBN 0 419 14930 9 (HB)
     0 442 31166 4 (USA)

British Library Cataloguing in Publication Data

The maintenance of brick and stone masonry structures.
  1. Buildings. Stone masonry. Maintenance & repair
  2. Buildings. Brickwork. Maintenance & repair
  I. Snowden, A. M. (A Maurice)
  692.1

  ISBN 0-419-14930-9

Library of Congress Cataloguing-in-Publication Data

The Maintenance of brick and stone masonry structures/edited by A. M. Sowden. — 1st ed.
    p.   cm.
    Includes bibliographical references and index.
    ISBN 0-422-31166-4
    1. Masonry – Maintenance and repair.   I. Sowden, A. M. (A. Maurice), 1922–
    TH5311.M33   1990
    693'.1—dc20

Contents

Contributors

Douglas J. Ayres
F.G.S., F.S.E.

Geotechnical Consultant; lately Soil Mechanics Engineer, British Rail Headquarters; former member of International Union of Railways' Tunnel Waterproofing Subcommittee; responsible for the development of the mechanical pointing technique and the B.R. gun.

A. Douglas M. Bellis
B.Sc., C.Eng., F.I.C.E., F.C.I.T.

Consulting Engineer; former Port and Estuary Engineer, Hull, for Associated British Ports; over 30 years' experience in port engineering on the Tyne, Mersey and Humber.

Michael A. Crisfield
B.Sc., Ph.D., C.Eng., M.I.Struct.E.

F.E.A. Professor of Computational Mechanics Department of Aeronautics, Imperial College, London; lately on the staff of the Bridges Division and Structural Analysis Unit of the Transport and Road Research Laboratory, undertaking research primarily involving the non-linear analysis of steel, concrete and masonry bridges. He has published over 50 technical papers and one textbook.

Geoffrey J. Edgell
B.Sc., Ph.D., C.Eng., M.I.C.E.

Head of Building Materials Division of British Ceramic Research Association; member of Council of British Masonry Society; member of Institute of Ceramics and various national and international committees; author/presenter of some 40 papers and lectures on masonry.

Ivor W. Ellis
C.Eng., F.I.C.E.

Consultant to, and former Managing Director of, Fondedile Foundations Ltd; past Chairman of the Federation of Piling Specialists; member of CIRIA Steering Committee on the State of the Art of Ground Anchors.

Richard O. Hall
B.Sc., M.Sc., C.Eng., M.R.P.R.I.

Deputy Head of Coatings, Polymers and Corrosion Unit, British Rail Scientific Services, Derby.

Colin J.F.P. Jones
B.Sc., M.Sc., Ph.D., F.I.C.E.

Professor of Geotechnical Engineering, University of Newcastle upon Tyne; former Assistant Director, West Yorkshire Metropolitan County Council, responsible for design, construction and maintenance of bridges, civil engineering structures and buildings, and Technical Director, West Yorkshire Civil Engineering and Technology Ltd; member of British and International Geotechnical Societies; member of various national bridge and technical committees and author of 3 textbooks and over 50 publications.

W. Brian Long
B.Eng., C.Eng., M.I.C.E.

Consultant to Balvac Whitley Moran Ltd and former Managing Director of Whitley Moran & Co. Ltd; founder member and first Chairman of the Sprayed Concrete Association; lately Chairman of the Federation of Resin Formulators and Applicators.

Ralph L. Mills
M.Sc., C.Eng., F.I.Struct.E.

Consulting Engineer, specializing in Historic Buildings and Ancient Monuments; formerly Chief Structural/ Civil Engineer with English Heritage for 20 years; previously with the Building Research Station and the National Coal Board.

John Morton
B.Sc., Ph.D., C.Eng., M.I.C.E.,
M.I.Ceram., M.Inst.M.

Senior Engineer with the Brick Development Association, responsible for the Engineering Function; member of the Technical Drafting Committee for the Structural Masonry Code and responsible for input to the Institution of Structural Engineers Continuing Professional Development courses on masonry; author of numerous papers and articles on engineering masonry, on which subject he also lectures widely.

John Page
B.Sc.

With the Road Research Laboratory (now Transport and Road Research Laboratory) since 1962; in the Bridges Division since 1970 he has studied aspects of loads and load effects on bridges. From 1984 he has had responsibility for the programme of tests designed to lead to a revised assessment technique for masonry arches.

John Powell
C.Eng., M.I.Struct.E., M.I.C.E.

Principal Bridge Engineer, British Waterways Board, having responsibility for all aspects of the maintenance and reconstruction of the Board's bridge stock throughout England, Scotland and Wales; formerly engaged on the design and maintenance of highway bridge structures in the bridge offices of the West Riding of Yorkshire and South Yorkshire County Councils; member of British Standards Technical Committee CSB/59, responsible for BS 5400.

Geoffrey F. Read
M.Sc., C.Eng., F.I.C.E., F.I.Struct.E., F.I.H.T., F.I.W.E.M., M.I.L.E., F.B.I.M.

Consulting Engineer; Research Fellow and Visiting Lecturer, Department of Civil and Structural Engineering, University of Manchester Institute of Science and Technology; former City Engineer and Surveyor, Manchester, and Joint Engineer to Manchester Airport Authority. As such, he became known worldwide as one of the leading campaigners for the renewal of sewerage infrastructure. He has written many technical papers on the subject as well as directing, among other things, a large programme of sewerage rehabilitation in Manchester; previously held appointments with six other local authorities and was an Advisor on Highways and Sewerage to the Association of Metropolitan Authorities; past Chairman of the City Engineers' Group and Founder President of the Association of Metropolitan District Engineers; author of numerous technical papers dealing with highways, sewerage, civil engineering at airports, multistorey car parks etc.

Barry A. Richardson
B.Sc., F.I.W.Sc., A.C.I.Arb.

Consulting Scientist, Expert Witness and Arbitrator; former Chairman of the Council of the Association of Consulting Scientists; currently Chairman of the Government Liaison Committee and representative of the Association on the Parliamentary and Scientific Committee at Westminster; former Chairman of the Construction Materials Group of the International Bio-deterioration Research Group; Fellow of the Ancient Monuments Society; author of books and many papers on the deterioration, preservation and decoration of structural materials.

A. Maurice Sowden
M.A., B.Sc., C.Eng., F.I.C.E., F.P.W.I.

Consulting Engineer; retired in 1984 as Assistant Director of Civil Engineering, British Rail Headquarters; 40 years' experience in the design, construction and maintenance of railway structures.

Contributors

V. Keith Sunley
C.Eng., M.I.C.E., M.I.Mech.E.

Head of Track Research, British Rail Research; former Head of Civil Engineering Structural Research, Research of Development Division, British Rail.

Robert C. de Vekey
Ph.D., C.Chem., M.R.S.C., D.I.C.

Head of Masonry Construction Section at the Building Research Establishment, concerned with the structural performance and safety of brick and stone masonry buildings; founder member of the British Masonry Society; associate member of the Ceramic Society/ Institute of Ceramics; over 30 years' experience of research into building materials technology, including work on the manufacture of bricks, blocks, ties, mortars and rendering and on the development of BS Codes of Practice in this field; author of some 70 papers, articles, guidance notes, etc. on masonry.

Preface

This book is intended as a practical guide for all who may be required to undertake or manage the maintenance of masonry structures; it is aimed particularly at those whose training and experience have been directed more to design and construction than to maintenance and more to steel and concrete than to masonry. It deals with all the tasks which may fall to the Maintenance Engineer, from the initial identification of defects and their diagnosis to their treatment and the monitoring of its cost effectiveness. Accordingly it seeks to bring together as much information as possible concerning proven techniques and to illustrate these with actual case histories. It also details some of the latest relevant developments. Most chapters conclude with a bibliography, to facilitate more detailed study of individual topics.

It relates to civil engineering structures – bridges, viaducts, tunnels, culverts and sewers, retaining walls, wharves and quays, towers and shafts – but not to buildings. The major part is written in the context of bridges and their associated retaining walls, which numerically predominate among masonry structures in the United Kingdom; separate chapters have therefore been included in respect of other types of structure, such as tunnels, sewers and maritime works, where special considerations apply and distinctive techniques of maintenance are often needed.

Each chapter is written by a specialist in the particular subject. Such plural authorship inevitably results in occasional duplication; little attempt has been made to eliminate this, as to do so would usually detract from the logical development of the arguments of the individual contributions. In most cases the overlapping points are of sufficient importance to warrant the emphasis of repetition.

The statements and opinions of each contributor are given in good faith, on the basis of his own experience and without any claim to being dogmatic or exhaustive. The views expressed are the contributors' own and there is no intention to imply that they are endorsed by present or former employers; neither are they necessarily shared by all contributors, although there is very considerable unanimity among us.

References to materials, products or processes are not exclusive. In all cases, it is implicit that they are used only by suitably experienced operatives and in accordance with current safety regulations and manufacturers' recommendations. Where appropriate, reference has been made to specific Regulations, Standards, Codes of Practice and the like; these should be construed as relating to the latest version or edition of the reference in question.

Although the book is written largely on the basis of British experience, relating courses of remedial action to British Standards and practices, the philosophies and principles described should be of universal application. Engineers in countries where structures are generally younger than those in the United Kingdom may yet have to face the full impact of the problems of deterioration and so may be able to take advantage of the details given as to successful countermeasures.

As Editor, I must record my appreciation of the cooperation I have received from all who have contributed chapters, illustrations or information and express my gratitude to innumerable British Rail colleagues (past and present) who have shared with me the fruits of so much experience.

Maurice Sowden

Glossary

Terms relating to Maintenance Management

Except where noted, the following definitions of keywords are taken from British Standard BS 3811:1984, *Glossary of Maintenance Management Terms in Terotechnology*. Those marked with an asterisk accord with the recommendations of the European Organization for Quality Control.

Maintenance	The combination of all technical and associated administrative action intended to retain an item in, or restore it to, a state in which it can perform its required function*.
Maintenance management	The organization of maintenance within an agreed policy.
Planned maintenance	The maintenance organized and carried out with forethought, control and the use of records to a predetermined plan based on the results of previous condition surveys (ex BS 8210: 1986).
Preventive maintenance	The maintenance carried out at predetermined intervals or corresponding to prescribed criteria and intended to reduce the probability of failure or the performance degradation of an item*.
Condition-based maintenance	The preventive maintenance initiated as a result of knowledge of the condition of an item from routine or continuous monitoring.
Scheduled maintenance	The preventive maintenance carried out at a predetermined interval of time, number of operations, mileage etc.
Corrective maintenance	The maintenance carried out after a failure has occurred and intended to restore an item to a state in which it can perform its required function*.
Emergency maintenance	The maintenance which it is necessary to put in hand immediately to avoid serious consequences.

Repair	To restore an item to an acceptable condition by the renewal, replacement or mending of worn, damaged or decayed parts.
Overhaul	A comprehensive examination and restoration of an item, or a major part thereof, to an acceptable condition.
Restoration	Maintenance actions intended to bring back an item to its original appearance or state.
Rehabilitation	Extensive work intended to bring an item up to current acceptable functional condition, often involving improvements.
Remedial work	Redesign and work necessary to restore the integrity of a construction to a standard that will allow the construction to perform its original function (ex BS 8210:1986).
Feedback	A written or oral report of the success or failure of an action to achieve its desired result, which can be used to influence design, performance and costs.
Diagnosis	The art or act of deciding from symptoms the nature of a fault.
Cost control	The regulation by executive action of the costs of operating an undertaking, particularly where such action is guided by cost accounting.
Budgetary control	The establishment of budgets relating the responsibilities of executives to the requirements of a policy, and the continuous comparison of actual with budgeted results, either to secure by individual action the objective of that policy or to provide a basis for its revision.
Life cycle costs	The total cost of ownership of an item, taking into account all the costs of acquisition, personnel training, operation, maintenance, modification and disposal, for the purpose of making decisions on new or changed requirements and as a control mechanism in service, for existing and future items.
Durability	The quality of maintaining a satisfactory appearance and satisfactory performance of required functions (ex BS Code of Practice CP 3).
Competent person	Person having sufficient professional or technical training, knowledge and experience to enable him or her: 1. to carry out allotted duties at the level of responsibility laid down; 2. to understand fully any potential hazards when carrying out those duties (ex BS 8210:1986).

Terms relating to masonry and its maintenance

The definitions marked with an asterisk are taken from British Standard BS 6100, *Glossary of building and civil engineering terms*, Sections 5.1 (1985), 5.2 and 5.3 (1984).

Masonry	Construction of stones, bricks or blocks*
Masonry unit	Component of masonry*
Abutment	End support of a bridge (Fig. 1)
Aqueduct	A water-carrying bridge or, more generally, an artificial channel for carrying water over long distances
Archivolt	A projecting moulding following the curve of the extrados on the end face of an arch

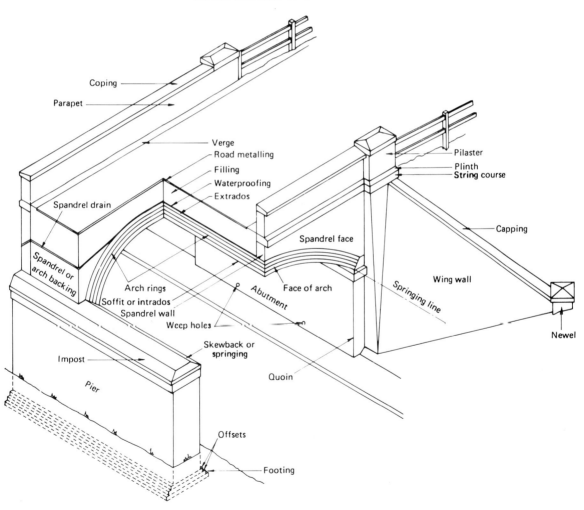

FIGURE 1 The component parts of an arch bridge.

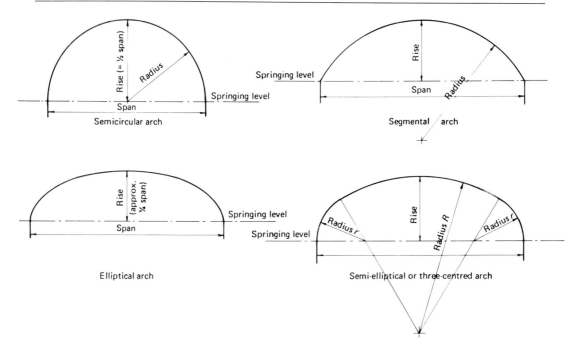

FIGURE 2 Arch profiles.

Arch profiles	Fig. 2
Arch rings	The curved courses which make up an arch (Fig. 1)
Ashlar	Squared stone blocks, finely dressed and jointed, laid regularly in parallel courses
Backing	Material (usually of lower quality) used to fill in or give support behind a structure
Balustrade	A series of short decorative columns, surmounted by a coping, to form a parapet
Band course	Plain contrasting course or courses, flush, projecting or recessed, carried horizontally along the face of a structure*
Bat	A portion of a brick, cut across its width
Batter	The backward slope of the face of a wall (specified as a one unit horizontally to *n* vertically) (Fig. 3)
Bedding plane	The plane of stratification of a stone
Bed joint	Horizontal joint in masonry
Bedstone	Large flat stone upon which a structural member is mounted or bedded

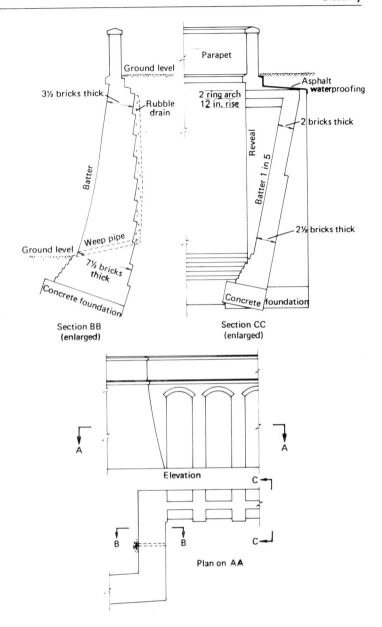

FIGURE 3 Typical brick retaining walls.

Block bonding
Joining one part of a wall to another (especially a facing wythe to its backing) by arranging for a group of masonry units in the former to extend as a block into, and be built into, a corresponding pocket in the latter

Bond	Arrangement of masonry units so that the vertical joints of one course do not coincide with those of courses immediately above and below*
Buttress	A projecting structure built against the exposed face of a wall to give it greater strength and stability
Byatt	A horizontal member to support centring or a working platform
Camp sheeting/shedding	A continuous line of sheet piles (usually of timber) to protect a bank or structural foundations against scour by water
Capping	The uppermost (finishing) course of a wall, which protects it but does not shed water clear of the surfaces of the wall below it (cf. Coping below and see Fig. 1); may be brick-on-edge, brick-on-end or purpose-made units
Cast stone/ Reconstructed stone	Precast concrete made to resemble natural stone
Centre/Centring	Temporary structure to support an arch during its construction or repair
Closer	A portion of a brick used to complete the bond pattern at the end of a wall
Cofferdam	Temporary dam forming an enclosure from which water can be removed to allow dry access to parts normally submerged
Collar joint	Curved bed joint between courses of masonry units in an arch*
Coping	The uppermost course of a wall, which protects it and sheds water clear of the surfaces of the wall below it (cf. Capping above and see Fig. 1)
Corbel/Corbelling	A series of courses each projecting beyond that below it, to form a bracket-like support or ledge
Core	The internal filling of a wall OR a sample of masonry or rock cut out by a diamond-tipped cylindrical drill
Cornice	Prominent horizontal architectural feature that projects from a wall*
Counterfort	Strengthening fin to a wall, built on the side of the retained material
Course	Single layer of masonry units of uniform height, including the bed joint*
Cramp	A shaped metal or slate tie placed across a joint between two masonry units (usually set into small recesses cut into

	adjoining stones), to hold them in their correct relative positions
Cross joint	Joint, other than a bed joint, usually at right angles to the face of a wall*
Crown	The highest part of an arch (Fig. 1)
Culvert	A covered channel or large pipe for carrying a watercourse below ground level, usually under a road or railway (NB There is a lack of unanimity as to the dividing lines between bridge/culvert/drain, but commonly a culvert is regarded as being of up to about 2 m clear span)
Cutwater	The end of a pier in a watercourse, shaped to divide the flow and deflect floating objects
Dentils	Individual headers projecting from the face of a wall to give a toothed effect, often under a cornice
Docking	Dipping bricks into water shortly before laying, to reduce their suction rate*
Dog toothing	Bricks laid with their corners projecting from the face of a wall*
Dowel	A short bar used to link adjacent units by being inserted into matching holes in them
Draw-down	Lowering of water level which, if it occurs rapidly (as on a falling tide), may adversely affect the stability of a quay wall by taking away the support afforded by the water at a time when the retained soil is saturated
Dressing	Surface finish to a stone produced by working rather than naturally
Drip moulding	A moulding on the underside of a (projecting) unit, designed to prevent water flowing back across it; a downward projection is known as a drip, a longitudinal groove as a throat
Dry stone wall	Stone wall constructed without mortar*
Efflorescence	Crystalline deposit on the surface of masonry after evaporation of water that has carried soluble salts from within*
Extrados	The outer (convex) curve of an arch (Fig. 1)
Face	Exposed surface of a structure or of a masonry unit as laid*
Fair faced	Work built with particular care, both to line and with even joints, where the finished work is to be visible (ex BS 5628)

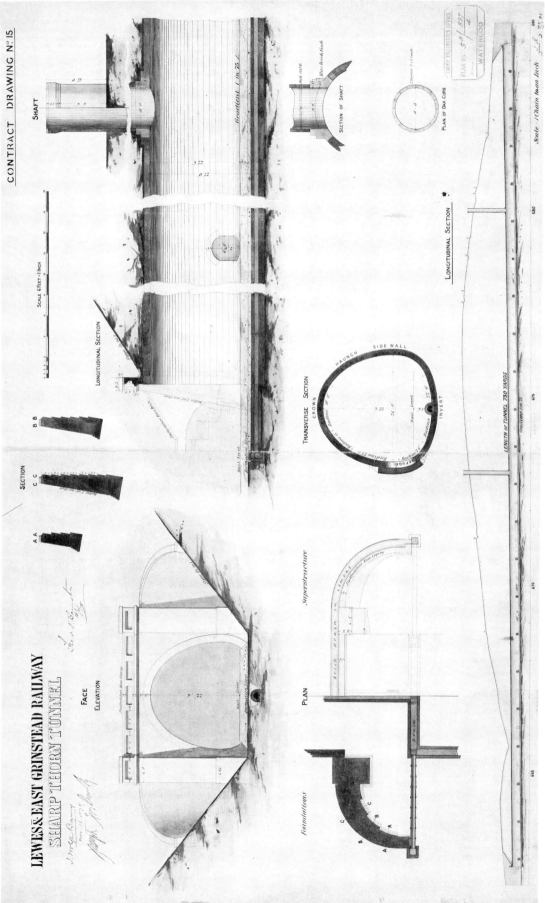

FIGURE 4 A typical railway tunnel (including air shaft). (*Courtesy of British Rail.*)

Footing	An enlargement of the bottom of a structure, in masonry or concrete, to increase its bearing area on the ground
Gabion	Unit formed by filling a wire basket with hand-packed stones; usually used to form a retaining or protective wall
Garland or Launder	A ring drain around the inner periphery of a shaft
Gauged brickwork	Brickwork built to fine tolerances*
Grout	A cementitious slurry or liquid chemical formulation injected to fill joints and cavities
Haunch	The part of an arch between springing and crown
Header	A masonry unit laid with its longer dimension normal to the face of the wall
Headwall	Retaining wall at the end of a culvert
Hearting	Infilling of broken stone*
Hydraulic (lime) mortar	Mortar containing which constituents enable it to set and harden under water
Impost	The upper element of an abutment or pier which supports an arch or other superstructure
Intrados	The inner (concave) curve of an arch (Fig. 1)
Invert	An inverted arch (i.e. with its concave face uppermost), commonly constructed between the toes of abutments for structural reasons or to prevent erosion of the ground (Fig. 4)
Jack arch	A short-span arch between I beams, springing from their bottom flanges
Joggled arch	Arch in which adjacent voussoirs are interlocked by means of visible rebates or steps*
Jointing	Finishing mortar joints as work proceeds, without pointing*
Jumper	A chisel-pointed steel bar which is used to form a hole in masonry, being driven by repeated blows from a heavy hammer
Kerf	Groove made by a saw*
Keystone	Central voussoir at the crown of an arch
Label	Drip moulding over an aperture
Lacing course	A course bridging a wall joint or collar joint to improve bond and integrity
Lagging	Timber battens laid on top of centres to support masonry units during construction or repairs

Launder	See Garland
Needle	A beam inserted through a wall to give it temporary support OR a support built into a wall and cantilevered from it to carry falsework etc.
Newel	A pillar forming the termination of a wing wall or stairway balustrade (Fig. 1)
Offset	A continuous recessed course that sets back the upper face of a wall from that beneath*
Overbreak	Rock excavated (usually unintentionally) in excess of that needed to accommodate construction to the stipulated dimensions
Overbridge	A bridge *over* the facility in question, e.g. a canal overbridge is one over a canal, irrespective of ownership
Overhand work	Work in masonry where the fair-faced side is finished by reaching over completed work*
Oversailing course	Continuous projecting course that carries forward the upper face of a wall from that beneath* (cf. Offset above)
Padstone	Masonry unit incorporated in a structure that distributes concentrated load*
Parapet	Upward extension of a wall, intended to prevent falls over a sudden change in level (Fig. 1)
Parging/Pargetting	A thin coat of mortar used to smooth over rough faces of masonry, often in preparation for waterproofing
Perpend	Vertical cross joint, appearing in the face of a wall between courses*
Pier	An intermediate support between adjoining bridge spans OR a thickened section located at intervals along a wall to strengthen it
Pilaster	A rectangular pillar formed by projecting masonry from one or both faces of a wall, as a termination or feature (Fig. 1)
Plinth	Projecting masonry at the base of a wall, to thicken it and give an appearance of additional strength
Pointing	Filling a partly raked back mortar joint to provide a finish*
Pozzolana	Volcanic dust (natural or artificially produced) which can be mixed with lime to impart hydraulic properties
Puddled clay	Clay mixed with water (and sometimes sand) until it is plastic and impermeable

Quoin	An external corner (Fig. 1) OR one of the stones forming it
Recess/Refuge	A recess in the face of an abutment, wall or parapet to allow pedestrians to stand clear of moving traffic (Fig. 4)
Reconstructed stone	See Cast stone
Relieving arch	Arch built into a wall to relieve that part of the wall below the arch from loads above*
Rendering	Mortar or concrete surface finish applied to a wall
Return	Visible surface at a change of direction in a wall, member or moulding*
Reveal	Return at each side of a recess or opening in a wall* (Fig. 3)
Revetment	A protective covering to a surface to prevent scour by water or weather
Rip-rap	Relatively heavy stones used to provide a revetment
Rise (of an arch)	Vertical height from springing level to the crown of the intrados (Fig. 2)
Rock-faced stone	Stone having a face produced naturally*
Rubber	A soft brick, especially made to be easily sawn, ground or rubbed with an abrasive, to accurate size or shape
Rubble	Stone of irregular shape and size*
Rustic brick	Brick with a rough textured face
Rusticated stone	Stone with its edges deeply dressed (bevelled or sunk) to make joints conspicuous
Saddle	A concrete slab cast over an arch to strengthen it or distribute the loads upon it
Saddleback coping	Coping weathered both ways from the centre of the section*
Scarcement	(The amount of) the setting back of a face of a wall from that below it, especially at the level of the top of the foundation
Skew arch	Arch in which the two horizontal axes are not at right angles*
Skewback	Surface of an inclined springing. . . .* (Fig.1)
Sleeper wall	Low load-bearing wall to provide intermediate support . . .*
Snap header	Half brick, appearing as a header on the face of a wall
Soffit	The underside of a member

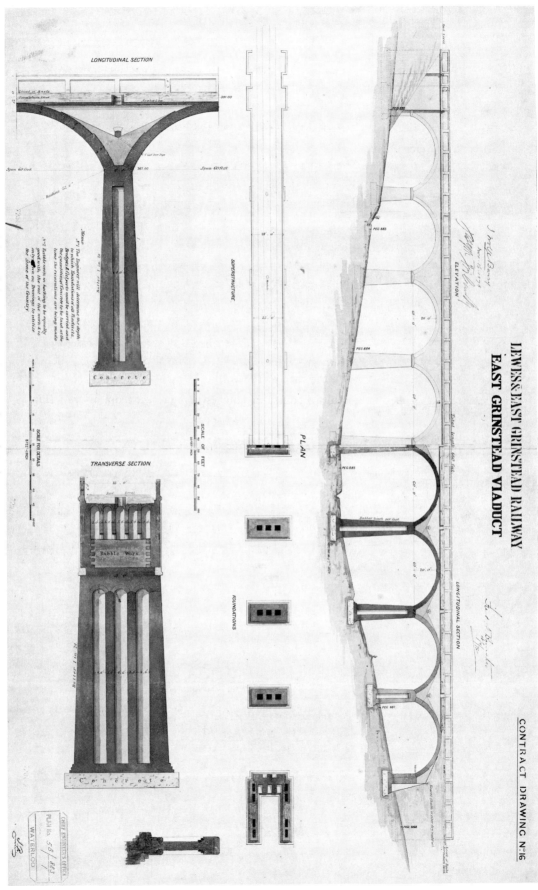

FIGURE 5 A typical viaduct (note the voided spandrels and piers). (*Courtesy of British Rail.*)

Soldier	Masonry unit laid with its longer dimension upright and parallel with the face of the wall, i.e. bedded on its smaller face
Spalling	Flaking from the face of a masonry unit caused by frost, crystallization of salts or mechanical action*
Spandrel	The part of an arched structure behind the extrados or between adjacent arches
Spandrel face	The exposed part of the spandrel, visible on the elevation of the structure (Fig. 1)
Springing	Plane from which an arch springs* (Fig. 1)
Squint	Special masonry unit used at an oblique quoin*
Starling	Protection to a pier in water made by enclosing its base with piling
Steeple coping	Saddleback coping whose upper surfaces are steeply inclined and meet to form an acute-angled ridge
Stretcher	A masonry unit laid with its longer dimension parallel with the face of the wall
String course	Moulded course or courses that project from a wall* (Fig. 1)
Underbridge	A bridge *under* the facility in question, e.g. a road underbridge is one under a road and a railway underbridge is one carrying a railway, irrespective of ownership
Underpinning	Strengthening the foundations of an existing structure by some form of piling or by the provision of additional supporting material to increase the bearing capacity
Vault	Arched ceiling over a void OR any space covered by arches
Viaduct	A bridge comprising a number of spans (usually four or more) (Fig. 5)
Voussoir	Wedge-shaped masonry unit in an arch*
Wall joint	Vertical joint within the thickness of a wall, parallel to its face*
Weathered	Sloped to throw off water*
Weep hole	A drainage hole through a wall (Fig. 1)
Window	A hole broken through a wall of sufficient size to allow direct visual observation of the condition of the wall and of what is behind it

Wing wall	Side-wall extending from the end of an abutment to retain an embankment or an excavated face (Fig. 1)
Wythe	A continuous vertical section of masonry, one unit in thickness

The above extracts from British Standards are reproduced by permission of the British Standards Institution. Copies of all British Standards mentioned in this book may be obtained from that Institution at Linford Wood, Milton Keynes, MK14 6LE.

PART ONE

Introduction

The approach to maintenance

A.M. Sowden

Inevitably, inexorably, all structures deteriorate towards a state of unserviceability and collapse. Maintenance is the art of controlling the rate at which they do so. Its objective is to ensure that, throughout their required life span, they are preserved in a satisfactorily functional condition with a proper regard for safety and economy. Proficiency in this field relies upon the exercise of skills and judgement derived from experience rather than theory; above all, it calls for an understanding of the long-term behaviour of structures and materials and for ingenuity based on a knowledge of a wide range of remedial techniques.

1.1 Maintenance - a neglected art

Maintenance rarely seems to receive the attention it deserves. Structure owners are preoccupied with the initial capital costs of their assets, usually without sufficient appreciation of the implications of ill-judged economies at this stage. There is a common failure to recognize the extent to which a structure represents an investment

requiring protection throughout its life. The fact that expenditure on maintaining structural fabric can often be deferred without apparent loss or harm leads to a lack of concern for signs of incipient deterioration, with the result that available finance is diverted from preventive maintenance to activities with greater obvious, but only cosmetic, impact. Yet neglect in the early stages is likely to lead to costs which will escalate rapidly the longer action is deferred until a stage is reached where essential remedial works make irresistibly urgent demands on resources; this rarely allows the best or most economic use of those resources.

As a result, the maintenance function has tended to be an under-rated activity – unsung and unappreciated, disregarded until some failure occurs (and then damned as mis-managed!). To engineers, it lacks the attraction and glamour of new construction, which is esteemed as exciting and innovative with visible results; there is little apparent appeal in a task which is seen as a tedious struggle against deterioration and decay, achieving no more than the perpetuation of what already exists. Too often it is taken for granted.

In reality, maintenance involves technical and managerial skills as great and as varied as those required for any other civil engineering activity and can be more challenging, demanding and fulfilling than most. Maintenance Engineers probably need a wider appreciation of the behaviour of a greater range of structures than do many designers; the materials they deal with have less reliable characteristics than their modern equivalents; there is more frequent need for the exercise of engineering judgement in fields where there are no established codes of practice and little authoritative guidance. The successful repair of an operational structure with minimal interference with its normal use may call for resourcefulness and ingenuity unmatched in new construction.

Maintenance practitioners cannot therefore be regarded as second-class engineers, with lower skills, for whom professional qualifications are irrelevant and who can be remunerated accordingly. The expenditure for which they are responsible is massive (even if, in practice, inadequate) and there is vast scope for its inefficient or wasteful application on inappropriate, ineffective or even unnecessary repairs.

There is therefore a critical need for a better appreciation of the importance of maintenance as an essential and integral part of sustaining infrastructure assets and for establishing appropriate policies for the care of those assets. This must involve suitable recognition of the role of its practitioners – professional, technician and artisan – and the provision of adequate resources to enable them to discharge that role properly. There is currently a general perception that levels of financing and resourcing of maintenance activities in the United Kingdom are inadequate, albeit to a degree on which there is little consensus. Whatever may be the extent of such shortfall, it is unquestionable that available resources could be used more effectively and efficiently if maintenance efforts were given more regard and attention and more purposeful direction by more experienced engineers, better provided with appropriate information and support.

In the latter respect, there is a dearth of printed material giving positive maintenance advice and guidance; what exists is generally of a fairly basic and limited nature, scattered over various publications and often hard to find. The published work of researchers on long-term performance is largely restricted to reports on accelerated ageing tests on materials, which may not accurately reflect the real interaction of natural and human agents of deterioration. Few textbooks or codes of practice deal with the interrelationship between cost, reliability and durability under practical conditions of exposure and use. The training given to aspiring engineers at most academic institutions fails to imbue even an awareness of the importance of maintenance or of the principles of exercising control over it; the qualification requirements of the professional institutions are only just beginning to encourage a different approach. Pleas for a more formal recognition of professional maintenance management as an integral part of the province of the civil engineer (e.g. Robertson, 1969) have gone largely unheeded.

In respect of techniques and equipment, maintenance has generally used traditional labour-intensive methods, with little mechanization. Maintainers' requirements differ in many respects from those of their designer colleagues, and their problems have tended to be neglected by development organizations and manufacturers. In many cases it has been left to Maintenance Engineers themselves to devise and develop their own solutions, as occurred, for example, in the technique of the mechanical pointing of masonry. There are still unfulfilled requirements for more effective and more economic methods in most fields of maintenance, and not least in that of brick and stone masonry.

To focus attention and efforts on such problems, Maintenance Engineers must make their needs better known and may well have to work together more closely to demonstrate to potential innovators and suppliers that there is a worthwhile market for cost-effective solutions.

1.2 Maintenance in context

Deterioration of a structure starts as soon as it is built; consideration of its maintenance should be – but seldom is – invoked long before that.

For the **conceptual design** of a new structure,

the engineer needs to establish what is the sponsor's policy (if indeed he has one) with regard to that structure's maintenance. What is its role in securing the objectives of the business? What is its required life expectation? How soon can it become obsolescent and require modification? What standard of maintenance can be regarded as acceptable and economically viable? What are the criteria or limits for the assessment of serviceability? To what extent can the disruption and indirect costs arising from maintenance operations be tolerated? It is probable that such considerations will not have occurred to the sponsor, and so the engineer may well have to convince him of their relevance and make him realize their economic implications. He must give appropriate advice and information. He must then ensure that the decisions he receives exert their proper influence on his basic design proposals. He should now be drawing on his knowledge of the long-term behaviour of different materials and structural forms and making design decisions in the light of the treatment they are likely to receive in the particular service for which they are intended; he must select them and arrange for their application in a manner appropriate to their individual characteristics and to the environment, in its widest sense, in which they will be used. He must be constantly aware that the maintenance performance of a structure, throughout its life, will be a reflection of his original design decisions.

In the **detailed design** stage such considerations continue to apply with equal force. The basic materials may have been settled, but it is still necessary to determine and define the appropriate qualities or grades of each and to ensure that incompatible materials are not placed in juxtaposition. There may be a case for the adoption of a 'mendable' design, in which elements likely to have a relatively short life or to be vulnerable to damage are arranged so as to be readily renewable, possibly in some pre-assembled form; for parts which could be difficult or costly to repair or replace, a degree of overdesign could well be

viable. There must be constant awareness of the risk of introducing features which will adversely affect durability or long-term appearance or which will restrict the required usage. In particular, since moisture, especially if charged with aggressive chemicals, is the greatest single cause of deterioration of most structural materials, there must be effective provision of waterproofing and/or drainage to ensure that water is shed quickly and led away in a controlled manner; ledges and crevices where moisture can lie and dirt accumulate should be ruthlessly designed out. The effects of weathering and surface staining on appearance must be carefully considered, as must the possibility of differential or thermal movement. A great deal of experience is needed to foresee the many factors which may accelerate deterioration. Equally, considerable expertise is required to counter them without increasing costs.

It is just as important, at this stage, to ensure ready access to all parts of the structure, initially to allow it to be properly examined and any defects observed, and subsequently to facilitate remedial operations. Not only does this mean the avoidance of inaccessible gaps between elements and fixtures (such as cladding and advertisement hoardings) which would obscure signs of trouble, but also it means the provision of safe means of reaching all parts of the structure – access platforms, manholes, catwalks and/or permanent fixing points for the attachment of ladders, scaffolds or safety lines. Masonry structures built by the Romans were frequently provided with stones which protruded from the face to accommodate scaffolding; it is curious that, 2000 years later, this lesson seems so often to have been forgotten!

At the **construction stage** the quality of materials and workmanship will have a major effect on subsequent maintenance. It is therefore necessary to define acceptable standards clearly and precisely and to ensure adherence to them by the employment of a reliable workforce, competently monitored and supervised. In weighing up the advantages of different methods of construction, the effect of each upon ageing characteristics ranks high among the relevant factors. Thus, durability might be improved by maximizing the amount of off-site construction, which can be undertaken under the more controlled and protected conditions of a factory rather than on a site exposed to the vagaries of weather; however, offsetting this could be a susceptibility to leakages at site joints. Even the programming of the work can have an effect; it is regrettably common to fail to allow adequate time for waterproofing operations to be undertaken with the care and thoroughness which they need if troublesome and unsightly leaks are to be avoided.

In support of the foregoing arguments, it is relevant that investigations of 510 defective buildings by the Building Research Establishment (1975) revealed that almost 50% of the defects were due to faulty design, 30% to faulty construction and 11% to faulty materials. Although these statistics relate to buildings in the United Kingdom, other sources such as Grimm (1980) suggest that those for masonry structures and for other countries are similar. The target areas for improving durability and reducing maintenance liabilities are obvious.

The inspection of a structure on a regular basis throughout its lifetime plays a key role in ensuring its continued safety and fitness for its purpose. Only through such inspection and detailed recording of the results can the Maintenance Engineer be provided with the critical information needed to enable him to be in control of the situation for which he carries responsibility.

Design, construction, inspection and repair are therefore interrelated activities, which rely for their successful accomplishment upon good communication between each of them. Improved standards of performance depend, above all, upon active and effective 'feedback' at every stage. A good deal of lip-service is paid to the concept of feedback, but in practice it takes place only to a very limited extent. There is a natural reluctance to publicize, or even admit to, shortcomings, with the result that valuable learning opportunities are lost. A few large undertakings are in a position to

compile case studies or manuals of guidance and good practice which draw upon their accumulation, over many years, of a considerable fund of experience, but these are issued only within the organization concerned and are not generally accessible. Little is readily available by way of published material. What there is tends to be greeted with a degree of unconcern and of failure to recognize its relevance to the reader's own activities. Such disinterest allows instructive experience to go to waste.

There is general acceptance of the value of construction experience to the designer, and cross-fertilization in these fields is relatively common. Therefore it is strange that the same argument is seldom extended to maintenance experience and even more rarely put to practical application. There can surely be no more reliable way of ensuring that the lessons of experience are absorbed than by direct personal involvement. There is therefore a strong case for rejecting any functional split between those engaged on design, construction and maintenance, and for the integration, within a single organization, of all aspects of the provision and management of structures. Within such an organization there should be the opportunity for the regular interchange not merely of information but, more importantly, of personnel between the various functions. There could be no better way of training complete craftsmen, whether at professional, technician, supervisory or artisan level. Career plans should allow staff to progress through the whole range of activities involved and no one should be considered for senior positions without such comprehensive experience.

Obviously, this is not possible in an organization which does not undertake the full range of those activities; it is suggested that standards would be raised if organizations which now specialize almost exclusively in new work were given the opportunity to broaden their experience by being retained to take responsibility for maintaining the structures they have designed and so discover for themselves the built-in shortcomings. The adoption of such a system on a much wider basis than currently prevails in the United Kingdom is worthy of serious consideration.

The individual designer whose duties do not give him such opportunities can nevertheless heighten his awareness of what can be done to improve the durability and long-term behaviour of structures. He can begin by recognizing that his own designs may be less than perfect in this respect and by being prepared to submit them to critical examination by experienced maintenance engineers; he should have due regard to constructive criticism from such sources and be willing to amend his designs or details accordingly. He should seek out local knowledge as to the durability and consistency of locally obtained materials. Participation in structural examinations of existing works is invariably instructive. He should certainly make a practice of revisiting 'his' structures some time after their completion to observe how they have performed and how their shortcomings could have been avoided, and he should seek to discuss their behaviour with the engineer responsible for their maintenance. He should develop the habit of looking at all structures with a critically discerning eye, taking note of apparent deficiencies in an attempt to identify their cause and how they could have been overcome. In short, he cannot expect the term 'feedback' to signify that he can rely upon being spoon-fed with the relevant information; he must take the initiative and make some effort on his own behalf.

The most effective acquisition of the required skills comes from personal participation. By working on inspection, assessment and the identification and rectification of defects, engineers come to understand the behaviour in use of different materials and forms of construction; they become familiar with the mode and rate of deterioration of each and learn to recognize those features and details which are prone to cause trouble; they acquire a more reliable measure of the significance of signs of incipient distress and of the probable rate of development of defects;

thus they are better able to appraise the severity of those defects and the urgency for remedial action. When such engineers are engaged on new work, the quality of their design and detailing and their concern for adequate standards of construction should be markedly improved. Similarly, Maintenance Engineers who have an appreciation of design concepts and constraints can more reliably decide on the relative importance of different defects, and the repairs they initiate are more likely to be cost effective.

Ideas inculcated by direct involvement are more assured of being applied in practice – possibly intuitively rather than consciously – than anything tenuously acquired at second-hand. This is not to decry vicarious experience, but bitter is undoubtedly better (even if initially unwelcome!)

1.3 Brick and stone masonry in context

The Maintenance Engineer working with brick and natural stone is relatively fortunate. These are among the oldest construction materials and have clearly demonstrated their superior durability. Properly used, they are capable of providing structures with a very long life, which are tolerant of significant abuse and neglect and which can successfully accommodate substantial deformation; they are available in attractive colours and textures, which can blend harmoniously into most environments, and they age gracefully to become integral, even cherished, elements of the local landscape. Since they are more chemically inert than most other constructional materials, their maintenance is correspondingly less demanding. For a variety of reasons, masonry structures tend to be generously proportioned, with considerable reserves of strength, enabling them to sustain without distress loadings much greater than those for which they were originally intended. As they are basically massive, the proportion of the stresses in them due to dead loading is usually so much greater than that due to live loads that they are less susceptible to variations in the latter.

The amount of brick and stone construction

FIGURE 1.1 A typical example of rough work behind a fair face (in a spandrel wall). (*Courtesy of British Rail.*)

which has been, and still is, undertaken has had the effect of developing and honing skills which incorporate generations of experience; traditions of quality workmanship have been preserved, so enhancing the standards achievable. Even so, the potential for further development may not have been exhausted; the relatively recent advent of the technique of prestressing opens up fresh possibilities of extending the use of the inherent compressive strength of these materials.

Masonry construction is not, however, without its potential deficiencies. Natural stone may vary greatly in its characteristics even when taken from different parts of the same quarry; it may have been badly selected or carelessly used in so far as it may have been incorrectly bedded in relation to its plane of stratification and the direction of the applied loading upon it. Bricks are a man-made product and subject to the vagaries which that implies; in the past they were commonly manufactured on or adjacent to the construction site, using locally available clay of varying suitability; often the degree of firing was insufficient, resulting in bricks which were soft and porous. There was little in the way of quality control, and the bricklayers themselves naturally favoured materials with high water absorption rates as these could be laid more easily and quickly. The mortar is a vital constituent of masonry construction and it too may vary in strength and durability. Old lime mortar, in particular, loses its cementitious nature with time and reverts to a sand fill in the joints; this may be washed out, so destroying the integrity of the structure.

In most older constructions little or no effort was made to prevent the entry of water into the masonry; such waterproofing as was provided is unlikely, after so long, to be still effective. Moisture in brickwork causes it to expand, and so a wet patch in a loaded area will tend to attract load to itself. The common practice of undertaking patch repairs using harder bricks and mortar than the original is another source of uneven stress distribution.

During periods of intense constructional activity, such as the Industrial Revolution and the age of 'railway mania', the amount and widespread distribution of building output were such that it was hard to find sufficient supervisory staff to maintain the degree of surveillance necessary to ensure a consistent quality of work. Masonry is particularly susceptible to poor workmanship concealed behind a fair face. There is plenty of evidence that past generations were as adept at sharp practices – and getting away with them – as any latter-day 'cowboys' (Fig. 1.1; see also Chapter 21, Fig. 21.4 and Section 21.11).

The visible faces of brickwork and stonework should therefore be regarded as unreliable indicators of what lies behind them. The latter could comprise materials of significantly lower quality, roughly laid and poorly mortared in; the thickness of an arch at its portal does not necessarily accord with the thickness elsewhere in its length. Such

original drawings as exist for the majority of older structures which were not of a large or complex nature seem to have been intended only as a rough guide to construction and so they are frequently inaccurate in detail; what is shown on the drawings as solid construction may, in fact, incorporate empty or rubble-filled voids or be thinner, but provided with counterforts. Thus, in the absence of reliable data as to homogeneity, physical properties and geometry, existing masonry structures are obviously not amenable to the precise calculation of factors of safety or to refined considerations of design adequacy.

Despite advances made over the years in the organization and mechanization of the processes leading up to the actual placing of masonry units, their erection is still a labour-intensive operation and hence masonry construction is only occasionally viable for current new work (Harvey, Maxwell and Smith, 1988). It is still extensively used to provide a more aesthetically acceptable face to concrete. Its bulk may increase costs of excavations, foundations and falsework, as well as detracting from the aesthetic appeal of modern construction. However efficient the arch form may be structurally, it is often not economical in providing conveniently usable space below it; it is probable that more arch bridges are renewed because of their failure to provide adequate headroom clearance over their full span than because of their structural inadequacies.

During most of the current century, therefore, construction using brick and stone has been limited, except in buildings. It follows that the engineer responsible for the maintenance of masonry structures will find that he has in his care mainly old assets – probably over 100 years old. They will have demonstrated a fine disregard for the accountant's current concepts of 'book life'. They may well seem indifferent to conventional theories of stability and capable of resisting almost any natural vicissitudes, other than undercutting by floodwater. However, the engineer can make no reliable assumptions about their reserves of strength. He must regard those parts which he can see and measure with a modicum of reserve because they may not be representative of the whole. His best guide to continuing satisfactory performance is likely to lie in the observation and recording, over a period of time, of the behaviour of each structure – of the visible and quantifiable evidence of deterioration, movement, distortion or cracking. Ultimately, he has to rely on his own engineering judgement.

1.4 The management of maintenance

Maintenance management has been authoritatively defined (see the Glossary) as 'the organization of maintenance within an agreed policy'. On this basis, it is evident that the Maintenance Engineer must ensure that such a policy has been defined and must so organize his activities that he can work effectively within its constraints. In particular, he must establish the standards of performance which he is required to achieve. These must at least be compatible with safety and comply with relevant legal and contractual obligations, but additionally should show concern for the operating and commercial strategies of the owner, for cash flow, for reliability and appearance and for the longer-term future of the assets. The main constraints are likely to be those of resources (finance, suitable labour and specialized plant) and opportunity (the extent to which the normal functioning of the asset can be interrupted while work is done on it).

The most effective results will be achieved through **planned maintenance** – work undertaken in accordance with a soundly based system of priorities, each operation properly planned and organized in advance, with the necessary labour, plant and materials assembled ready for use when required. This calls for a systematic and disciplined approach. Properly applied, it leads to a greater margin of safety against failure and a reduced risk of having to resort to **unplanned maintenance** – unforeseen or emergency action to prevent imminent unserviceability or collapse. Such 'breakdown maintenance' is likely to be wasteful of repair capacity, expensively disruptive to the normal utilization of the asset and destructive of staff morale and public confidence. The more that planned maintenance has to be cut back in response to stricter constraints, the greater is the emphasis needed on vigilant and frequent examination to ensure against being taken unawares by a crisis situation.

Planned maintenance can be categorized as either preventive (to forestall significant damage) or corrective (to remedy established defects).

Preventive action comprises the routine removal of features which could initiate trouble and the rectification of incipient defects before they cause real harm. In the case of masonry structures it includes curbing overloading, removing vegetation growth, clearing drainage outlets, taking precautions against scour or mining subsidence, applying preservative or protective treatment and repointing. Theoretically, it should include waterproofing, but in practice that is invariably a major operation which, for that reason, is rarely done as promptly as it should be. Intuitively, all engineers would acknowledge that 'prevention is better than cure', but when money is in short supply the temptation is strong to spend what is available on corrective work with more immediate claim and more obvious beneficial effect; when preventive maintenance would unduly disrupt commercial activities or involve costly temporary works, it may indeed be preferable to defer it.

Corrective action is the work needed to restore the integrity of a damaged or deteriorated structure and includes the repair or replacement of defective elements. Often it cannot be undertaken without first establishing, and dealing with, the cause of the trouble. The virtues of a 'stitch in time' approach should not be allowed to obscure the risk that premature action may have only a cosmetic effect and achieve no more than a temporary palliative unless the basic deficiency has been correctly diagnosed and overcome. Fortunately, most defects in masonry, if due to natural rather than human causes, become critical only slowly; they are evidenced by visible signs whose development can be monitored over a sufficiently long period of time to enable their origin to be correctly identified.

Even when the cause has been established, it is still a matter of judgement as to when to embark upon remedial work. Unless the defects are unacceptable on practical or aesthetic grounds, it may be advisable to do no more than continue to monitor their progression in the hope that they may converge upon a new state of equilibrium and so require only superficial making good or even no action at all.

The Maintenance Engineer must be able to identify, reliably, key points in the deterioration process at which steps should be taken to ensure that structures remain suitable for their intended function. Particularly significant are:

1. the critical point of disrepair, which CIRIA (in press) recognises as 'the point at which the onset of progressive failure occurs. If the structure is repaired before this point is reached, its life may be extended indefinitely. If repairs are not carried out, major loss of use and reconstruction will be inevitable if the structure is to be returned to its functional level';
2. the point at which the rate of expenditure needed to maintain serviceability escalates to such an extent as to be uneconomic by comparison with partial or complete replacement.

However thorough the investigation of defects may have been, however accurate the diagnosis of the cause and however detailed the proposals for overcoming it, it is often impossible to foresee the extent of the remedial work required until the defective area has been opened out and exposed to examination in depth. By the time that stage is reached there is no turning back and repairs have to proceed, but fine judgements then have to be made as to whether to adjust the work to suit the resources allocated to it or vice versa. In these circumstances, technical considerations ought to prevail; in case of doubt, extending the work by more than is apparently necessary is better than curtailing what ought to be done. Uncertainty over the extent of a repair may make inappropriate the conventional approaches to the reimbursement of contractors.

1.5 Financial considerations

The management of maintenance depends not only upon such technical aspects as the nature, extent and timing of work, but equally critically upon the proper planning and control of resources, especially finance. This normally revolves around an annually allocated budget.

Prior to any particular budget year, decisions have to be made regarding the relative claims of preventive and corrective work and between various alternative repair and renewal options; priorities have to be assessed in respect of the various items of work vying for attention, estimates have to be prepared and a programme must be built up, from which a budget is compiled for at least the coming year but preferably for a rolling programme extending some five years ahead; much of the latter may be provisional, but it will nevertheless allow sensible advance planning, reduce the likelihood of unacceptable peak loads and foster the establishment of a well-motivated workforce. Economic judgements have to be made at every stage, and when budget proposals have been formulated the owner has to be convinced, before he will allocate the necessary finance, that they are fully justified and that they represent value for money. A cost–benefit analysis may be called for.

Provided that safety is not jeopardized, it is possible to make a conscious choice between the quality levels of work undertaken and the costs incurred. Lowering the quality may reduce immediate expenditure but is likely to increase subsequent recurring costs; a reduction in preventive maintenance will increase the amount of later corrective work; improved maintenance can set back the time when the structure must be replaced. The options may be many and sensible economic decisions can be made only by comparing total costs over an extended period – preferably 'whole-life' costs. The timing of future operations assumes considerable significance since, in economic terms, the present value of work which is deferred is less than if it were carried out now. The technique of discounting (see Institution of Civil Engineers, 1969) allows a proper financial comparison between, for example, reconstruction now and repairs which would allow that reconstruction to be deferred for a specific time; similarly it could demonstrate the viability of preventive maintenance which prolonged the period before corrective action became necessary. Its application depends upon realistic estimates of costs, both direct and indirect, throughout the remaining life of the structure or at least for many years ahead. However, these costs are very difficult to assess because their nature is dependent upon so many imponderables and their extent is so unpredictable.

No two structures, however apparently similar, behave identically, and so history and experience can be no more than guides as to the cost of repairs. They are, however, virtually the only guides there are. There is an almost total absence of any reliably derived or generally accepted norms as to what ought to be spent on the maintenance of structures.

In 1981 the Organization for Economic Cooperation and Development (OECD) issued a report on Bridge Maintenance prepared by a panel of experts from 12 European countries, Canada, Japan and the United States. On the strength of a survey of current strategies and practices in respect of road bridges, it was recommended that 'at least 0.5 per cent of the replacement value should be devoted yearly to maintenance expenditure in order to achieve a satisfactory standard'. The UK view, as expressed by the Department of Transport engineers on the panel, was to the effect that such a rate of expenditure is 'sufficient to cope with essential maintenance, but not adequate

to prevent some long term deterioration'.

A further report on Bridge Rehabilitation and Strengthening, prepared on a similar basis, was issued by the OECD in 1983. This recommended that 'an annual sum equivalent to 1.5 to 2 per cent of the replacement value of the total bridge stock should be ear-marked for rehabilitation or the replacement of "below standard" bridges'. It was made clear that this figure excluded the cost of maintenance as covered by the earlier OECD report.

In 1982, an eminent British consulting engineer put forward the view that 'for most ordinary structures, which have been well designed or built, normal regular maintenance costs lie within a fairly narrow band, generally approximating to around one per cent per annum of the original capital cost, excluding inflation and financing charges' (Brown, 1982).

These disparate figures derive from broad-brush approaches which fail to distinguish between such variants as types and ages of structure, the materials and forms of their construction, the usage to which they have been subjected or the maintenance standards which have been applied to them.

Somewhat more specific are the figures agreed between the Department of Transport and British Rail for assessing the annual costs of maintaining and ultimately renewing road bridges of different forms of construction, expressed as percentages of current new construction costs; for brick and masonry, these percentages are as follows:

in substructures, 0.375%;
in superstructures, 0.50%.

It should be noted that these figures include provision for ultimate renewal. They were determined some 50 years ago and there is no extant record of the criteria on which they were based but they are patently inconsistent with the more recent recommendations quoted above.

Equally, there are no established measurable scales against which to assess quantitatively the severity of defects or the degree of improvement achieved by their repair. The Department of Transport alpha-numeric scales (see Chapter 5, Section 5.5) rely entirely on subjective judgements of deterioration. The condition factor of the 'modified MEXE' method of arch assessment (see Chapter 7, Section 7.3.1) is of limited scope and is empirically based on equivocal estimates of the significance of defects observed visually; it pays scant regard to the effect of such imperfections as ring separation, which may not be apparent and is rarely determinate in extent, but evidence is emerging that this defect can cause serious reduction in carrying capacity.

There is therefore a need to refine our ability to quantify, by measurement and testing, the significance of different types and degrees of deterioration in a wider range of structures and to institute a comprehensive and systematic examination of maintenance standards and costs, with the objective of establishing criteria against which the cost effectiveness of different practices and strategies can be reliably measured. Not only would this assist in providing data and a rational basis on which to make economic choices between alternatives, but it would also afford a more credible justification for the level of funds which should be allocated to each year's maintenance budget to achieve the required standards of serviceability and safety. Failing this being undertaken on a national scale, there is scope for individual organizations and even individual engineers to keep records of their maintenance operations which will enable cost effectiveness to be assessed, on however limited a scale.

During the budget year there are still judgements to be made, which are influenced as much by financial as technical considerations. On the question of choosing between undertaking the repair work by direct labour or by contractors, political considerations may supervene, but, these apart, there are many maintenance operations which require a degree of specialist expertise or equipment such that the employment of skilled contractors is clearly indicated. For more common operations and small-scale patching a proper comparison must balance the advantages of the familiarity of direct labour staff with the type of work and the environment concerned against the possible greater efficiency of contractors.

Financial control of expenditure is constantly necessary, but it can be thrown into disarray by short-term changes in budget allocations which are certainly counter-productive in the longer term. Reductions mean cutbacks in preventive work, wasteful temporary palliatives in place of permanent remedies and situations in which the engineer finds himself at the mercy of events rather than in control of them; increases often lead to work being undertaken primarily because it can absorb the additional resources quickly rather than on the basis of need or effectiveness. Similarly, if monies unexpended for any reason in one financial year or on one asset cannot be carried over to another, this may encourage work to be undertaken in an attempt to avoid the loss of the available allocation, irrespective of whether it represents value for money.

Maximum flexibility in the allocation of funds and other resources allows optimum effectiveness to be achieved.

On the expiry of the budget year, an audit or review should be undertaken to determine whether the maintenance effort has been effective and whether the resources devoted to it have been used efficiently. Honest critical appraisals of performance of workers and machines should be used constructively as a basis for improvement.

1.6 Environmental considerations

Aesthetics are frequently disregarded in repair work, with results that are only too regrettably obvious. Cases of 'unsympathetic' rehabilitation abound. Ugly patching, using dissimilar materials in an entirely haphazard manner, is particularly prevalent in masonry structures. The perceptive Maintenance Engineer should be as concerned to harmonize the appearance of repair materials with those existing as he is to match their physical properties. Because of the small areas usually involved, the additional costs, in terms of money and effort, in finding bricks or natural stone of similar size, colour, surface texture and weathering characteristics to the original materials and using them on at least the readily visible faces are usually modest and well justified by the results. The existing materials are likely to have been obtained locally, and so this is the most promising source of further supplies, but it must be recognized that many local brickworks and quarries have been closed. It may be possible to take advantage of recovered materials, either by cleaning and re-using them or by employing them to make reconstructed stone units. Where it is impracticable to avoid a mismatch, the limits of the new facework should be as regular and symmetrical as possible and desirably should coincide with some feature (for example, a horizontal band course or a vertical return face). Existing aspects of coursing, bonding and pointing, as well as decorative features, should be reproduced in new faces exposed to public view. Attempts to disguise incongruities by the application of coatings of paint or rendering are likely to be short lived and cannot be recommended.

Prominent new features such as the ends of anchor- or tie-bars should be symmetrically located, with pattress plates no larger than necessary, and properly protected against corrosion. Where strengthening of brick or stone masonry involves visible thickening in a different material (for example, underlining an arch or 'corsetting' a pier), the contrast in materials can be made an acceptable distinctive feature, but care is still necessary to maintain compatible proportions in the structure as a whole.

Surface staining (including efflorescence) is unsightly and unfortunately common in new facework. Advice on how to minimize its incidence is given in Chapter 8, Section 8.3.5; the most effective countermeasure is to prevent water from percolating through the brickwork.

Percolating water also has the effect of encouraging the accumulation of dirty and slimy deposits which are equally undesirable environmentally, especially on faces with which the public may come into contact. In such circumstances counteractive measures should have a high priority and should not be regarded as purely cosmetic, since the removal of dirt can help to preserve the face.

The true measure of the Maintenance Engineer's success may well lie in the effectiveness with which he has used his available resources, but few people will be aware of what he has achieved in that respect and neither will they know the extent to which the repairs he has undertaken on a structure may have extended its economic life. The general public is more likely to judge his efforts primarily by the concern he has shown for the sympathetic preservation of what they may well have come to regard as an integral and indispensible part of the local scene. In fact, on occasions he may be constrained by environmental or historic considerations to undertake restorative maintenance when engineering or economic factors would suggest demolition or renewal; he has then to exercise his professional responsibilities on the broadest basis.

1.7 The achievement of effective maintenance

It is worthy of emphasis that the proficiency of the Maintenance Engineer relies fundamentally on his ability to differentiate between defects requiring attention and those which are benign or tolerable, in order to establish reliably the urgency for remedial action; equally, it demands the ability to devise and implement practical, effective and economic solutions. Whatever the defect, the objective of the repair must be to remedy it and not just to patch; damage should not merely be made good superficially but should be prevented from recurring. Diagnosis of the cause, from the observation of the symptoms, is therefore critical.

Skill in these areas depends to a great extent on professional judgement derived from knowledge and experience, but such judgement is supported and enhanced by ensuring that the maximum amount of back-up information is available, so that all deductions and decisions are soundly based. In all but the most simple and straightforward problems, the recommended approach is as follows.

1. Compile a dossier of everything that is known about the structure – its original construction, its subsequent history, the usage to which it has been subjected, previous repairs and their effectiveness, the results of earlier investigations, records of defects and their development.
2. Examine regularly and thoroughly, to a strictly disciplined routine. Record the results in as much detail as possible and add them to the dossier (see Chapter 5).
3. Investigate any significant deterioration; monitor signs of distress, sample and test (see Chapter 6); open out to reveal hidden information (such as geometric anomalies, voids, unrecorded previous repairs) and carry out surveys of relevant factors (subsoil conditions, sources of water percolation, location of public utility services etc.).
4. Analyse the results of these investigations, in combination with the history from (1) and the record of progressive deterioration from (2).
5. Identify the cause of the deterioration and assess the urgency for preventive or corrective action in order to meet currently required standards of performance. This may involve the assessment of the residual capacity of the structure (see Chapter 7) and the implications of deferring action.
6. Consider the various possible options in the light of defined maintenance policy and available resources. Decide upon what to do, and when and by whom it is to be done, having regard to technical, economic and operational criteria and constraints. Ensure that the remedial action addresses both the defect and its cause.

7. Plan, assemble the necessary resources and implement that decision, exercising careful control and supervision over the quality of materials, workmanship and finished appearance.
8. Record what has been done and at what cost, adding this information to the dossier.
9. Monitor and record the effectiveness of what has been done.
10. Feed back any relevant information to designers, examiners and fellow Maintenance Engineers.

There are many sources of background information which can be used to compile the dossier (point 1) or to build up one which is deficient. The most unlikely sources are often surprisingly fruitful. Possibilities include the following:

1. the files, records and muniments of the owner's organization
2. the archives of local authorities
3. local libraries (for historical books and past issues of local newspapers and periodicals)
4. local historical societies and museums
5. local geologists, geological records and societies
6. the Institution of Civil Engineers library and Panel for Historical Engineering Works
7. water and other service authorities
8. Institute of Geological Sciences
9. National Coal Board and the bodies concerned with the extraction of other minerals
10. Public Records Office, Kew
11. British Library Newspaper Library, Colindale Avenue, London NW9
12. aerial photographs (held by the Royal Commission on Historic Monuments).

The search for forgotten or overlooked data can be a fascinating activity which appeals to anyone with a bent for historical research. The building up of a coherent picture, in the manner of a jigsaw puzzle, from an aggregation of discrete pieces of information is rewarding, particularly when it validates a theory derived from observations on the structure itself.

1.8 References and further reading

Building Research Establishment (1975) *Digest 176, Failure Patterns and Implications.*

British Standards Institution

Code of Practice CP 3, Chapter IX (1950), *Code of Functional Requirements of Buildings: Durability.*

BS 5390: 1976. (1984) *Code of Practice for Stone Masonry.*

BS 5628 *Code of Practice for the Use of Masonry.*

Part 1: 1978. *Structural Use of Unreinforced Masonry.*

Part 2: 1985. *Structural Use of Reinforced and Prestressed Masonry.*

Part 3: 1985. *Materials and Components, Design and Workmanship.*

Brown, C.D. (1982) Life cycle costs for structures, particularly bridges. *Proc. Conf. on the Challenge of the 80s,* Paper d.4, paragraph 16, Institution of Civil Engineers, London.

Construction Industry Research and Information Association (in press) *The maintenance and rehabilitation of old waterfront walls.* E. & F. N. Spon, London.

Curtin, W. G., Shaw, G., Beck, J. K. and Parkman, G. I. (1984) *Structural Masonry Detailing,* Granada, London.

Grimm, C. T. (1980) Masonry failure investigations. *Proc. Symp. on Masonry: Materials Properties and Performance,* Orlando, FL, December 1980, Special Technical Publication 778, pp.245–60, American Society for Testing and Materials, Philadelphia, PA.

Harvey, W. J., Maxwell, J. W. S. and Smith, F. W. (1988) Arch bridges are economic. *Proc. 8th Brick and Block Masonry Conference* (ed. J. W. de Courcy). Sept, Dublin. Elsevier Applied Science Publishers, London, vol. 3, pp.1302–10.

Institution of Civil Engineers (1969) *An Introduction to Engineering Economics,* Institution of Civil Engineers, London.

Ireson, A. S. (1987) *Masonry Conservation and Restoration.* Attic Books, Builth Wells.

Knight, T. L., Hammett, M. and Baldwin, R. J. (1989) Achieving aesthetically acceptable brickwork.

Part I The role of the specifier.

Part II Site management.

Part III The craft of bricklaying.

Proc. British Masonry Society – Masonry (3), Workmanship in Masonry Construction. Pp. 15–17, 26–7 and 28–30. Building Research Establishment.

Narayanan, R. S. (1989) Workmanship: the designer's role. *Proc. British Masonry Society – Masonry (3), Workmanship in Masonry Construction.* Pp. 7–10. Building Research Establishment.

Nash, W. G. (1986) *Brickwork Repair and Restoration.* Attic Books, Builth Wells.

O'Malley, R. E. (1951) *Economics of the Maintenance of Highway Bridges: Concrete and Masonry Bridges.* Road Paper No. 33. Institution of Civil Engineers, London.

Organisation for Economic Cooperation and Development, Road Research. (1981) *Bridge Maintenance.* pp. 30–32. OECD, Paris.

Organisation for Economic Cooperation and Development, Road Transport Research. (1983) *Bridge Rehabilitation and Strengthening.* p.97. OECD, Paris.

Roberts, J. J., Edgell, G. J. and Rathbone, A. J. (1986) *Handbook to BS. 5628, Part 2.* Viewpoint Publications, London.

Robertson, J. A. (1969) The planned maintenance of buildings and structures. *Proc. Inst. Civil Eng. Suppl.* Paper 7184 S, pp.483–96.

Sutherland, R. J. M. (1981) Brick and block masonry in engineering. *Proc. Inst. Civil Eng. Part 1,* 70, 31–63 (Paper No. 8411).

Sutherland, R. J. M. (1989) Load bearing masonry in *Civil Engineer's Reference Book,* 4th edn, (ed. L. S. Blake), Butterworth, London.

Turton, F. and Cox, N. S. (1953) *The Design of structures in relation to Maintenance and Inspection,* Railway Paper No. 48, Institution of Civil Engineers, London.

Turton, F. (1972) *Railway Bridge Maintenance,* Hutchinson Educational, London.

Bricks

J. Morton

2.1 Brick making

Brick making was one of civilization's first steps towards self-preservation, once man moved out of the caves. Techniques have, of course, progressed dramatically from the early days, some thousands of years ago, but the basic principles of manufacture remain the same today as they were then – namely, the art of combining earth, fire and water. The process involves four main stages, which are described below.

2.1.1 Winning the clay and preparing it

After the clay has been excavated, it is either stored in a spoil heap in or near the quarry or taken to the preparation plant. A spoil heap is usually used when the various bands or layers of clay need to be carefully mixed together for colour consistency or when the clay needs to be weathered over a winter period. In the preparation plant it is usually crushed and ground to a fine powder before being mixed with water.

2.1.2 Shaping the brick

The three main modern processes used to shape the green brick are as follows.

(a) Extrusion

The prepared clay is forced through a rectangular die and the resulting column is cut into 1–2 m lengths. The column may be of solid clay or it may have perforation patterns contained within it to reduce the energy consumption during drying and firing. Each length is then converted to a number of individual bricks by cutting it with a series of wires; in some instances it is pushed through static wires while in others the wires rotate through the stationary clay body. In this process, before the clay is extruded it is passed through a de-airing machine which removes as much entrapped air from within the clay as is required for successful production.

(b) The semi-dry pressed method

In the semi-dry pressed method the clay powder is fed in a fairly dry form (i.e. with low water content) into moulds where it is pressed; each brick is individually formed by the pressing process and then taken for drying and firing. Perhaps the most prolific use of the pressed brick is in Fletton brick production, which uses bands of Lower Oxford Clay. The relative volume of production of the semi-dry pressed bricks known as Flettons makes them an important element in the brickwork scene in Britain today, although they are used mainly in housebuilding as opposed to civil engineering work.

(c) Handmaking

Handmade or soft mud bricks are still made today by the method used centuries ago. When the clay has been won and after any preparation that may be required, it is brought to the brickmaker's table. Just enough clay for one brick is rolled and kneaded to remove excess air and to form it into the rough shape for the mould. This warp of clay is then thrown downwards into the mould, any excess being removed with a wire cutter. The brick is removed from the mould, the sides of which have been sanded to aid this process, and taken for drying and firing. While some bricks are still produced in this way, in many 'handmade brickworks' the clay is formed into the shape of a brick using a machine which simulates the way in which the human brickmaker throws the brick. In all other respects the process is virtually identical.

2.1.3 Drying the green brick

The method of drying depends upon the method of manufacture. Broadly speaking, extruded bricks are dried in special drying rooms into which hot clean exhaust air from the kiln is fed until the water content has been sufficiently reduced. In the case of handmade bricks, traditionally these have been dried in long open rows; the period of drying depends on the weather but is normally two to three weeks. Where handmade bricks are made using more modern production facilities, it is common to find special drying rooms being used.

The semi-dried pressed process does not involve a separate drying operation. Of course, in its green state the clay has a much lower moisture content than have the extruded or handmade bricks, but indeed a drying process does exist within the heating, firing and cooling cycle of the continuous Hoffman kiln operation.

2.1.4 Firing the brick

(a) Clamp fired

Historically, handmade brick production has been fired in a traditional brick clamp. The clamp is merely a large quantity of dry green bricks carefully constructed on a foundation of already fired bricks on which a layer of fuel, usually coke, has been spread. The fuel is lit at one end and the fire

is allowed to travel along the clamp. However, this traditional way of firing bricks is being replaced by more automated forms of kiln firing which improve efficiency and provide improved levels of quality control.

(b) Tunnel kiln

Many handmade bricks are fired in tunnel kilns, as are the majority of bricks produced by the extrusion process. Bricks pre-loaded onto kiln cars travel through the kiln, and the firing zone is stationary at or near the centre. Travelling towards the firing zone, the bricks are further dried and heated until, in the firing zone itself, temperatures of some $1000 - 1100°$ C convert the brick earth into the hard ceramic we know as brick. The remainder of the journey down the kiln is through the cooling zone. Some $20 - 30$ h after entering, the brick leaves the kiln – still very hot to the touch.

(c) Hoffman kiln

In the Hoffman kiln, which is used to fire all semi-dried pressed bricks, the fire moves round the various cell compartments. This first requires the bricks to have been warmed and dried as part of the firing cycle. An interesting aspect of the Lower Oxford Clay is the combustible microscopic globules of oil contained within it, which comes from microscopic sea creatures trapped in the clay when it was laid down in geological time. It is this content of organic material which provides the energy efficiency of the Fletton brick-making process.

2.2 Calcium silicate bricks

The foregoing is concerned with clay brick production. Another type of brick – made of calcium silicate – is also common in Britain, although far fewer of these are produced than those made of clay. They are manufactured by mixing sand with lime or crushed flint, and the green brick is produced by a pressing process. Unlike clay bricks, they are not fired; instead they are cured in high pressure steam within an autoclave. They have distinctively sharp arrises. As a result of recent production developments it is difficult to distinguish some of the modern calcium silicate facing bricks, particularly the multicoloured variety, from similar clay bricks unless they are viewed from a short distance.

2.3 Material characteristics

Clay and calcium silicate bricks are available in strengths from 10 to 15 N/mm² upwards. In the case of clay, brick strengths of perhaps 150–180 N/mm² are possible, although the normal maximum is about 100 N/mm²; therefore the range is quite large. In the case of calcium silicate, strengths up to 35–45 N/mm² are available.

In modern design the strength of the brick can be used to determine the strength of the brickwork if the mortar designation is known. British Standard BS 5628: Part 1 gives brickwork strengths as 'characteristic' values, since it is a limit state code. However, it does not give any values for brickwork with other than modern mortars. In assessing the strength of old brickwork made from cement-free mortars or those with a very weak cement content, the tables of permissible stresses contained in CP 111 (now withdrawn) are often more useful, although some conversion is required to give the equivalent of characteristic compressive strength values. Once the compressive strength is known, it is possible to estimate a value for the modulus of elasticity E. BS 5628: Part 2 suggests a figure of 900 times the characteristic strength of the masonry.

Another important characteristic is water absorption. This and the mortar designation control the flexural strength of the masonry or, as perhaps it is better understood, the characteristic tensile strength in bending. It has been found that this value is higher when the water absorption of the brick is lower for any given mortar. Values are given in BS 5628: Part 1, Table 3.

2.4 Types of brick

Engineering bricks are characterized by high strength and low water absorption, which give them high durability and the ability to prevent upward capillary movement of water. They can therefore be used in damp-proof courses. Engineering A bricks are defined as having water absorption of less than 4.5% and a strength of more than 70 N/mm², while engineering B bricks have less than 7% water absorption and a strength of more than 50 N/mm². Engineering bricks demonstrate a greater resistance to abrasion than others and are widely used as protection in the inverts of culverts; engineering A bricks (usually bull-nosed) are used for water steps where abrasion may be high.

One type of engineering brick is the blue variety, typically the Staffordshire Blue. The colour comes from the method of manufacture combined with the type of clay used and its chemical composition. Engineering bricks, do not have to be blue; similarly, not every blue brick is necessarily of engineering quality.

Among the other types commonly encountered in the structural maintenance field, London Stock bricks are handmade in the Home Counties. Their characteristics are high water absorption, relatively low strength (15–35 N/mm²) and an attractive appearance; they are highly durable.

Gault bricks were once common but now they are available from only one brickworks. Distinguishable by their buff off-white yellowish colour, their properties are low strength (typically 10–15 N/mm²), high water absorption and moderate frost resistance. They are made from Gault clay which is bleached during the manufacturing process by interaction with the high proportion of free lime.

Red Rubbers are oversized soft red facing bricks specially made to be cut and finely rubbed down on all faces; they are used for very high quality arch work and other ornamental brick features. The soft clay deposits used for manufacture are very carefully screened and mixed with a high proportion of fine sand to facilitate the hand-rubbing techniques used to obtain the 0.5 mm lime jointing which is typical of this work. Owing to the high cost of these bricks and the skills needed to work with them, demand is practically restricted to conservation work.

Glazed bricks are characterized by their glass-like finish, which gives a tile effect. They are manufactured using a strong dense extruded wire-cut brick on which is sprayed a glaze based on feldspar, which may also contain metal oxides to obtain a selection of different colours. They are available in all standard sizes and in special shapes; however, they are a specialist product with only one or two manufacturing units in Britain. In addition to their ability to reflect light in locations where natural lighting is poor, they are readily cleaned and less prone to permanent defacement by graffiti.

2.5 Special shapes

British Standards BS 3921 and BS 187 deal with the specification of clay and calcium silicate bricks with normal shapes. There are often requirements for other shapes – e.g. plinth bricks to effect a setting back in the face of a wall, squint bricks for use at acute or obtuse quoins, bull-nosed bricks and coping bricks. It is possible to have virtually any practical shape purpose made and so avoid the need to cut bricks to fit. Those most commonly sought are listed in British Standard BS 4729 and are often referred to as 'standard' special shapes; they are more readily available than 'special' specials which are not included in the British Standard.

2.6 Accepting delivery

Although it seems obvious, it is important to check bricks for compliance with what has been specified and ordered **before** they are built into a structure. Visual inspection for colour, finish,

size, shape and freedom from defects, while being very useful, has its limitations. All subjective views must be subservient to the quantitative tests defined in the British Standards.

As Vitruvius wrote, some 2000 years ago, 'Whether the baked brick itself is very good or faulty for building, no one can judge off-hand . . .' But then, of course, quality assurance was less well developed in the Roman Empire!

2.7 References and further reading

Brick Development Association (1979a) *Building Note 1, Bricks and Brickwork on Site.*

Brick Development Association (1979b) *Design Note 3, Brickwork Dimensions and Tables.*

Brick Development Association (1985) *Special Publication 13, Specification for Clay Bricks.*

British Standards Institution

BS 187: 1978. *Specification for Calcium Silicate (Sandlime and Flintlime) Bricks.*

BS 3921: 1985. *Specification for Clay Bricks.*

BS 4729: 1990. *Specification for Dimensions of Bricks and Special Shapes and Sizes.*

BS 5628. Code of Practice for the Use of Masonry.

Part 1: 1978. *Code of Practice for the Use of Masonry. Structural Use of Unreinforced Masonry.*

Part 2: 1985. *Structural Use of Reinforced and Prestressed Masonry.*

Part 3: 1985. *Materials and Components, Design and Workmanship.*

BS 5642: Part 2: 1983. *Specification for Copings of Pre-cast Concrete, Cast Stone, Clayware, Slate and Natural Stone.*

Code of Practice CP 111: 1970 (withdrawn) *Structural Recommendations for Load Bearing Walls.*

Building Research Establishment (1973) *Digest 157, Calcium Silicate (Sandlime, Flintlime) Brickwork.*

Building Research Establishment (1974) *Digests 164 and 165, Clay Brickwork.*

Construction Industry Research and Information Association (1986) *Structural renovation of traditional buildings.* Report No. 111, 3, pp. 19–20, CIRIA, London.

Everett, A. (1986) *Mitchell's Building Construction: Materials,* Batsford, London.

McKay, W. B. (1971) *Building Construction,* Longman, London.

Vitruvius (1960) *The Ten Books of Architecture,* Dover Publications, New York.

Chapter three

Natural stonework

B.A. Richardson

The natural stone used for masonry is known to geologists as rock, and it is convenient to classify it according to its origin into primary or igneous rock, secondary or sedimentary rock, and metamorphic rock.

The igneous rocks, which are formed by solidification of molten material, include basic rocks such as basalt and dolerite, often known as whinstone, and acid rocks such as granite. Whilst whinstone is most commonly crushed for road-stone, granite is widely used as a structural material but it varies greatly in its nature. It is composed largely of silica deposited as crystals which may be small, granular and open textured, with an appearance of a rather coarse sandstone, or large, compact and virtually impermeable.

The sedimentary rocks are formed from water-borne or airborne debris. In the case of sandstones, deposits of silica granules, usually from granitic rocks, are consolidated with an amorphous silica or calcium carbonate cementing matrix, often in conjunction with aluminium or iron oxides. The properties of sandstones depend largely upon the cementing matrix as the silica granules are virtually inert. The limestones constitute the second major group of sedimentary rocks and are formed either from accumulations of animal shells in calcareous cement or by crystallization from solution to give a characteristic oolitic structure. In either case the resulting limestone is usually predominantly calcium carbonate, although other carbonates such as magnesium carbonate sometimes occur. Sedimentary stones of intermediate structure also occur and are termed arenaceous limestones or calcareous sandstones, depending upon whether calcium carbonate or silica predominates.

The metamorphic rocks are developed from all the preceding types under the influence of heat and pressure. Thus the schists and gneisses are formed from igneous rocks, quartzite from sandstones, marbles from limestones, and slates from mud and shale.

Suitability for structural purposes depends upon a number of factors, the most important being availability and workability. Obviously local stones have been preferred but they have not always been ideal, and the progressive improvement in communications over the years has caused stones to be obtained from ever-widening areas. Anyone concerned with the maintenance of natural stone masonry must therefore possess an exceptionally broad knowledge, even if his activities are confined to a particular locality.

The way in which natural stone is used in masonry will depend very largely upon the properties of the material that is used. In most natural stone masonry in the British Isles porous sandstones or limestones are used because they are most easily worked. In a building or any other construction enclosing a space, resistance to rain penetration will depend upon the balance between the water absorbed, the capacity of the wall and the rate of subsequent evaporation. Heavy showers alternating with bright and windy periods are usually less likely to lead to major water accumulations than continuous drizzle with exceptionally humid conditions which obstruct evaporation, but dampness will usually be apparent on areas of thinner wall section, such as window and door reveals, whatever the prevailing weather conditions, whenever solid porous masonry is used. In the nineteenth century natural stone was often selected for its denseness and low porosity as this resulted in high compressive strength and low permeability, but such stone was often particularly susceptible to crystallization deterioration, i.e. damage caused by the presence of salts or by freezing, as such stones were often predominantly microporous. (The mechanisms of deterioration will be explained in Chapter 8, Section 8.2). It was also believed that dense stone with high compressive strength inevitably possessed a high modulus of rupture, but this was not necessarily the case and fractures in stone lintels could often be attributed to poor stone selection in this respect. Unfortunately, the normal approach to the selection of natural stone has not significantly improved during the twentieth century. Indeed, there has been a tendency for the use of natural stone to be confined to a few readily available stones with well-known properties, such as Portland, Bath and Clipsham limestones, but the most marked development has been a profound reluctance to use natural stone at all and to crush the stone instead for use as an aggregate in concrete which, often mistakenly, is considered to be a much more reliable and predictable structural material.

Whilst erosion and spalling deterioration of limestones and sandstones is very obvious when it occurs and suggests that the use of these sedimentary stones should be avoided, despite the excellent performance of properly selected stones, problems are also encountered with igneous and metamorphic stones. With porous sedimentary stones any water penetration through defects in mortar joints will be absorbed into the adjacent stone, but where impermeable granite, marble or slate is used, any defects in the mortar joint become particularly apparent as there is no adjacent porous stone to absorb water penetrating through leaks in the mortar joints. In addition, it may be difficult to establish adhesion between the mortar joints and the adjacent stones, and special attention to structural bonding may be necessary at all stages in the design, construction and maintenance of such masonry.

Obviously this description of natural stone masonry is considerably simplified. There is, in fact, an enormous range of building stones of many different types but, even within a single quarry, stones vary from bed to bed and also within a bed. Often the physical properties and durability of a particular stone have been established in the past through its reputation in service, and the obvious fact that currently available stone may possess entirely different properties, even if it is obtained from the same quarry, is often entirely ignored. Thus reputation cannot be considered a reliable method for stone selection for either new construction or repair purposes.

Clearly, appearance is of special importance when endeavouring to match repair stones to the masonry of an existing structure. Information on the original quarry will be helpful, but the quarry may be entirely worked out and it may be necessary to seek an alternative source of stone, usually from a quarry working the same geological deposits as this is the most reliable method for ensuring a close match. It is particularly important to appreciate that the colour and texture match must be correct after weathering; fresh new stone will usually weather or 'mellow' to an entirely different colour. Samples of stone must be tested to confirm probable durability, compressive strength and perhaps modulus of rupture; the last property is particularly important when stone is used as a thin cladding rather than as a load-bearing solid structure.

References and further reading

Ashurst, J. and Dimes, F. G. (1977) *Stone in Building: Its Use and Potential Today,* Architectural Press, London.

British Standards Institution BS 5390: 1976 (1984). *Code of Practice for Stone Masonry.*

Building Research Establishment (1983a) *Digest 269, The Selection of Natural Building Stone.*

Building Research Establishment (1983b) *SO 36, The Building Limestones of the British Isles.*

Building Research Establishment (1986) *Report BR 84, The Building Sandstones of the British Isles.*

Building Research Establishment (1988) *Report BR 134, The Magnesian Limestones of the British Isles.*

Building Research Establishment (1989) *Report BR 141, Durability Tests for Building Stone.*

Davey, N. (1976) *Building Stones of England and Wales,* Bedford Square Press, London.

Shadmon, A. (1989) *Stone: an introduction.* Intermediate Technology, London.

Shore, B. C. G. (1957) *Stones of Britain,* Leonard Hill, Glasgow.

Stone Industries (1985) *Natural Stone Directory,* Ealing Publications, Maidenhead, Berkshire.

Warnes, A. R. (1926) *Building Stones,* Ernest Benn, Tonbridge, Kent.

Mortar

R. C. de Vekey

4.1 Introduction

Mortar is a material that is plastic when fresh and then stiffens rapidly when in contact with pieces of stone or brick and sets to a hard adherent mass over a period of days, weeks or longer. The primary purpose of mortar is to fill the gaps between blocks of stone, brick or concrete that result from their imperfect size and shape and thus to make possible the construction of masonry which is structurally stable and weatherproof. If the masonry units are accurate to fractions of a millimetre or selected to dovetail together, as in the case of dry stone walling, it is possible to build structurally stable masonry without mortar. In modern slender brickwork cladding and lightly stressed load-bearing brickwork there is also a requirement for a minimum bond to impart a resistance to lateral loading (flexural resistance).

The basic properties required in all mortars are as follows:

1. good workability, to minimize the labour of applying and spreading it and to allow a short but dependable time interval in which the unit can be positioned and adjusted without disrupting the bond
2. a relatively rapid rate of stiffening, so that following work is not unduly delayed

3. sufficient compressive, tensile and shear strength in the hardened state for the application
4. a good bond to the units to minimize rain penetration and maximize the tensile and flexural strength of the masonry
5. some capacity to creep and yield, to allow for movements during the settling in and drying out of a structure and, in some cases, for small amounts of movement during the lifetime of the structure caused by environmental effects, imposed loads, settlement, changes of use, etc.
6. durability when subjected to any form of environmental effect such as wind, water, abrasion, freezing, chemicals and pollutants
7. constituents that do not reduce the durability or the appearance of the masonry by attacking the units or causing staining.

Although, in principle, mortar could be one of a huge range of possible compositions, most of the mortars used in both old and contemporary structures are based on a mixture of a filler, usually sand, a hydraulic binder (e.g. cement) and water. To make some compositions sufficiently plastic for easy positioning of the masonry units an additional workability aid or plasticizer must be included. Generally the binder constituent is finer than the sand filler and the optimum plasticity characteristic is obtained for a mixture of 1 part binder to 3 parts filler. This is achieved with modern mixes based on ordinary Portland cement by keeping a ratio of binder to sand of 1:3 but diluting the cement with a non-setting fine powder, usually lime, to obtain weaker mixes. Mixes with lower proportions of binder or fines are made more plastic by the addition of surfactants which increase plasticity by entraining air.

4.2 Mortar constituents

4.2.1 Fillers (sand)

Sand suitable for mortar consists of rock particles in the size range between 5 mm and 75 microns in diameter. Commonly it consists of naturally occurring layers derived from recent alluvial deposition (e.g. river beds and sea beaches) or older deposits due to past alluvial or glacial action. In some areas it may be derived from dunes or by crushing quarried rocks. The chemical and geological composition will reflect the area from which it is derived. The commonest sands are those based on silica, partly because of its wide distribution in rock such as sandstones and the flint in limestones and partly because it is hard and chemically resistant. Other likely constituents are clay derived from the decomposition of feldspars, chalk or limestone from shells in some marine sands and crushed rocks, sands from weathered granites which contain micas and crushed basalts and granites. Very flaky materials such as slates and micas are not suitable as it is difficult to make them workable. Very absorbent materials are also unsatisfactory for dry-mixed mortars since they cause rapid falls in workability by absorbing the mixing water. They may be suitable for mixes based on 'coarse stuff'.

In order to be at all usable, mortar sand must contain no particles with a diameter greater than about half the thinnest joint thickness. Additionally it should be mostly free of clay particles (around 75–30 microns) which cause unsatisfactory shrinkage characteristics and chemical interactions with the binder. Ideally it should also have a good range of particle sizes from the largest to the smallest (a good grading) since this leads to good packing of the particles to give a dense strong mass resistant to erosion, permeation and chemical attack. Many

naturally occurring alluvial deposits fall naturally into the required grading and can be used as dug or with a few coarser particles screened off. These are usually termed 'pit sands'. Sands that are outside the normal range must be sieved to remove coarse fractions and washed to remove excess clay particles. For particular purposes sand can be sieved into fractions and regraded but this is rarely done for a mortar sand. Figure 4.1 shows the grading curves for the sands allowed under the current version of BS 1200 (amended in 1984). This gives two allowed grades, S for structural use and G, with slightly wider limits, for general purposes.

Most of the normal constituents of sand are virtually chemically inert to normal environmental agents, i.e. water, dissolved salts, carbonic acid, humic acids, other organic compounds and 'acid rain', but chalk or limestone particles will be dissolved slowly by mild acids and clays may react in time with acids or alkalis. Most sand constituents are also fairly hard and are resistant, in themselves, towards mechanical abrasion and erosion by dust in wind and water.

4.2.2 Binders

Binders are the components which cause the mortar to set to a hard mass and thus allow the masonry to resist high stresses. Because they must be finely divided to be able to penetrate the spaces between sand grains and to react in some way to give the change from soft to hard, they must be inherently more chemically reactive than the other components. Their reactivity is their Achilles' heel in that they often react with chemicals in the environment with resultant deterioration.

Currently, the most popular binder for general purposes is **ordinary Portland cement** (OPC) which is made by heating limestones containing silica with clay to give a complex mixture of calcium aluminates, calcium silicates and aluminoferrites together with a little uncombined lime (calcium oxide). The resulting 'clinker' is ground

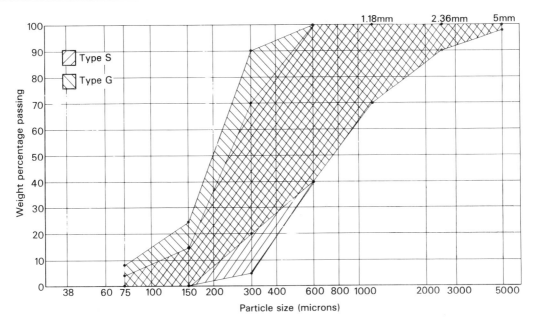

FIGURE 4.1 Grading envelopes for mortar sands.

to a fine reactive powder with the addition of about 5% calcium sulphate which slows down the otherwise flash set of the aluminate.

Sulphate-resisting Portland cement (SRPC) is similar to OPC but has less of the aluminate component which is susceptible to a deleterious reaction with water containing dissolved sulphates.

Portland blast furnace (super sulphated) cement, which is largely (around 60%) ground blast furnace slag – a reactive glass by-product of iron smelting – activated with OPC and lime, is also used.

Masonry cement is a factory prepared mixture of OPC with a fine inert filler – plasticizer (around 20%) and an air-entraining agent to give additional plasticity. It is intended solely for making bedding mortars mixed with sand and water. Currently the fine powder is normally ground chalk but the British Standard allows a wide range of inert and semi-inert materials. In the future Portland cements blended with pulverized fuel ash (PFA) may well be used for mortars either as masonry cements or as blended cements under new standards.

Hydraulic lime was widely used in the past but is seldom used now because it is not generally available. It is basically a quicklime (calcium oxide) produced by heating impure limestone to a high temperature. The impurities, usually siliceous or clay, lead to the formation of a proportion of hydraulically active compounds such as calcium silicates or aluminates. The mortar is made as normal by gauging with sand and water but there will be some emission of heat while the lime is slaking.

Pure quicklime (CaO) has also been used widely for making mortars in the past. It is made up as for hydraulic lime mortar, but since it does not have any setting action in the short term it can be kept for days or weeks provided that it is covered and prevented from drying out. The wet mix with sand is termed 'coarse stuff'. Contemporary lime mortar may be made from hydrated lime ($Ca(OH)_2$) but is otherwise similar. The initial setting action of this mortar depends only on drying out in contact with the units, and so it is not suitable for the construction of slender structures which require rapid development of flexural strength. Over long periods of months or years this mortar hardens by carbonation of lime to form calcium carbonate, but it is never as hard or durable as properly specified cement mortars. Most cements are blended with pure hydrated lime in various proportions to make hybrid binders which give mortars with a lower strength and rigidity but still maintaining the plasticity of the 1:3 binder:sand ratio.

4.2.3 Other constituents

Binder, sand and water are the only ingredients of basic mortars, but many mortars, particularly modern formulations, contain other ingredients.

(a) Plasticizers

The traditional plasticizer is hydrated lime, which is used as a substitute binder, but many organic compounds can be used to induce plasticity by other mechanisms. All the classic plasticizers for mortar operate by causing air to be entrained as small bubbles. These bubbles fill the spaces between the sand grains and induce plasticity. Typical materials are based on Vinsol Resin, a by-product of cellulose pulp manufacture, or other naturally available or synthetic detergents. They are surfactants and alter surface tension and other properties. The super-plasticizers used for concrete plasticize by a different mechanism which does not cause air entrainment, but they are not used widely as mortar plasticizers.

(b) Latex additives

A number of synthetic copolymer plastics can be produced in the form of a 'latex', i.e. a stable finely divided suspension of the plastic in water. When combined with hydraulic cement mixes these materials have a number of effects; in particular, they increase adhesion of mortar to all substrates and also increase the tensile strength, reduce the stiffness, reduce the permeability and increase the durability. Because of these effects they are widely used in flooring screeds and renders but they are also used to formulate high

bond and waterproof mortars. The better polymers are based on copolymerized mixtures of butadiene, styrene and acrylics. Polyvinylidene dichloride (PVDC) has also been marketed for this application but there can be some loss of chlorine which can attack buried metals. These materials should never be used with sands containing clay particles. The manufacturers' advice should be followed with regard to dosage (usually in the range 5%–20% of the cement weight).

(c) Pigments

For architectural reasons mortar of a particular colour may be required. If this cannot be attained by selection of suitable natural sands and binders then pigments may be added. These are in the form of powders of inert coloured compounds of a similar fineness to the binder; thus they tend to dilute the mix and reduce strength. The contemporary Codes of Practice limit the content of most pigments to 10% and carbon black to 3% maximum of the weight of the cement.

(d) Retarders/accelerators

Retarders are used to delay the initial set of hydraulic cement mortars to allow their supply as 'readymix'. Accelerators usually based on calcium chloride have been marketed. None of the current codes permit the use of any form of chloride because of its effect on buried metals. Alternatives such as calcium formate may be satisfactory. Accelerators should not be used as an alternative to proper precautions when building in frosty weather as they are ineffective for this purpose.

4.3 Mortar formulations and uses

Mortar is called upon to cope with a huge range of requirements, which are sometimes in conflict, and to work with a variety of materials in a range of climatic conditions. To obtain optimum performance the composition must be tailored to the aplication. The broad principles are as follows.

Increased contents of hydraulic cements give stronger, denser and more impervious mortars

Increasing ability to tolerate movement but decreasing strength and durability →

← Increasing strength, durability, resistance to percolation

TABLE 4.1 Strength of mortars and brickwork

Mortar Designation	Type of mortar (ingredients) (proportion by volume)			Mortar strength range (N/mm²)	Brick strength (N/mm²) Characteristic compressive strength of brickwork (N/mm²)				
	Cement: Lime: Sand	Masonry cement: Sand	Cement: Sand + Plasticizer		7	20	35	50[a]	70[b]
(i)	1:0-0.25:3	-	-	11-16	3.5	7.5	11	15	19
(ii)	1:0.5:4.5	1:2.5-3.5	1:3-4	4.5-6.5	3.5	6.5	9.5	12	15
(iii)	1:1:5-6	1:4-5	1:5-6	2.5-3.6	3.5	6	8.5	11	13
(iv)	1:2:8-9	1:5.5-6.5	1:7-8	1.0-1.5	3	5	7	9	11
(v)	1:3:10-12	1:6.5-7	1:8	0.5-1.0	2	4	6	7.5	8.5
(vi)	0:1:2-3[c]	-	-	0.5-1.0	2	4	6	7.5	8.5
(vii)	0:1:2-3[d]	-	-	0.5-1.0	2	3	3.5	4.5	5

Increasing ability to resist frost attack while green; decreasing water:cement ratio →

Increasing bond and hence rain resistance; increasing long-term durability →

[a] Class B engineering brick.
[b] Class A engineering brick.
[c] Hydraulic lime.
[d] Pure lime.

which are more durable, bond better to bricks under normal circumstances and harden rapidly at normal temperatures. They also increase the drying shrinkage and their use with low strength units may lead to cracking owing to movement.

Decreased contents of hydraulic cements give weaker, more ductile, mortars, which are more tolerant of movement and matched better to low strength units, but at the cost of a reduction in strength, durability and bond. There is a corresponding reduction in shrinkage and hardening rate.

Sharp well-graded sands give very high compressive strength, low permeability and generally good bond but poor workability, while fine 'loamy' sands give high workability but generally with reduced compressive strength and sometimes reduced bond.

The addition of lime confers plasticity and, particularly for the wet stored mixes, water retentivity – the ability of the mortar to retain its water in contact with highly absorbent bricks – which facilitates the laying process and makes sure that the cement can hydrate. Lime mortars perform poorly if subjected to freezing while in the green (unhardened) state, but are very durable when hardened. Lime is white and thus tends to lighten the colour of the mortar. In some circumstances it can be leached out and may cause staining.

Air entrainment tends to improve the frost resistance of green mortar and allows lower water: cement ratios to be used, but such plasticized mixes are less durable and less water retentive than equivalent lime mixes. Air-entrained mixes also need careful manufacture and control of use since very high air contents or retempering (i.e. reworking of the mix usually after the addition of further water) can lead to very poor performance.

Pigment addition weakens mortar and the content should never exceed the limits quoted in section 4.2.3 (c) above. Polymer additives can markedly improve some properties, such as bond, flexural strength and impermeability, but they are costly and should only be used where there is a particular requirement.

Table 4.1 gives the common formulation of both contemporary mortars and the lime mortars likely to be encountered in most of the civil engineering structures built before 1920 where rebuilding has not taken place. The table gives some estimate of the performance of the mortars in terms of the compressive (cube) strength of mortar and brickwork since this is the most relevant factor to the design and performance of civil engineering structures such as columns and arches. The characteristic strength is quoted since it gives a better idea of the true behaviour than does the permissible stress which embodies an unquantified factor of safety. This has meant that some interpolation of the figures in Code of Practice CP 111 has been necessary since the lime mortars are not dealt with by current Codes of Practice. Other important performance factors are indicated as trends at the foot of the table but cannot generally be quantified. The brickwork strength values given are for standard format bricks but differ very little for squat concrete blocks and hollow concrete blocks. Solid concrete blocks with a ratio of height to least horizontal dimension of between 2 and 4 give approximately twice the strength in masonry. Dressed stone blocks can normally be designed as though they were solid concrete blocks of the same strength and similar shape.

4.4 Selection of mortars for repairs to civil engineering structures

Current recommendations for the specification of mortars for engineering brickwork (Table 4.2) give much stronger mixes with a higher cement content than the traditional lime mixes used in the past. These have the advantage of greater durability but do not tolerate so much movement without cracking. Larger self-contained structural repairs can be carried out with modern mortar formulations. Where repairs are of small areas of existing structures built with lime mortars, it is preferable to try to match the repair mortar strength and stiffness to the old mortar. This avoids problems

TABLE 4.2   Minimum recommended mortar designations for new works and large repairs

Exposure condition	Example applications	Clay bricks, normal salts	Clay bricks, low salts	Cal. sil. bricks	Concrete units
Internal	Internal walls of buildings	OPC (iv)	OPC (iv)	OPC (iv)	OPC (iv)
External sheltered	Main external walls of buildings	OPC (ii)	OPC (ii)	OPC (iii)	OPC (iii)
External exposed	Parapets of buildings and bridges, boundary walls, drained/waterproofed earth-retaining walls	SRPC (ii)	OPC (ii)	OPC (iii)	OPC (iii)
Buried (no frost)	Footings, linings to long tunnels	SRPC (ii)	SRPC or OPC (ii)	SRPC or OPC (ii)	SRPC or OPC (ii)
External in contact with ground	Bridge/tunnel linings, unwaterproofed earth-retaining walls	SRPC (i)	SRPC or OPC (i)	SRPC or OPC (ii)	SRPC or OPC (ii)
Extreme exposure	Cappings, copings and sills to external walls	SRPC (i)	OPC (i)	OPC (ii)	OPC (ii)
External in contact with moving water	Wet bridges, canals, surface water drains, culverts, manholes etc.	SRPC (i)	SRPC or OPC (i)	SRPC or OPC (ii)	SRPC or OPC (ii)
In contact with foul or acid water	Industrial sewers, drainage from peat or woodland	SRPC (i)	SRPC (i)	–[a]	SRPC (ii)

SRPC is necessary where leachable sulphates are present either in the units or in groundwater in contact with the masonry. Manufacturers' advice should be taken on the suitability of a particular unit for any application.
[a]Sandlime units are not normally recommended for foul drains.

caused by 'hardspots' in the structure that have different movement and impact characteristics which can lead to failures due to differential movement. Mortars with a low cement content can be made more durable and adherent by the addition of polymer latexes based on butadiene–styrene or styrene–acrylic copolymers.

4.5 References and further reading

British Standards Institution
BS 12: 1978. *Specification for Ordinary and Rapid Hardening Portland Cement.*
BS 1014: 1975 (1986). *Specification for Pigments for Portland Cement and Portland Cement Products.*
BS 1200: 1976. *Specification for Building Sands from Natural Sources. Sands for Mortar for Brickwork, Block Walling and Masonry.*
BS 4027: 1980. *Specification for Sulphate-resisting Portland Cement.*
BS 4551: 1980. *Methods of Testing Mortars, Screeds and Plasters.*
BS 4721: 1981 (1986). *Specification for Ready Mixed Building Mortars.*
BS 4887. *Mortar Admixtures.*
   Part 1: 1986. *Specification for Air-entraining (Plasticizing) Admixtures.*
   Part 2: 1987. *Specification for Set-retarding Admixtures.*
BS 5224: 1976. *Specification for Masonry Cement.*
BS 5628. *Code of Practice for the Use of Masonry.*
   Part 3: 1985. *Materials and Components, Design and Workmanship.*
British Ceramic Research Association (1985), *Special Publication 109, Achieving the Functional Requirements of Mortar.*
Building Research Establishment (1973) *Digest 160, Mortars for Bricklaying.*
Construction Industry Research and Information Association (1986) *Structural renovation of traditional buildings.* Report No. III, 3, 20–2, CIRIA, London.
National Building Studies (1950) *Bull. 8, Mortar.*
Water Research Laboratory (1986) *Inform. Guidance Note 4-10-01; Bricks and Mortar.*

PART TWO

Inspection, investigation and assessment

Inspection and recording

A.M. Sowden

Inspection is the basic means by which the Maintenance Engineer is kept aware of the condition of the assets entrusted to his care. Without a system of regular periodic examination of those assets he is effectively blindfold in the discharge of his responsibilities. Such a system is an integral part of the consideration of safety and serviceability in that it provides the link between the conditions to which a structure is subjected and the manner in which it performs (OECD 1976). Furthermore, its disciplined application is a pre-requisite to the economic timing, planning and implementation of remedial works and fully justifies its cost on those grounds. Rather than add to the overall expenditure on maintenance, it may well effect economies by demonstrating that defects are in fact stable or that their rate of development is such that repairs can be deferred; it is likely to aid the diagnosis of the cause of any trouble, so making possible more cost-effective repairs. Relatively inexpensive inspections carried out on a regular basis may, in the long term, lead to substantial cost savings, since, in certain circumstances, they may eliminate the need for other costly investigations. *Ad hoc* or crisis inspection is no substitute for a well managed inspection procedure.

5.1 Objectives

Although there are various types and levels of inspection, they all have the objectives of observing the current condition of assets and of providing factual information as a basis for the following:

1. affording assurance as to structural safety
2. providing for the economic management and control of operational serviceability by
   (a) identifying the need for preventive action
   (b) detecting incipient defects at an early stage
   (c) monitoring the development of those defects in order to determine the urgency for and the nature of corrective action
   (d) compiling quantitative records of deterioration on which to base maintenance planning
3. providing feedback to improve standards of durability in design and construction and of cost effectiveness in maintenance
4. providing data to assist in formulating future maintenance strategies
5. checking on changes in service conditions
6. ensuring continuing compliance with social and legal obligations and guarding against political or professional embarrassment

5.2 Scale and frequency

In terms of scale and frequency of inspection, most of the established systems comprise three broad categories, which are described below using the terminology of the OECD report on bridge inspection (1976).

The aim of **Superficial inspection** is to bring to early notice fairly obvious defects and deterioration which, if not dealt with, could escalate to a condition needing costly repairs. Examples include new cracks or water emission, fallen or displaced masonry, and signs of vandalism or impact by vehicles.

Special structural expertise or technical knowledge is not needed to observe such indications. Checks of this nature can be made on a casual or regular basis. It should be incumbent upon all the owner's employees who are familiar with a structure or who work in its vicinity to heed and report any developments which they consider untoward. Those who regularly patrol the road, railway or premises should make a cursory examination of structures as part of their task of overall surveillance. Depending on the length of the intervals between more detailed examinations, regular superficial examinations should be undertaken intermediately (preferably annually) by trained personnel, who should observe the structure from all viewpoints and levels without the aid of ladders but using binoculars where appropriate. They should report, in writing, 'by exception'. In particular, structures such as bridges should be observed under traffic, and any excessive vibration, deflection, looseness or variation in the width of cracks should be reported.

Principal inspections involve thorough examination by trained personnel, of all parts of a structure at prescribed intervals. The intervals may range from two to as much as ten years, depending upon the type of structure, its age and condition, and the difficulty of gaining access to it. The examination should be based on a prepared check list and made at close range (i.e. within touching distance). In the United King-

dom, Department of Transport recommendations for highway bridges (1983) subdivide principal inspections into two categories, referred to as general (at intervals not exceeding two years) and principal (at intervals not exceeding six years). The former are predominantly visual, with some recourse to simple monitoring equipment such as movement gauges and plumb lines. Principal inspections are more intensive, possibly involving the use of more sophisticated equipment; they may therefore necessitate the employment of technically qualified staff or specialist personnel with skills in accurate measurement or non-destructive testing.

British Rail relies primarily upon a single type of regular Detailed Inspection by trained artisan examiners at intervals normally not exceeding six years; depending upon what these reveal, decisions are made as to whether to embark on more technical monitoring or specialist investigations in a limited number of cases.

There are other variants, which should reflect the maintenance management policy of the undertaking concerned, but all have the objective of ensuring that every part of a structure is closely examined by suitably knowledgeable persons at least once during the period in which degradation could render any part unfit for its purpose.

Whatever maximum interval between inspections is prescribed, it is important that the responsible Maintenance Engineer is given absolute discretion and the resources to increase the frequency of examination for the whole or any part of individual structures which invoke concern, such as might be occasioned by their construction, their poor condition, their susceptibility to scour, mining settlement in the vicinity, persistent disregard of weight restrictions or vulnerability to vandalism.

For all principal inspections, detailed written reports must be prepared to record the condition of the structure, expressed in quantitative terms and illustrated by dimensioned sketches and/or photographs, as described in Section 5.7.

Special inspections are occasioned by some

unusual circumstance, such as exceptional loading or accidental impact, flooding or high seas, the discovery of a major weakness, or the need to reassess the capacity of a structure to cope with a change of use. They should be thorough and detailed, but may be limited in scope to particular critical elements. For example, after severe flooding, all substructures potentially affected should be specially examined for the effects of scour or for any accumulation of debris. If evidence is needed in support of such a policy, it is to be found in studies by Smith (1976) into 143 bridge failures dating from 1847; by far the largest single cause of these (in fact, of nearly half of them) was scour of foundations during flood conditions; 3 subsequent instances of the failure of masonry structures on British Rail have been similarly attributable. Should a specific abnormal defect be discovered in one structure and there is perceived to be a risk of its being repeated in others of similar construction, that risk should be investigated by examining them all.

5.3 Recruitment and training of examiners

The essential role of examiners is:

1. to observe and record the condition of each structure and, in particular, the changes in that condition which have occurred since the previous examination;
2. to transmit that information, clearly, accurately and promptly to the responsible engineer, so that he may use it as a reliable basis to decide on appropriate action.

At its simplest, therefore, an inspection team may comprise no more than a single examiner, who is basically an artisan, accompanied by an unskilled 'mate'; the latter assists with the necessary equipment, 'foots' the ladder and may act as 'look-out' on roads or railways. At the other end of the range, a qualified and experienced engineer may lead a team of skilled specialists, graduates and technicians. Such a team might well be necessary

for a Special Inspection or for a Principal Inspection of a complex or problematical structure, but would be unjustifiable for the great majority of cases, especially those in masonry. The inspection organization should certainly be supervised by a professional engineer and the ultimate decisions on remedial action should be made by an experienced Maintenance Engineer, who would normally initiate work only after making a personal visit of inspection; but the bulk of the routine examination can be most economically undertaken by artisan staff who have been trained for this task but whose other technical skills are limited. So long as the essential requirement is only for accurate factual reporting, rather than technical judgements or quantitative assessments of Condition Factors (see section 7.3.1), the prime attributes of a good examiner are that he should be:

1. healthy, physically fit, capable of climbing around structures and with a reasonably good head for heights;
2. conscientious, responsible and temperate;
3. self-motivated, interested and capable of working with minimal direct supervision;
4. observant and able to describe clearly, in writing and by sketches, what he observes.

On this premise, it is not necessary for those who undertake routine examinations to have any craft or technical skills at all, but, in the light of the arguments in section 1.2, as to the need to train complete craftsmen, it is desirable to recruit examiners from among skilled artisans who have the background of practical skills which will enable them to take advantage of subsequent promotional opportunities into the ranks of works supervisors. In fact, this is the usual source of recruitment. Undoubtedly, there is some merit in having masonry structures, for example, inspected by former bricklayers or masons, who therefore have some understanding of the characteristics of the materials with which they are dealing and sufficient appreciation of the significance of the defects they observe to be able to draw immediate attention to any which are in a critical condition or in need of early expert scrutiny. However, since so many structures are constructed of more than one basic material, it is more common to employ 'all-purpose' examiners.

There is an arguable case for examiners to have technical knowledge, such that they can make informed engineering judgements and actionable recommendations. A degree of technical expertise may well be necessary in the examination of complex structures or when decisions have to be taken regarding remedial action, but the great majority of structures need no more than to be kept under regular observation so that incipient defects are identified promptly and monitored. It would be extravagant for normal routine examinations to be undertaken by more highly qualified personnel than is necessary to provide such a basic screening service. Furthermore, staff of the calibre of incorporated engineer or higher might soon lose interest in routine reporting on structures in generally satisfactory condition.

Bearing in mind the precept that it is the rate of development of defects which is all important, there are obvious advantages in allowing examiners to acquire some familiarity, over a period of time, with the particular structures that they are required to cover. Too rapid promotion and too frequent turnover of personnel militate against this, but continuity must be balanced against the need to maintain enthusiasm and to provide career advancement opportunities which will attract staff with the right motivation.

If examiners are recruited from the ranks of artisans, there must be provision for prompt technical back-up as necessary, and, where special access arrangements have to be made for inspection, opportune advantage should be taken of these for appropriate technical involvement jointly with the examination. Inspection is an integral part of the maintenance function, but it is preferable for the examination role and the implementation of remedi measures to be kept as separate activities within the maintenance orga-

nization; each should be undertaken full-time by different personnel to ensure objectivity and avoid the predominance of personal interests. This should not preclude the interchange of staff and information between them. There is also merit in a system whereby a period of attachment to a structural examiner is included as part of the professional training of would-be chartered engineers – with potential advantage to both.

Whether examiners are drawn from the ranks of artisans, technicians or graduates, they will benefit from training to ensure the achievement of competent and consistent standards. (It was an untutored novice who is alleged to have reported on a window to the effect that four panes were 'cracked both sides', while two others were 'severely missing'!) Training should be aimed at instilling competency in the following:

1. how to plan and organize inspection activities
2. what to look for
3. how to record the results using standard terminology, unambiguous phraseology and informative sketches
4. how to measure and monitor defects
5. how to use cameras and any special equipment
6. how to operate in safety
7. what action to take in an emergency

To ensure an adequate supply of able examiners, British Rail have long considered it essential to mount regular formal training courses which incorporate practical work and conclude with an assessment of each trainee's ability and aptitude for the work. These courses are of up to eight weeks' duration, and comprise two weeks basic training, three weeks bridge and structures examination option, and three weeks building examination option. There is also provision for periodic revision and updating courses.

At the time (1976) when the OECD report on Bridge Inspection was compiled, few countries apart from France and the United States had formalized inspection training. Several recognized the need but relied on on-the-job training. However, to quote the OECD report, 'maintaining an efficient and competent bridge inspection force goes much beyond initial training or even on-the-job training. The whole inspection organisation must be of such a nature that it invites competent young people to include bridge inspection in their professional careers. This means pay scales and advancement opportunities above, or at least equal to that of other sub-divisions . . . (such as design, construction, etc.) must be established and maintained.' Such sentiments can only be strongly endorsed, even if the subdivision responsible for maintenance is, in the above quotation, apparently relegated to the category of 'etc.'!

5.4 The organization of a structural examination

Before commencing a principal or detailed inspection on site, the examiner must acquaint himself with the structure and any relevant background information about it which may be available in the records. He should study the last inspection report, so that he can direct particular attention to the condition of defects noted previously and he should ascertain whether there are any elements or features which need special observation or follow-up check measurements. In the case of a large or complicated structure, he should be supplied with a key plan identifying individual members in a consistent manner and possibly with outline drawings of critical elements. He must take action to deal with any problems of accessibility which might inhibit comprehensive examination with a proper regard for safety; there may be a need to arrange for the temporary closure of a road or a railway, for look-out men or traffic regulation, for the current in electrical conductors to be temporarily switched off, for lighting, for scaffolding to be erected or mobile access equipment to be supplied, for cladding panels or advertisement boards to be removed, for excessive vegetation growth or marine fouling to be cleared, for special arrangements to be made to secure entry into land owned or occupied by third parties, such as locked tenancies under arches, or for special precautions to be taken to permit safe entry into confined spaces. Unless he is already familiar with the structure, a preliminary site inspection in daylight is therefore a virtual necessity.

If he is thwarted in any way in completing an examination, he should draw special attention to this fact. It is incumbent upon the Maintenance Engineer to give whole-hearted support to his examiners in insisting upon the removal (wherever practicable, the permanent removal) of all impediments to proper examination. The random removal of individual cladding panels and the like may suffice if all the parts of the structure so revealed prove to be in good condition, but it is preferable for cladding and hoarding to be erected sufficiently clear of the structure to allow access behind them.

This may be an opportune point at which to discourage any ideas of avoiding the need to maintain redundant arches merely by infilling them. However thoroughly the filling material is compacted and even if it is finally grouted up to the underside of the vault, settlement will assuredly occur over a period of time, removing support from the structure and leaving voids too small to allow access for examination. Deterioration of the masonry can therefore proceed unchecked; in fact, on occasion, this has led to cases of unexpected settlement which could have had serious consequences. Avoidance of such an eventuality can be ensured only by breaking the arch as part of the abandonment operation. For similar reasons, 'blind' vaults should be opened up to permit regular examination or filled in.

To gain access to other parts which are normally obscured, the examiner must take advantage of any opportunity which may arise as a result of opening out or the erection of scaffolding for repairs, and periods of drought or exceptionally low tides which reveal scour or the condition of foundations. He must also plan the inspection operation in such a way as to minimize the interruptions which it may cause to traffic and the use of costly special equipment. To this end, he may be able to employ time-saving techniques such as photography and a tape recorder.

The examination itself should be undertaken systematically, preferably following an established routine, to ensure that no part of the structure is missed. Observations should be recorded in a field book or on a tape recorder as soon as they are made, never trusting to memory, and should be promptly transferred to the appropriate record form. Defects should be described and, wherever possible, indicative dimensions should be quoted to allow comparison with those noted in previous and subsequent years, to enable their development to be monitored.

In the particular cases of very high structures or parts in deep water, it may be unreasonable to expect either the examiners or the Maintenance Engineer to inspect these personally, but that does not afford an excuse for neglecting them. Assistance should be enlisted from specialist steeplejacks or divers, but, however competent they may appear to be, it would be unwise to rely upon their having the necessary skills in examining and reporting unless closely supervised or to expect them to make the engineering judgements which are the engineer's responsibility. They must therefore be very carefully briefed as to what they are to look for and how they are to report on what they find. The areas to be examined should be broken down into elements, to be dealt with in a pre-planned sequence, and there should be a clear understanding as to how defects are to be identified, described, measured and recorded. Good communications between the responsible engineer and the person making the examination are critical and should normally be continuously maintained by telephone or radio link. Photography is certainly a desirable adjunct, while closed circuit television and video recording may well justify their cost, insofar as they provide something approaching first hand observation even from a remote position, as well as a permanent record. It is obvious folly to schedule such inspections other than at times when weather and

temperature conditions, water levels and visibility are likely to be at their most favourable.

The statistics quoted above highlight the importance of underwater surveys covering not only the condition of waterfront structures but also the extent of scour in their vicinity. Bed levels may change quickly due to current or wave scour or to littoral drift and it is essential to be alive to the possibility that such structures may be put at risk thereby. In such circumstances, Special Inspections of vulnerable assets assume particular importance. Under water, poor visibility is likely to make structural examination more difficult, despite the use of lamps, and it is often possible to work only by touch. Planning then needs even greater attention than for work in clear water. Rather than make direct measurements, it may be better to transfer shapes to templates and sizes to marks on a rod or to calipers, which can be accurately scaled above water level. In critical cases, it will be necessary to dam off and lower water levels by pumping; this may be the only way of gaining access to some culverts.

In short, an attitude which neglects parts difficult to examine properly is wholly unacceptable.

5.5 The observation and measurement of defects

The common indications of actual or potential distress in masonry include the following.

Cracks running through the bricks/stones and the mortar joints between them may be seen. The examination record should note their position and orientation (e.g. vertical, horizontal or diagonal; parallel, normal or oblique to the axis of an arch or tunnel), their length and their displacement (i.e. the amount by which their sides have moved apart). Displacement should be measured on three axes, to record movement vertically, sideways and forwards. If a crack has opened more at one end than at the other, this is a useful pointer as to its cause and therefore should be noted. To assist in the observation of any future growth in the length of a crack, its current extent can be indicated by a small mark in waxed chalk or paint at each end. Cracks in the road surface over a bridge or in the ground behind a wall are equally relevant signs.

Distortion of regular shape, **misalignment** and **tilting** usually indicate that movement has occurred without necessarily giving rise to cracking because of the inherent flexibility imparted by the mortar joints. However, the distortion could have occurred during, or immediately following, construction; to determine whether it is stable or progressive, measurements of the amount of distortion should be taken periodically from a fixed point or base line.

Bulging and **drumminess** (a dull hollow sound, as opposed to a 'ring' when struck with a light hammer) are normally indicative of separation of the face from the masonry behind it. The amount of forward movement at the bulge and the extent of the area affected should be measured using a line or plumb bob or possibly a microlevel (see Section 5.6).

Undue movement under live load is often discernible, but accurate measurement of it is beyond the scope of the examiner, who can only rely on a subjective view as to its severity. It may be possible to gain a rough idea of its magnitude by observing the movement relative to a fixed point, such as a mark on the stile of a ladder. (It should be noted, however, that more accurate measurement of the crown deflection of a masonry arch is of dubious value, whether expressed in absolute terms or as a fraction of the span. It will be affected by any settlement or spreading of the abutments under load. It will also depend upon the modulus of elasticity of the masonry, which can be expected to vary widely between structures and possibly within the same structure. It will be influenced by the degree of saturation and so may vary with changing weather conditions. Indeed, the deflection: span ratio of one bridge, under identical test loading, was found to vary between 24×10^{-5} and 52×10^{-5} during the course of a single winter. In such

circumstances, it is apparent that absolute values of deflection provide an insensitive yardstick of condition and their measurement is unlikely to afford reliable criteria of serviceability. Nevertheless, in default of better criteria, engineers tend to use such values empirically as indicators of condition.) From the examiner's point of view, the relative movement which he can detect by placing his hand across a fracture is likely to be more significant.

Mortar loss is evidenced by deep open joints between masonry units or by loose or displaced units. Individual stones in an arch may have dropped to a position where they protrude from the soffit; sometimes loose bricks in a wall move back when struck with a hammer. The depth from the face to which mortar is missing or perished should be recorded, as should the location and extent of the area affected. (NB Loose masonry units in an arch soffit are potentially hazardous if there is a risk of them falling out, and so it may be necessary for the examiner to forestall this by temporary measures such as driving small timber wedges into the joints.)

The depth and area affected by **spalled** or **eroded** surfaces should be quoted.

Water percolation through the masonry cannot easily be measured, but an indication can be given by a broad classification such as 'damp' (discoloration of the surface, moist to the touch, but not dripping), 'wet' (drops falling regularly) or 'running' (a trickle or jet of water). A report of water percolation should also give information as to the recent incidence of rainfall, as this may indicate whether the source of the water is natural or is a leaking water pipe or sewer.

A limited amount of information on **erosion of foundations in water, undermining or scour** is obtainable in shallow water by using a probe and by measuring depths of water and soft silt with a graduated rod. However, generally, if trouble of this nature is suspected as a result of routine examination, a survey using specialized staff or equipment is likely to be necessary to determine its extent.

Some authorities require defects to be assessed as to their severity and extent; for example, the Department of Transport Bridge Inspection Guide (1983) suggests alpha-numeric scales as follows.

Extent
A. No significant defect.
B. Slight, not more than 5% affected (of area, length, etc.).
C. Moderate, 5% to 20% affected.
D. Extensive, over 20% affected.

Severity
1. No significant defects.
2. Minor defects of a non-urgent nature.
3. Defects of an unacceptable nature which should be included for attention within the next two annual maintenance programmes.
4. Severe defects where action is needed (these should be reported immediately to the engineer) within the next financial year.

The G, F and P ratings on the Department of Transport Report Form (Fig.5.6) are regarded as corresponding to categories 1, 2 and 3 respectively on the above Severity scale.

In France, a series of documents has been issued, each of which is an illustrated catalogue of defects occurring in a particular type of structure, with the aim of achieving greater uniformity in the designation of observed defects and of classifying them according to their degree of severity. In the document dealing with masonry structures (Ministère des Transports, 1982), five indices of severity are defined as follows:

B. defects without important consequences other than aesthetic
C. defects which indicate the risk of abnormal development
D. defects which indicate developing deterioration
E. defects which clearly show a change in the behaviour of the structure and which bring its durability into question
F. defects which indicate the approach to a limit

state, necessitating a restriction of use or removal from service.

Such assessments of severity do, of course, involve a significant element of judgement, calling for a degree of technical knowledge which may be beyond the type of examiner with which we have mainly been concerned in Section 5.3. Indeed, it can be argued that this judgement cannot always be made on purely visual evidence and often should only be made by an experienced engineer in the light of all the information available from inspections, investigations and records.

It is indisputable, however, that every maintenance organisation needs some kind of severity rating system which enables items of remedial work to be scheduled in priority order, so that limited resources can be methodically allocated to the most demanding cases. It can be relatively simple, perhaps no more than an indication of the need for action urgently ('this year'), 'next year', 'sometime' before the next examination or 'never' (or, at least not before further examination and reconsideration in x years' time). Where the organization is of any size, such a system should be capable of computerization.

5.6 Equipment

The examiner engaged on routine inspections of brick and stone masonry needs only basically simple and unsophisticated equipment with which to make the required observations and measurements. It is preferable that he should not be encumbered by heavy or damage-prone devices. Recourse to more complex instruments and advanced techniques should be left to technical staff or specialists engaged on follow-up investigations.

Thus, in the case of masonry structures the examiner should normally be able to rely on little more than a testing hammer and a probe, a rule and a line. However, he may first have to establish reference points by fixing non-corrodible pins in the face of the masonry or by attaching 'tell-tales'

FIGURE 5.1 Means of monitoring cracking: mortar pad. (*Courtesy of British Rail.*)

accurately machined reference face on to which the level fits, but it is expensive to install.

A simpler approach being developed by staff at British Rail is to form pads of rapid-setting material which can bond to any surface of concrete, rough stone or metal in wet or dry conditions. Called Clinopads, they have formulations of lightweight thixotropic resin or hydraulic cement base which nevertheless permit a precise planar impression to be made to house a demountable, pocket sized microlevel without the likelihood of loss of particles of filler, cement etc. which would alter the accurate reading. Depending upon the microlevel used, precision tests have indicated an accuracy of 0.02°. The Clinopad is unobtrusive and cheap to replace in case of vandalism, and several can be installed in an hour. It is stable over years so that long-term movements can be monitored.

It used to be common practice to monitor the growth of cracks by applying pads of cement mortar to the surface of the masonry across the crack at suitable locations, with the date of so doing being scribed on each pad (Fig. 5.1). Further development of the crack fractured the pad and the resulting gap could be measured; reasonable accuracy was achievable by using a simple crack-width gauge (Fig. 5.2). Not infrequently, however, the part of the pad on one side of the crack became detached, thus precluding further measurements. The more recently developed Avongard tell-tale consists of two matching plates of acrylic plastic which are fixed across a crack so that they overlap for part of their length (Fig. 5.3); on one plate is a calibrated grid, while the other is transparent and marked with a hairline cursor. As the crack opens or closes, the two plates move relative to each other and the extent of the movement can be measured by the movement of the cursor over the grid. The latest technique involves installing small stainless steel discs or brass screws on each side of the crack and using vernier, dial or digital electronic calipers to measure the distance between them (Figs 5.4 and 5.5). It requires no great skill to achieve an

to it, in order to ensure repeatability of readings.

To measure bulging, tilting and loss of verticality, he could use a string-line and some convenient heavy object to make a plumb bob. (A useful tip for stabilising a plumb line in windy conditions is to suspend the bob in a bucket of water). Alternatively, the recent availability of low cost electronic microlevels that give a digital display of the angle of a horizontal or a vertical surface to a claimed accuracy of 0.5°, 0.1° or even 0.01° has led to their use in monitoring structures. One system involves the use of a metal frame which is fixed to a structure by means of stainless steel screws bonded into holes drilled into the masonry. This special frame provides an

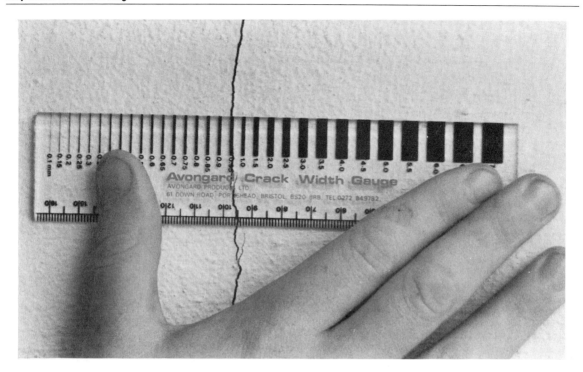

FIGURE 5.2 Means of monitoring cracking: crack-width gauge. (*Courtesy of Avongard.*)

accuracy of 0.02 mm with this system. Furthermore, the only elements permanently fixed to the structure, the discs or screws, are sufficiently unobtrusive to be markedly less prone to vandalism than any form of tell-tale.

The next steps in the monitoring of movement are usually beyond the scope of artisan examiners and are undertaken by technical staff. They may involve the measurement of settlement and/or distortion using accurate surveying techniques. For determining the distortion of an arch profile, the use of a steel tape is unlikely to achieve sufficient accuracy; it may be necessary to use a tensioned Invar wire Distometer or a photographic or laser measuring technique. These and other more refined and specialized methods of investigation, measurement and non-destructive testing are described in Chapter 6. At this point, however, it may be opportune to stress the value of plotting all measurements on a time base to facilitate the identification of trends and changes

in the rate of movement.

Returning to the requirements for routine examination, there is sometimes a need for special equipment to enable the examiner to reach, in safety, the parts of the structure he has to inspect and this should always be readily available to him. Examples include scaffolding or a mobile access platform (see Section 5.8), safety harness and line, boat, lighting, protective clothing, head/eye protection, waders and safety equipment for entry into confined spaces (see Section 5.9.4). For underwater work, there is need for rods to check for scour, scrapers to remove marine growth, probes for investigating cavities, lights, measuring devices and means of communication (see Section 5.4). Where visibility permits, a diver's slate is useful for making notes and sketches, and where it is poor, a plastic bag can be used to provide a clear water lens to facilitate the close inspection of surfaces and irregular shapes.

FIGURE 5.3 Means of monitoring cracking: plastic tell-tale. (*Courtesy of Avongard.*)

5.7 The use of report forms

Examiners must compile fully detailed written reports as a permanent record of what they observe. To facilitate this, most major owners of structures use specially devised forms intended to encourage clear, concise and comprehensive coverage. That used for motorway and trunk road bridges in England and Wales is the Department of Transport's Inspection Report Form BE 11/86 (Fig. 5.6).

From the experience of many decades British Rail have evolved the Bridge and Structure Examination Report Form shown in Fig. 5.7. Like the Department of Transport form, this lists the elements generally found in structures, against which the examiner is required to give a broad indication of the condition of each part by recording G (good), F (fair) or P (poor). Even

with a separate schedule which records a further 60 items for inclusion as appropriate in the blank spaces under 'Name of part', the list is not exhaustive, but it constitutes a check list which helps the examiner. Similarly, the rating assessments by the examiner are not regarded as conclusive, but they help the recipient to pinpoint the apparent worst features of the structure. Ample space is left for detailed descriptions and sketches or photographs of defects, and these are obligatory for every part rated P; continuation sheets are used as necessary. There is space for general comments and for drawing special attention to any critical feature or significant new development. There is also provision for follow-up recommendations/comments to be made by the examiner's supervisor and for decisions/instructions for action to be added by the Maintenance Engineer; these instructions may stipulate a

FIGURE 5.4 Means of monitoring cracking: dial caliper used between locating holes in targets. (*Courtesy of Avongard.*)

reduced interval before the next routine examination, call for a further inspection by more technically qualified staff or require an investigation by specialists.

Whatever the layout of the report form, its whole emphasis must be on recording the nature and extent of all defects in a quantitative way (see Section 5.5) so that changes can be monitored. Sketches should be approximately to scale and dimensioned; photographs should incorporate a reference scale and, to identify the location, reference marks can be chalked on the structure in a position where they will appear in the photograph. (Such marks and any used to highlight defects are unsightly and so should subsequently be rubbed off.) A note should be included in the report as to the condition of any maintenance work previously undertaken, since this will afford an indication of its effectiveness and hence give

guidance as to materials and techniques for dealing with similar defects in the future.

Forms such as those illustrated in Figs 5.6 and 5.7 are not, of course, suitable for all types of structure. Tunnels, for example, need a type of report which allows defects to be plotted on a developed plan of the intrados – a kind of 'worm's eye view'. For this purpose, it is customary to divide the tunnel itself longitudinally into sections of about 10 m (indicated by markers fixed to the tunnel wall) and peripherally into at least five sectors (crown, two haunches and two side-walls). The report then takes the form of a grid in which each rectangle represents one sector 10 m long to a suitable scale (1/100 to 1/200) and the defects found in that area are plotted diagrammatically thereon in their correct relative positions.

Other types of structure may need other vari-

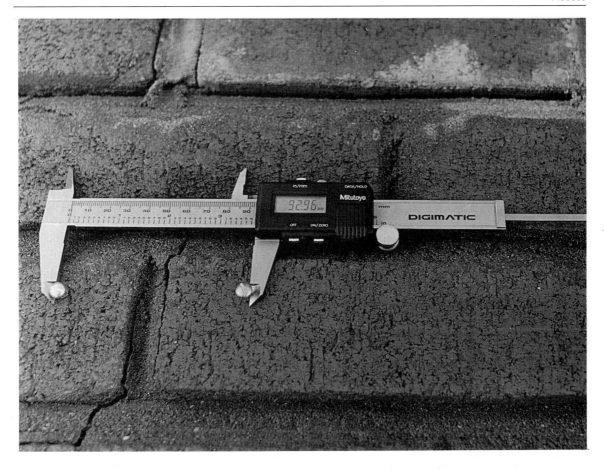

FIGURE 5.5 Digital electronic caliper used between brass screws. (*Courtesy of Avongard.*)

ants; the precise form of the report is important only in so far as it encourages clear, comprehensive, accurate and informative recording.

5.8 Access

Access to masonry structures for the purpose of detailed inspection may present particular difficulties because of the scarcity of adequate ledges and places where a man can stand on the structure or which can be used to support a ladder; indeed, many structures such as viaducts are so high as to be beyond the reach of ordinary ladders, while fixed ladders are a rarity. Scaffolding, whether built up from the ground or suspended, is seldom economic if erected solely for examination

purposes and not used jointly for other major operations.

There is therefore considerable potential for the use of mobile platforms (Figs 5.8–5.19) ranging from simple ones which extend only in an upward vertical direction (such as those of the scissors lift type) to self-propelled hydraulic machines with booms which are articulated, telescopic and capable of rotational movement. Although there is a definite place for the former type, most of them have relatively limited scope, whereas the latter are available over a wide range of capabilities (up to over 60 m working height and more than 25 m out-reach) and all have ample payloads for inspection purposes. Upward-operating hydraulic units are common as they can be used for so many other

Department of Transport

BE11/86

Trunk Road / Motorway Structure Inspection Report

Structure No. `_ _ _ / _ _ _ _ _ _ _ _ _ _ / _ _ _ . _ _ / _ _ _`

Grid Ref. `_ _ _ _ _   _ _ _ _ _`

Jnno   Road No.   R.Type   Kilom.   Spurl/r   E   N

Agent Authority

Structure name

Date of Inspection `_ _ . _ _ _ . _ _ _ _`

Type of Inspection *(Please tick)*   G ☐   P ☐   S ☐   Inspected by

(eg. 15 - JUN - 1987)

Defect Assessment

	Estimated Cost	Extent	Severity	Work recc. & Priority	Comments
1. Foundations					
2. Inverts or Aprons					
3. Fenders					
4. Piers or Columns					
5. Abutments					
6. Wing walls					
7. Retaining walls or Revetments					
8. Approach Embankments					
9. Bearings					
10. Main beams / Tunnel portals / Mast					
11. Transverse beams / Catenary cables					
12. Diaphragms or bracings					
13. Concrete Slab					
14. Metal deck plates / Tunnel linings					

Defect Assessment (cont.)

BE11/86

	Estimated Cost	Extent	Severity	Work recc. & Priority	Comments
15. Jack arches					
16. Arch ring / Armco					
17. Spandrels					
18. Tie rods					
19. Drainage systems					
20. Waterproofing					
21. Surfacing					
22. Service Ducts					
23. Expansion Joints					
24. Parapets / Handrails					
25. Access gantries or walkways					
26. Machinery					

Reasons for priority allocation

FIGURE 5.6 Department of Transport Inspection Report Form. (*Courtesy of Department of Transport.*)

FIGURE 5.7 (*facing page*) British Rail Bridge and Structure Examination Report Form. (*Courtesy of British Rail.*)

.... Region
.... Area
Line ....
Stns. between ....

Type of Over/Under/Bridge/Structure
....
Carrying ....
Over ....

Bridge No. .... No. of spans ....
Name ....
....
At .... (Mileage)
Route Code ....

Name of Part	G-good F-fair P-poor	Remarks (Refer to parts by name) Sheet of
Main Girders		
Cross Girders		
Rail Bearers		
Floor		
Rivets & Bolts		
Arch Ring		
Spandrels		
Abutments		
Piers		
Wing & Retaining walls		
Pointing		
Parapets & Pilasters		
Columns & Cylinders		
Trestles & Crossheads		
Bedstones & Cills		
Bearings		
Ballast plates/Boards		
Longitudinal timbers		
Waterproofing		
Drainage		
Gutters & Downpipes		
Handrails		
Painting		
Track/Road Condition		

General Comments

Tick as appropriate	
Change of construction	
Closed line	
C.W.R.	
Rail Joints	
25I Axle/Abnormal Rd., Loads	
Weight Restriction Plates	
Inaccessible Parts	
Tell Tales	
Plumbing Points	

Examined .... (Examiner) .... (Date)   Date of previous detailed examination

Recommendations   Action   Next detailed examination due

Signed .... (Date)   .... A.C.E. .... (Date)

FIGURE 5.8 Rail-mounted access platforms: the forerunner (c.1900). (*Courtesy of British Rail.*)

access purposes, but the fact that they need a firm and reasonably level area from which to work creates a demand for similar machines able to work in the downwards mode.

British Rail, with responsibility for nearly 80,000 bridge spans, pioneered the 'goes-under' type in 1957 with a rail-mounted version and since then have had up to five units virtually fully occupied at any one time. The way in which the design has evolved shows what can be achieved by collaboration between a purposeful maintenance organization and enterprising commercial firms.

After initial in-house experiments in the early 1950s with a gantry suspended from a portal frame spanning the tracks, the first successful device was fundamentally a folding two-arm unit,

built by Simon Engineering from standard components, which was mounted on a flat rail wagon and inverted to operate in the downwards direction to a depth of 29 ft (8.8 m) below rail level, with an under-reach of 15 ft (4.6 m) (Fig. 5.10). For the second-generation machine, an American-built TU 37 unit was adapted by Elstree Plant Ltd (now EPL International) to fit onto a well-type bogie wagon; with a gross weight of 57 tonnes, it had an additional turntable positioned at the end of a slewing arm, thus greatly increasing articulation, while a telescoping outer boom extended its working range to 50 ft (15.2 m) vertically downwards and a maximum working under-reach of 40 ft (12.2 m) (Fig. 5.11).

The next development, in conjunction with

FIGURE 5.9 Rail-mounted access platforms: scissors lift platform. (*Courtesy of Permaquip.*)

Armfield Engineering Ltd, was a major breakthrough – a 'universal' machine, capable of operating in both upward and downward modes (Fig. 5.12). To achieve maximum extension, it was designed up to the capacity limit of British Rail's largest well-type bogie wagon and, with the counterbalance necessary to ensure stability, had a gross weight of 100 tonnes. However, the weight of the articulated structure was kept below 9 tonnes by using thin-walled aluminium box sections. The working envelope of this unit is shown in Fig. 5.13: vertically it extends to 24 m upwards and 16 m downwards, while in the upward configuration the horizontal out-reach is some 17 m

and in the downward one the under-reach is sufficient to cover a soffit width of up to 17 m. The slewing boom is capable of clearing a bridge parapet or other lineside obstruction up to 1.8 m above rail level and permits the operation of the articulating booms without the need to displace overhead electrification wires. It can thus be used for access to bridges over and under the tracks, as well as to station roofs, retaining walls and other lineside structures.

The latest model (Fig. 5.14), by Wickham Engineering, sacrifices the extremes of reach in favour of improved versatility, in that the articulated structure (based on the Topper hydraulic platform) is mounted on a lorry chassis which is capable of travelling and working on both road and rail and is self-propelled. It retains the upwards and downwards facility, with the working envelope shown in Fig. 5.15.

Each of these British Rail machines has had a working platform capable of accommodating three men or 600 lb (272 kg), and therefore of allowing either inspection or minor repair work. Railway track occupation for such activities must be confined to a single track and for only short periods, generally at night or at weekends. The maximum utilization must be achieved irrespective of the weather or other conditions prevailing at the time. Mobile access units currently used by British Rail are therefore required to go into service within minutes of arriving on site; they must be capable of reliable and safe operation in subfreezing temperatures and with wind speeds of up to 40 m.p.h. track on a wheel base with adverse super-elevation (cant) up to 6° without the employment of outriggers or rail clamps (as these would inhibit mobility). At the work platform there has to be a power supply and floodlighting as well as a console to control all movements; the latter is duplicated on the vehicle deck. The unit's power source and electrohydraulic services have to be extensively duplicated as a precaution against failure; electrical cut-outs and mechanical stops prevent encroachment into the structure clearances of adjoi-

FIGURE 5.10 Rail-mounted access platforms: Simon unit (1957). (*Courtesy of British Rail.*)

ning tracks/traffic lanes. It must be possible for the booms to be retracted in the event of a malfunction and for the operators on the working platform to be recovered even if totally incapacitated. These measures are precautionary; only safety is rated above reliability. In the stowed position, the machines must be within the railway loading gauge, able to travel at speed between working sites and able to withstand the effects of rough shunting.

For structures under or alongside roads, it seems that few owner authorities consider that they have sufficient demand to justify the purchase of special downward-operating machines, but an increasing variety of models is currently available on hire, to the extent that inspection of even the

most inaccessible parts of such structures need no longer present any significant problem. For example, there is a truck-mounted version of the well-tried TU 37 unit (Fig. 5.16) with a downward reach of 14.6 m, an under-reach of 12.2 m and a three-man cage capacity. A German-made (Cramer) unit available in the United Kingdom carries a telescopic platform which extends from 12.7 to 20 m under a bridge and to 10.6 m below road level; it has a maximum uniformly distributed payload of 900 kg or it can support a load of 450 kg at the tip of its telescopic arm without the use of outriggers (Fig. 5.17).

A simpler and less sophisticated unit (the Bridge Inspector by Pressings (Metal) Ltd) provides access to bridge soffits at a cost well below

FIGURE 5.11 Rail-mounted access platforms: Elstree Plant Ltd TU 37 unit (1974). (*Courtesy of British Rail.*)

that of conventional hydraulic machines. Trailer mounted and with a gross weight of less than 4 tonnes, it can be used on load-restricted footways, albeit with stabilizing jacks. As a cantilever supporting one man it has an under-reach of 13.2 m (Fig. 5.18), while twin units used together as in Fig. 5.19 can cope with bridges some 27 m wide.

Even simpler devices such as slung platforms, boatswain's chairs and suspended cradles find application in particular cases including retaining walls and tunnel shafts. There is also scope for the technique of abseiling, which is possibly the least demanding of metal equipment, if not of mettle!

5.9 Safety

Considerations of safety pervade every aspect of the Maintenance Engineer's function and this fact might justify a separate chapter covering the application of precautions generally to all activities. However, safety is not something to be regarded as separate or additional, superimposed on the expeditious accomplishment of a task; it is an integral part of every task. It is not the icing on the cake, but an essential part of the cake recipe. Safety in inspection operations will therefore be discussed here (but much of what follows is of general relevance to all maintenance activities).

FIGURE 5.12 Rail-mounted access platforms: Armfield Universal unit (1980). (*Courtesy of British Rail.*)

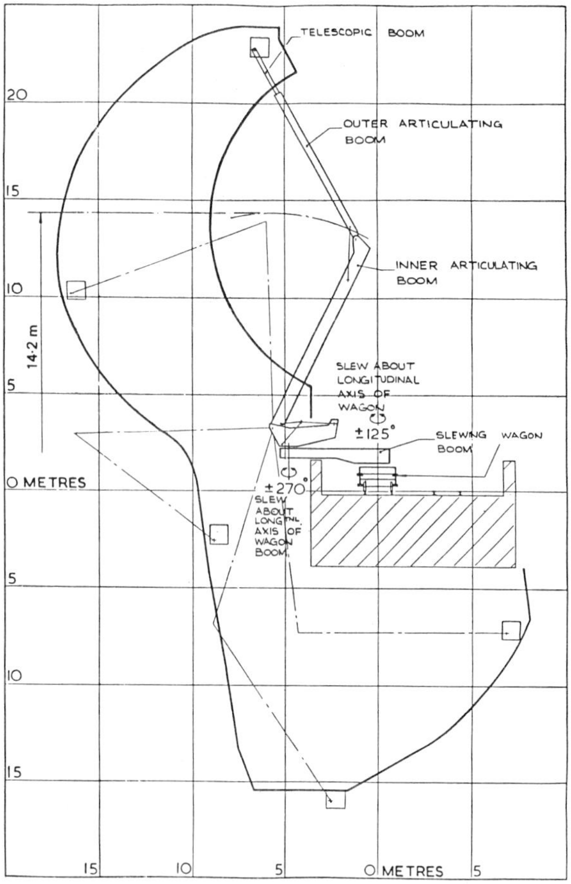

FIGURE 5.13 Rail-mounted access platforms: working envelope of the Armfield Universal unit. (*Courtesy of Armfield Engineering.*)

The nature of the examiner's task and the manner in which he has to undertake it make it potentially hazardous. It is vital that he should be afforded every assistance to overcome the hazards and that he should be temperamentally averse to taking risks. He must be supplied with the appropriate equipment, which must be maintained in good condition; he must be trained in its use and in safe systems of working; he must be imbued with a safety awareness, in respect both of his own activities and of their possible effect on others.

It is perhaps a mixed blessing that so much of his work involves the use of ladders. Research has shown (Shimmin *et al.*, 1980) that workers' subjective perceptions of risk situations rate work on ladders as the most hazardous of construction operations; they may therefore be disposed to exercise more care in this activity. In fact, for whatever reason, on the basis of accident statistics it ranks relatively low among risk situations. Nevertheless, the rarity of accidents to examiners should not be allowed to breed complacency. The requirement that all parts of a structure should be observed from within touching distance does pose some problems of safe access from a ladder,

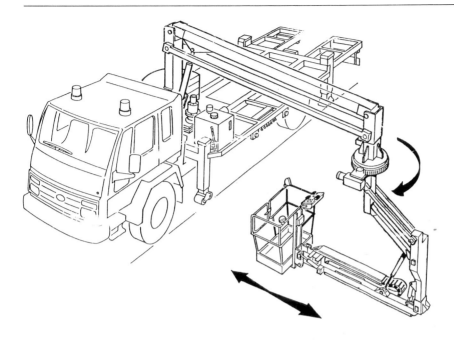

FIGURE 5.14 Road vehicle mounted access platform: Wickham unit (1989). (*Courtesy of Wickham Engineering Co.*)

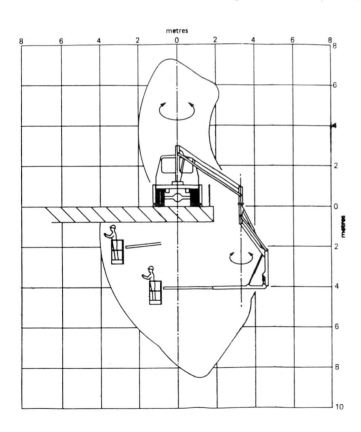

FIGURE 5.15 Road vehicle mounted access platform: working envelope of the Wickham unit. (*Courtesy of Wickham Engineering Co.*)

particularly in the case of masonry arches. One obvious and simple contribution to countering that problem is to fit the examiner's hammer with a long handle (at least a metre long), to increase his reach. At the other end of the scale, the more widespread use of mobile access units (see Section 5.8) is advocated.

At the head of the relevant statutory requirements stands the Health and Safety at Work Act, 1974.

This places a duty on

1. the employer, to ensure the health, safety and welfare of his employees, by providing and maintaining suitable workplaces, plant, systems of work, working environment and training and also to ensure that the health and safety of persons other than his employees are not put at risk by his activities
2. those in control of premises, to ensure that such premises and the access to them are safe and without risk to the health of anyone working there
3. the suppliers and installers of articles for use at work, to ensure that those articles are safe and without risks to health when properly used, and also to carry out such tests and examinations to ensure that they remain so and to make available adequate information as to their proper use
4. the employee, to take care for the health and safety of himself and anyone else who may be affected by his acts or omissions at work and to cooperate with his employer in enabling the latter to comply with his statutory duties.

The Act therefore imposes responsibilities upon a structure's owner (as employer and controller of premises), upon the Maintenance Engineer (as the owner's agent and employee), upon providers of equipment (scaffolding, access platforms, investigatory apparatus etc.) and upon examiners (as employees). It applies the criminal law to health and safety at work, so that those who fail to discharge the duties it imposes on them are effectively criminals.

It is couched in somewhat general terms however; for more specific guidance we need to refer to the Factories Act of 1961, especially, in the current context, to Part VII, Section 127 (Works of Building and Engineering Construction), and also to the Construction Regulations (Statutory Instruments). Among the latter those most relevant to the inspection of masonry structures are as follows.

S.I. 94 The Construction (Working Places) Regulations 1966, in respect of ladders, scaffolds, working platforms and gangways, boatswain's chairs, safety belts etc.

S.I. 95 The Construction (Health and Welfare) Regulations 1966, in respect of first aid boxes, shelter and protective clothing, washing facilities etc.

S.I. 399 The Diving Operations at Work Regulations 1981.

S.I. 1580 The Construction (General Provisions) Regulations 1961, in respect of supervision of safe conduct of work, ventilation, work adjacent to water, clearances, electrically charged cables or apparatus, lighting, construction of temporary structures etc.

S.I. 1581 The Construction (Lifting Operations) Regulations 1961, in respect of hoists, lifting appliances, suspended scaffolds etc.

Of particular relevance are the regulations and precautions applicable to the following.

FIGURE 5.16 Road vehicle mounted access platform: Elstree Plant Ltd TU 37 unit. (*Courtesy of EPL International.*)

5.9.1 Ladders

Ladders have to be of good construction, of suitable and sound material, and of adequate strength. They must be properly maintained in good repair and with no part missing, defective or inadequately fixed. It is recommended that every independent ladder (i.e. not fixed to a structure or forming part of a scaffold) be inspected by a competent person at six-monthly intervals and that a proper register of such inspections and of any repairs effected be kept. Additionally, the examiner should check his own ladder daily. (The term 'competent person' is not defined in the legislation, but the definition given in BS 8210, as quoted in the Glossary, should prove to be adequate and acceptable).

A ladder more than 10 ft long must not be used unless

1. it is equally and properly supported on each stile, on a level and firm footing, and is not standing on loose packing
2. it is securely fixed at or near its lower end or near to its upper resting place (a requirement once incorrectly paraphrased by an over-zealous safety officer as 'No one shall climb a ladder until the top has been tied'!) – alternatively, a person can be stationed at the foot of the ladder to prevent it slipping
3. it is secured where necessary to prevent undue swaying or sagging.

FIGURE 5.17 Road vehicle mounted access platform: Cramer unit. (*Courtesy of EPL International.*)

Long runs of ladders should be provided with intermediate landing places not more than 30 ft apart (measured vertically) and every landing place should be provided with guard rails. Particular care is needed in the vicinity of overhead or third-rail electrical conductors, especially when a metal ladder is being carried.

Ladders permanently fixed to structures must comply with BS 4211: 1967 and be provided with back hoops and rest platforms as appropriate. On tall ladders where, for any reason, back hoops are impracticable, consideration should be given to installing a fall-arrest system complying with BS 5062 and issuing the examiner with a safety harness and clips so that he can take advantage of it.

5.9.2 Scaffolds

Whatever the type of scaffold – static, suspended or mobile – in order to comply with S.I. 94 it must be of sound construction and erected under the immediate supervision of a competent person and by competent workmen. It must be properly maintained; to this end, it must be examined by a competent person at intervals not exceeding seven days and after exposure to adverse weather conditions, and a report must be made, in a specified form, of every inspection. An examiner seeking to take advantage of a scaffold erected for other purposes would be wise to consult these inspection reports to confirm compliance with the requirements of the regulations and so assure himself, as far as possible, of the adequacy of the scaffold before using it.

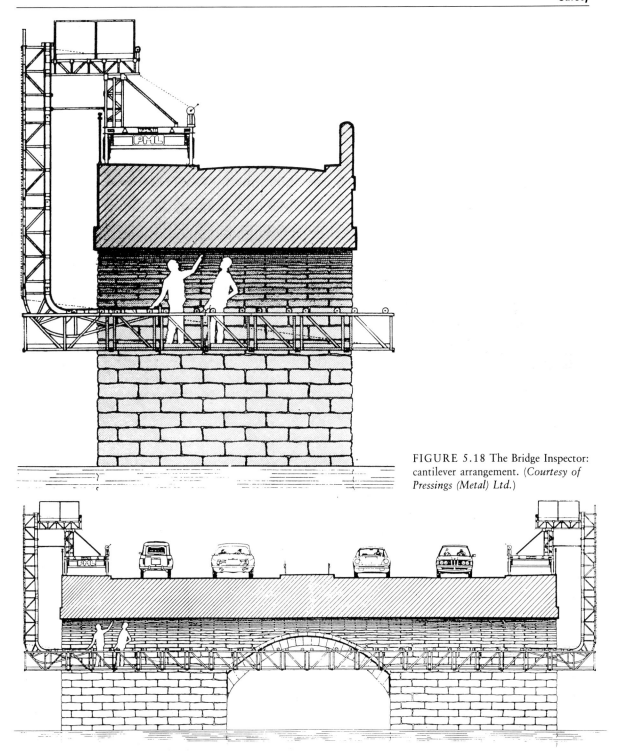

FIGURE 5.18 The Bridge Inspector: cantilever arrangement. (*Courtesy of Pressings (Metal) Ltd.*)

FIGURE 5.19 The Bridge Inspector: twin unit arrangement. (*Courtesy of Pressings (Metal) Ltd.*)

5.9.3 Traffic

Many of the areas requiring examination are over, under or adjacent to roads and/or railways and cannot be dealt with satisfactorily unless precautions are taken in respect of passing traffic. In all cases high visibility clothing should be worn. It may be possible to regulate road traffic by marshalling, coning off or diversion so that it passes well clear of the working space. This is best done in association with the police authorities. The appropriate signs, their siting and lighting are covered by the Department of Transport's *Traffic Signs Manual*, Chapter 8.

Adjacent to operational railways, it may be possible to undertake a limited amount of inspection in the intervals between passing trains, but only under the protection of a qualified look-out man. More commonly, it is necessary to arrange for the temporary interruption of railway traffic by taking 'possession' or 'occupation' of the relevant track(s), including the 'current isolation' of overhead or third-rail electrical conductors. This is usually possible only for limited periods, often at night, and must be organized by arrangement with the railway authority weeks or even months in advance; permission is granted only subject to strictly disciplined procedures.

5.9.4 Confined spaces

Confined spaces are relatively common in the types of structure under consideration, in the form of

1. 'blind vaults' or cavities within structures, which are virtually enclosed and poorly ventilated
2. abandoned tunnels
3. culverts, sewers and drains
4. other spaces with limited or no ventilation, such as shafts, wells, deep catchpits.

The atmosphere in a confined space may contain hazardous concentrations of gases, vapour or dust from sources such as a former work activity within the space or introduced from outside (e.g. from petrol escaping into sewers); they may result from fumes emitted when sludge is distributed during cleaning for inspection. Alternatively, oxygen deficiency may occur in drains as a result of the ingress of moisture or by absorption of oxygen by certain constituents of soils.

The Factories Act (Section 30) has the effect of precluding entry into a confined space for inspection (or any other purpose) until a responsible person has assessed the situation, ensured that any necessary cleaning and ventilation has been carried out, made the appropriate checks and specified what precautions should be taken to allow safe entry, either with or without breathing apparatus. A 'responsible person' in this context is one who is experienced in the work and in the hazards involved and who is familiar with the relevant chemistry. There must be strict adherence to the precautions he specifies, and the Health and Safety Executive (1977) recommends that they be set down in writing in the form of a 'permit to work' system.

To overcome the possibility of the statutory requirements becoming unnecessarily restrictive to structural examination work, it has been found advantageous to give examiners special training and then to authorize them, on their own initiative, to enter into a defined series of confined spaces – generally those which are known to be ventilated by a through airflow and in which noxious fumes or oxygen deficiency are unlikely. Even so, they must always be supported by a second person who has been briefed as to the appropriate emergency procedures and must remain at the entrance to the space throughout the time the examiner is inside. This arrangement covers the majority of structures requiring examination.

5.9.5 Infection

Anyone who has been exposed to possible infection, such as by coming into contact with sewage or contaminated water, should wash thoroughly with soap, water and disinfectant, particularly before handling food or drink. Contaminated clothes and boots should be washed and dried as soon as possible. Cuts, scratches and abrasions of the skin should also be washed thoroughly and an antiseptic dressing applied.

5.9.6 Diving

Whether underwater inspections are undertaken by contract or by direct labour specialist divers and whether subaqua or standard diving equipment is used, the Maintenance Engineer has a duty to ensure compliance with the proper safety precautions, as defined by the Diving Operations at Work Regulations and their associated Guidance Notes. There must be someone with overall responsibility for each diving operation and he must appoint a diving supervisor to take immediate control. The diving supervisor must have adequate knowledge of the techniques to be used and a certificate qualifying him to dive (but he must not dive whilst acting as diving supervisor). He will be responsible for ensuring that the necessary plant and equipment are available, properly maintained and tested, and that each diver is medically fit, competent and qualified; he must prescribe diving rules and monitor their observance. There must be someone to operate the plant and equipment and a standby diver ready to dive, unless there are two divers in the water able to provide immediate mutual assistance. Therefore the minimum team, including the supervisor, to undertake diving inspections comprises four persons. Considerations of cost and convenience must not be allowed to inhibit this vital operation.

5.10 References and further reading

British Standards Institution
BS 1129: 1982. *Specification for Portable Timber Ladders, Steps, Trestles and Lightweight Stagings.*
BS 2037: 1984. *Specification for Portable Aluminium Ladders, Steps, Trestles and Lightweight Stagings.*
BS 3572: 1986. *Access Fittings for Chimneys and Other High Structures in Concrete or Brickwork.*
BS 4211: 1967. *Specification for Ladders for Permanent Access to Chimneys, Other High Structures, Silos and Bins.*
BS 5062. *Self Locking Anchorages for Industrial Use.*
  Part 1: 1985. *Specification for Self Locking Safety Anchorages and Associated Anchorage Lines.*
  Part 2: 1985. *Recommendations for Selection, Care and Use.*
BS 6037: 1981. *Code of Practice for Permanently Installed Suspended Access Equipment.*
BS 8210: 1986. *Guide to Building Maintenance Management,* Appendix C, *Temporary Access Equipment.*
Construction Industry Research and Information Association (in press) *The Maintenance and Rehabilitation of Old Waterfront Walls,* E. & F. N. Spon, London.
Department of Transport (1983) *Bridge Inspection Guide,* HMSO, London.
Eccleston, B. C. and Tindall, B. C. (1985) Access techniques for inspection and repair. *Proc. Conf. on Structural Faults and Repair,* Engineering Technics Press.
Falcon, K. C. and Houghton, A. F. (1983) Design of a mobile articulated structure for railway bridge maintenance. *Struct. Eng.* **61** A (11).
Health and Safety Executive (1977) *Guidance Note GS 5, Entry into Confined Spaces.*
Institution of Structural Engineers (1980) *The Appraisal of Existing Structures.*
Ministère des Transports (1982) *Défauts Apparents des Ouvrages d'Art en Maçonnerie,* Laboratoire Central des Ponts at Chaussées, Paris.
Organization for Economic Cooperation and Development, Road Research (1976) *Bridge Inspection,* OECD, Paris.
Price, W. I. J. (1982) Highway bridge inspection: principles and practices in Europe. *Proc. Symp. on Maintenance, Repair and Rehabilitation of Bridges,* International Association for Bridge and Structural Engineering, Zurich.
Shimmin, S., Corbett, J. M. and McHugh, D. (1980) Human behaviour: some aspects of risk taking in the construction industry. *Proc. Conf. on Safe Construction for the Future,* pp. 13–22, Institution of Civil Engineers, London.
Smith, D. W. (1976) Bridge failures. *Proc. Inst. Civil Eng., Part 1,* **60,** 367–82.
Tucker, C. (1981) Moveable platforms for inspection and maintenance of railway bridges. *Colloq. Int. sur la Gestion des Ouvrages d'Art. Inspection, Maintenance and Repair of Road and Railway Bridges, Brussels/Paris, September 1981,* Presses de l'Ecole Nationale des Ponts et Chaussées, Paris and Editions Ancient ENPC, Brussels, Vol. I, pp. 253–8.

Chapter six

The experimental investigation of defects

V.K. Sunley

this type. Although new methods sometimes give inconclusive or even misleading information in early applications, the rate of development of equipment and knowledge is such that they cannot be dismissed for all time. Therefore, where possible, methods of checking the suitability of these methods will be discussed.

6.1 Introduction

Why should we investigate defects? A crack is a crack, and investigations take time and cost money. Why not just fix it and move on to the next pressing problem?

Stated boldly like that, such questions seem puerile. After all we would expect rather more treatment than a bottle of pain killers for a broken leg. It is obvious that the disease and not the symptoms should be treated. The difficulty is that when dealing with masonry structures the disease is often well hidden behind the symptoms and the symptoms are common to many diseases. The aim of this chapter is therefore to help engineers to identify the causes of defects so that relevant and long-lasting, as opposed to cosmetic, remedial action can be specified. This will be done by identifying and categorizing the defects and discussing methods of investigation which can be applied to establish their cause.

Not all the methods of investigation discussed are new and employ high technology, but those that are pose a particular problem for a book of

6.2 Types of defect

When confronted with a defective structure, a start to the process of establishing the cause of the defect can be made by asking the question: is it a structural defect or a material defect?

In this context a structural defect is caused by the inability of the structure to support the loads superimposed upon it. Examples of this type of defect are cracks, bulges, excessive deformation, missing bricks or blocks and, in the extreme, collapse. However, these defects are not exclusively structural.

A material defect is caused by the inability of the fabric of the structure to withstand the environment it is exposed to. Examples of this are cracks, spalling, erosion and visible signs of the loss of mechanical properties, e.g. perished mortar.

Even trying to make this, the broadest, classification presents difficulties. Cracks can be structural or material and so can spalling. The face of masonry can be spalled by excessive compressive loading, freeze–thaw action or extreme heat as was seen following the fire in British Rail's Summit Tunnel in 1984 (see Chapter 12, Section 12.5). However, clues can be found in the appearance and history of defects. Cracks in a brick or stone which are not reflected in its surroundings are not likely to be structural, but a crack passing through adjacent bricks, stones or otherwise sound mortar joints is likely to be structural. Even so, the inability of the outer visible parts of the structure to carry the load applied can be caused by material deterioration inside.

History is a good teacher, and if its lessons are studied attentively a good deal can be learned about the causes of defects in aged masonry structures. Questions can be asked, for instance, about the geological stability or the likelihood of mining subsidence, accidental impacts and the weathering capability of the probably local, construction material. These material factors are not usually recorded along with details particular to the structure. Another potential source of valuable information is details of previous remedial work and what was discovered about the construction during the execution of that work.

There is no purpose in trying to produce a list of all the possibilities as it would approach the infinite and readers are known to become bored when faced with considerably less than the infinite. The point is that in order to home in as quickly as possible on the cause of a defect all the information available should be used. As well as information obtained from measurements, which will be discussed later, there are clues in the appearance and history which contribute to answering the question: is the defect structural or material? The answer influences the type of investigation employed to establish the extent or seriousness of a defect.

6.3 Causes of defects

The causes and characteristics of defects are considered in detail elsewhere in this book. However, it is necessary to mention them here in the context of deciding upon the method of investigation and the instrumentation suitable for obtaining further evidence to relate the symptom to a cause.

For this purpose it is useful to classify the possible causes into four categories, each of which has a group of relevant investigation methods associated with it. The categories of causes considered here are construction, long-term loading, transient loading and the environment.

6.3.1 Construction

The cause of a defect may have its roots in the way in which the structure was built, and this is not always obvious from external appearances. The original drawings are not usually much help in this matter because frequently they do not exist, which is not surprising given the age of most masonry structures. When they are available they tend to be more a guide to the overall dimensions for the mason than an engineering drawing with construction details. Also, even in 'the good old days' the expedient which could be hidden was not universally frowned upon. Therefore it is necessary to probe deeper to find out how the structure was built.

The external dimensions are readily available but the internal dimensions are not. The thickness of a retaining wall should not be judged by its visible top or edge. The number of rings in an arch bridge or tunnel should not be assumed to be the same as that shown at the portal. There is frequently an increase in the thickness of a brick arch away from the crown and also a structurally significant backing to the haunch.

Other internal features can also affect the development of a defect. The presence of voids, either intentional or unintentional, is an obvious example. Also of interest is the degree to which the structure is interconnected. Headers are used to tie together layers of bricks or stones, and it is important to know whether the whole thickness is tied together or whether this is only partially the case. Solid-looking viaduct piers have been found to be made of a dressed stone outer shell filled with a rubble and mortar (in places) core with no connection between the two. Brick arches and tunnel linings have been found to comprise rings connected in pairs rather than throughout the full thickness.

The material from which a masonry structure is built is also not obvious from external appearances. A blue engineering brick exterior has been known to hide a softer brick interior. Also important, but not so obvious, is the way that the bricks and stones are laid. This affects the proportion of mortar which will generally deteriorate more quickly than the brick or stone, especially if the waterproofing has failed.

The reader may ask 'Is there nothing I can trust?', and the answer is, 'Not at face value without investigation'.

6.3.2 Long-term loading

The reason for making the distinction between long-term loading and transient loading is that different techniques are required to measure their effects. However, there is no precise dividing line between the two, and it is sufficient to say that events measured in hours or more rather than seconds are considered to be long term.

The most obvious case of long-term loading is the dead load. This is the weight of the structure and also the weight or the pressure exerted by whatever the structure is supporting. It may be the most obvious, but measuring the effects of dead load is probably more difficult than anything else.

Similar in many ways to dead load is the loading imposed by ground movement or foundation weakness, because when this has occurred historically it will have caused a redistribution of the dead-load forces within the structure as it moved to cope with its change in circumstances. It is important to consider this type of effect, which is built into the structure, because materials fail as a result of the total loading and not just the loading applied when the engineer is taking measurements.

Ground movements may also occur whilst the structure is under observation, and it is easier to determine the effects of these. A good example of this is mining subsidence. The progress of the subterranean activity is known and the ground strains and settlement can be predicted. Using these, it is possible to estimate if or when a structure will be affected and observations can be mounted to suit.

Superimposed on these long-term and usually irreversible loads will be cyclically varying long-term loads. An example of these is thermal loading, which has a diurnal as well as an annual cycle. Another is the change in pressure due to the changes in the level of the water-table, which will be an irregular cycle but may also have a long-term trend if something is done to alter drainage or drains become blocked. These cyclical changes do not usually have a major effect, but they should be considered when making long-term observations. If they are not, an observation made on a hot summer day, when compared with one made in winter, may give a misleading impression of what is happening.

6.3.3 Transient loading

Instead of trying to define the duration of a transient loading event, it is again more useful to think in terms of the type of equipment which would be used to record it. Transient loading needs to be monitored by equipment capable of recording continuously.

Traffic is the principal cause of transient loading. It can cause live loading due to the direct effects of vehicle axle loads superimposed on the structure or the usually second-order effects of a vibration transmitted to the structure via the ground or air. Impacts are another source of defects caused by traffic – usually high road vehicles hitting low bridges. The defects caused in this way are usually easy to diagnose and the most effective prevention known is someone in a yellow jacket with a clipboard looking intelligently at the structure. Other types of transient loading include wind loading and seismic loading. Fortunately, the latter is not common in the United Kingdom.

When investigating transient loading effects it is not sufficient to consider only the magnitude of the load. The frequency of its application is also important, because the effects can be amplified if the natural frequencies of the structure are excited. A good example of this is a long train with equally spaced axles crossing a bridge. At the speed of the train when the axle passing frequency coincides with the bridge's natural frequency, each successive axle will reinforce the motion excited by the previous one. This will not happen at speeds higher or lower than this critical value.

When attempting to diagnose the causes of defects it is important to recognize the possibility of incremental development. A long-term monitoring exercise may show a continuous development of a defect, whereas the truth may be that development is by means of a series of increments each caused by a transient loading event.

6.3.4 Environment

The last of the categories of defect causes under consideration is the environment within which the structure works. Two aspects of the environment have already been considered, namely wind and temperature. This is because they cause loading to be applied to the structure. What this category is meant to cover are the environmental effects which act upon the fabric of masonry structures.

The weather creates the most important aspects of a structure's environment, with water being the most unfriendly element. Much of the deterioration of the structural fabric has as its root cause the breakdown of the structure's waterproofing and drainage system. Once the water can permeate through the structure the way is open for damage caused by freeze–thaw, erosion, leaching out of the mortar and general reduction of the strength of the structural fabric. Rain can contain pollutants and ground water absorbs chemicals which can attack a structure by chemical action. An example of this is sulphate attack on

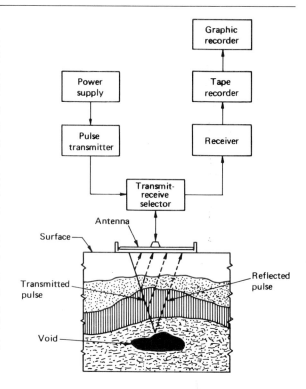

FIGURE 6.1 Schematic diagram of a ground-probing radar system.

cementitious materials like concrete and mortar. When a structure is exposed to flowing water, scouring can result. This is caused by the water alone to some extent, but mainly by the particles it carries with it.

Another severe, but fortunately rare, environment is caused by fire. The high temperatures applied to the structural fabric can cause spalling, and in the case of bricks prolonged high temperatures can lead to surface vitrification. The latter may not be a problem initially, but the vitrified layer may spall off later. This is a case where a knowledge of the structure's history can be useful, because after a while the effects of an isolated fire can look similar to freeze–thaw action.

6.4 Investigation methods

This part of the chapter poses an almost impossible task as there must be nearly as many

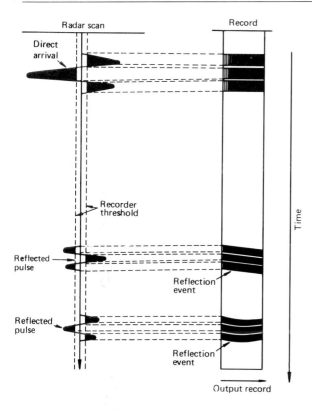

FIGURE 6.2 Idealized recording of a radar scan.

investigation methods as there are engineers investigating problems associated with masonry structures. However, with sublime optimism, an attempt will be made to discuss the available techniques whilst recognizing that derivatives and adaptations proliferate to suite particular circumstances. This is also an area where the rate of development of new techniques is high, and so new methods will appear and some which at present do not give satisfactory results may well develop into industry standards.

As with the causes of defects, the investigation methods are considered under the four headings of construction, long-term loading, transient loading and the environment.

6.4.1 Construction

There are two basic approaches to the investi-

gation of a structure's construction: non-destructive testing and direct visual or measurement methods. The former will be considered first.

(a) Radar

It has been found that high frequency radio waves can penetrate the earth to a depth of up to 20 m. As a result ground-probing radar is used for applications such as the detection of buried objects, the location of cavities and geological interfaces, and the estimation of their depth (Holt and Eales, 1987). Therefore it is a logical progression to attempt to use radar for the non-destructive examination of the internal construction of brick and masonry structures.

Modern commercial ground-probing radar systems use the pulse method. In this, a short pulse of radio frequency is emitted by an antenna in contact with the ground. There follows a period of quiescence during which the pattern of reflections can be observed. These reflections and the time taken for them to return indicate the presence and distance of a back-scattering surface in the line of propagation of the transmitted radio wave. The pulses are transmitted at a high repetition rate so that repeated reflections can be identified with confidence, which is important for detecting weak signals. The frequency of the transmitted pulse is a compromise, because the lower the frequency is, the greater is the depth of penetration but the poorer is the ability to discriminate small dimensions. For ground-probing radar the transmitted frequency is chosen between 50 MHz and 5 GHz, depending on the attenuation of the materials being probed. A schematic diagram of the radar system is shown in Fig. 6.1 and a representation of the recorded output is shown in Fig. 6.2.

Radar has been used with some success for investigating tunnel linings. Lining thicknesses have been confirmed, areas of ring separation within the lining have been suggested and the locations of blind shafts behind the lining have been determined. These results have been verified by direct methods, and this has boosted confidence in the technique to the extent that development work is being undertaken. This is particularly aimed at optimum antenna design and computer-enhanced signal processing for masonry investigations.

(b) Sonic methods

The non-destructive testing of steel components by ultrasonic methods is a well-established procedure, and they are also used successfully, albeit at lower frequencies, for concrete. However, these materials are relatively homogeneous compared with masonry structures, and the high resolution at shallow depths of penetration of the high frequency ultrasonic methods are inappropriate for masonry. The answer to this problem has been to develop sonic methods (Komeyli-Birjandi *et al.*, 1984).

The principle of operation of the sonic method is that one face of the structure is hit with a hammer and the impact is recorded by an adjacent accelerometer. Another accelerometer on the opposite face of the structure records the arrival of the transmitted compression wave, and the time between transmission and reception is calculated. If this procedure is repeated over a regular grid, variations in the transmission time of the compression wave over the structure can be plotted.

The transmission time of the compression wave depends on the density of the material and the presence of voids through which the wave will not travel. If a void is present, the wave will travel around it, therefore lengthening the transmission time. If the void is large or near either the transmit-side or the receive-side accelerometer, a signal may not be received at all. A study of the variation in transmission times over a structure will indicate changes in density or the presence and extent of voids.

Sonic methods have been used with some success on simple masonry structures where there is access to both sides of the structure, e.g. abutments and piers. Any variations in the length of the transmission path due to the width of the structure must obviously be taken into account.

More difficult to interpret, however, are internal construction features such as changes in wall thickness, internal arches and changes in fill material from concrete to rubble or earth. Visible features such as tie-bars, which could provide a preferential transmission path for the compression wave, may also affect the recorded transmission times.

(c) Thermography

Thermal-imaging cameras are used to record variations of temperature and to display these as different colours representing different temperature ranges. The magnitude of the ranges, or the sensitivity, depends upon the requirements of the application, but very small temperature differences can be detected. The cameras measure absolute temperature and use a reference such as liquid nitrogen.

The application of thermography to masonry structures is based upon the idea that the temperature of the surface will depend to some extent upon what is behind it. For example, the surface may be backed up by continuous conductive material, voids or running water, and this will affect the surface temperature. However, variations in surface emissivity will also affect the results.

This technique has been used on a tunnel lining, and variations of surface temperature were duly recorded and the locations of suspected voids were interpreted. However, when the lining was subsequently drilled the correlation between suspected voids and actual construction was found to be poor.

(d) Endoprobes

Endoprobes, sometimes called borescopes, are a means of seeing deep into existing fissures or specially cored holes. They are therefore not a completely non-destructive technique, but the cored holes can be filled after use and would generally only be a problem on a visually sensitive surface.

An endoprobe is a tube packed with optical

FIGURE 6.3 Endoprobe with power supply and light source.

fibres with a lens at one end and an eyepiece and light input at the other. The lens end of the endoprobe is placed in the hole, light is passed down it to illuminate the inside of the hole and the image is viewed at the eyepiece. A typical example of this equipment is shown in Fig. 6.3.

The many variations and features which can be obtained for endoprobes make them an almost universal tool for internal inspections. These include a wide range of lengths and diameters, solid tubular or flexible bodies, lenses for forward, sideways or retro viewing, eyepiece focus and angled viewing, still and video camera attachments, and mains or battery power supplies.

Endoprobes have been widely used for investigating the construction of masonry structures. A small-diameter hole is all that is needed to explore wall thickness, internal voids or brick ring separation. Their main problem is that they only allow discrete points to be inspected and so a systematic pattern of holes will be required. Even then, large features can easily be missed unless there are other signs indicating their presence.

In the present state of the art, the previously mentioned non-destructive methods which can be used for investigating the construction of a masonry structure, i.e. radar and sonic methods, should be checked using an endoprobe survey.

6.4.2 Long-term loading

In order to measure the effects of long-term loading the techniques used need long-term stability. If the instruments are designed for relative measurements, for example the movement of one part of the structure with respect to another, the instrument must be immune to the influence of all changes except the one being measured. In the case of absolute measurements, e.g. the movement of a structure with respect to a remote datum, the datum points must be truly fixed and a means of checking that this continues to be the case must be arranged.

The following techniques have been found to be suitable for measuring the effects of long-term loading such as strains and displacements and are discussed in that order.

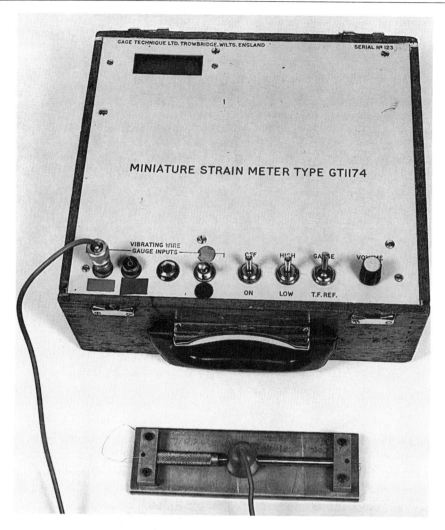

FIGURE 6.4 Acoustic strain gauge with recording equipment.

(a) Acoustic strain gauge

The acoustic or vibrating-wire strain gauge is essentially a thin wire tensioned between end blocks which is plucked electromagnetically so that it vibrates. The electromagnetic circuit then counts the number of cycles of vibration in a given time and thus obtains the frequency of vibration. This frequency is related to the tension in the wire, and so if the end blocks are rigidly attached to a body undergoing strain, the tension in the wire and therefore the frequency of vibration will change. For a given gauge length, the strain change can be found from the frequency change. Provided that the anchorages at the ends of the wire do not move and the wire does not corrode, long-term stability is assured. An example of this device is shown in Fig. 6.4 and further discussion can be found in Tyler (1974).

The gauge lengths generally obtainable are restricted to between 50 and 150 mm. Strains measured in masonry structures are usually small, unless they are being measured across a crack, and

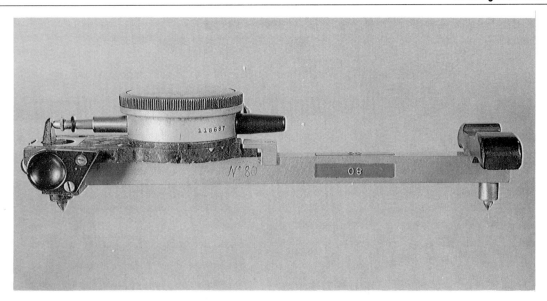

FIGURE 6.5 Demec mechanical strain gauge.

so the larger gauge length would generally be preferred. The resolution of a vibrating-wire strain gauge with a gauge length of 150 mm is $\pm 0.5 \times 10^{-6}$, which is well suited to the smaller strains generally found.

When dealing with such small quantities it is important to compensate for thermal strains. This can be done either by attaching a gauge to an unstrained piece of material in the same temperature environment and subtracting its strain from that of the active gauge or by using prior knowledge of the material's thermal strain characteristics and the temperature. Special vibrating-wire gauges can be obtained which also measure the temperature. In this case the coil in the electromagnetic circuit has a known resistance – temperature characteristic.

Vibrating-wire strain gauges suitable for attaching to the surface of a structure or embedding into concrete can be obtained.

(b) Mechanical strain gauges

There are many types of mechanical strain gauge, but the most useful must be the Demec because of its versatility and robustness (Morice and Base, 1953).

The Demec gauge is essentially a bar with a small conical point at one end and a spring-loaded pivoted lever at the other. On one side of the pivot is another small conical point and a dial gauge records the lever's position on the other side of the pivot. Small targets with central conical recesses are stuck to the structure, with their distance apart being equal to the Demec gauge length in the middle of its measuring range. A setting-out bar can be obtained to simplify the positioning of the targets. The conical points of the gauge can then be inserted into the conical recesses of the targets and the reading on the dial gauge observed. The relationship between the dial gauge reading and strain depends on the gauge length and lever arm ratio and is provided with the Demec gauge. An example of this device is shown in Fig. 6.5.

Demec gauges are available with gauge lengths from 50 to 2000 mm, but 200 mm is the most common; it is well suited to the measurement of strains in masonry structures and can accommodate the usually uneven surface. The accuracy of the 200 mm gauge is $\pm 6 \times 10^{-6}$; however, this is somewhat theoretical as much depends on the skill of the operator. It is important that the conical points of the gauge conform to the recess

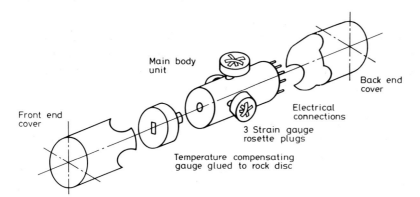

Main body unit

Front end cover

Back end cover

Electrical connections

3 Strain gauge rosette plugs

Temperature compensating gauge glued to rock disc

FIGURE 6.6 Exploded view of a plug gauge cell.

in the targets and so the targets must be mounted flat. In order to overcome small inaccuracies in the mounting of the targets, the same point should always be inserted into the same recess. A rule such as always having the dial gauge end to the right will achieve this, and the effect of the 'feel' of the operator can be minimized by always using the same operator.

The temperature compensation should take into account the thermal expansion/contraction of the gauge itself as well as that of the structural material. An Invar rod, which has a zero coefficient of thermal expansion, can be obtained for the former, and the techniques described for the vibrating-wire strain gauge can be used for the latter.

(c) Measurement of pre-existing stress or strain

When strain measurements are made on an existing masonry structure, they are relative to the state of stress at the start of the measurements. In order to find the total strain, it is necessary to measure the pre-existing strains relative to the stress-free state.

Several plug gauge cells have been developed to achieve this (Kim and Franklin, 1987). Basically they involve installing strain-measuring devices inside the fabric of the structure, recording the strains, and then cutting out a core from the structure, which concentrically encompasses the strain-measuring devices (overcoring) to create a stress-free state in them and re-recording the strains. The difference between the two strain readings is the strain due to the state of stress currently existing inside the structure.

One device of this type, shown in Fig. 6.6, is known as the CSIR or CSIRO cell, developed by the South African Council for Scientific and Industrial Research and the Commonwealth Scientific and Industrial Research Organisation of Australia respectively. These cells use a similar technique and employ electrical resistance strain gauges, which are described in a later section of this chapter, for measuring the strains.

The first step in the installation of the cells is to core a hole into the structure. If the core is unbroken and has no planes of weakness in the area where the strains are to be measured, then installation can proceed. The diameter of the cored hole depends on the particular cell; for the CSIR cell, it is 38 mm. Three strain gauge rosettes, each comprising three or four elements, are placed on the cell and pre-coated with adhesive. An installing tool is used to position the rosettes parallel to the bore and at intervals of 120° around it. When the bore in the area where the rosettes are to be stuck has been cleaned and dried, the installing tool and cell are inserted into the cored hole to the desired depth and rotated to give the correction orientation to the rosettes. Pressure is then applied through the installing tool to press the rosettes against the inside of the cored hole. After allowing time for the adhesive to set the pressure is released, the installing tool is withdrawn and the hole is plugged with electrical connections accessible from the outside.

The first set of strain readings can now be taken. The gauges are then concentrically overcored to produce a stress-free state in the material and another set of strain readings is measured. The principal stresses existing in the structure at that location can be calculated using these sets of strains and Young's modulus and Poisson's ratio for the material.

The accuracy of the method depends upon the accuracy to which the rosettes have been located and to which the material properties are known. However, the measured strains should be accurate to $\pm 5 \times 10^{-6}$.

This type of cell is restricted to materials which can be assumed to be homogeneous. Also, because of the use of electrical resistance strain gauges, wet conditions should be avoided and as short a time as possible allowed to elapse between the strain measurements before and after overcoring. To overcome these two problems, developments have been started using a derivative of the acoustic strain gauge mentioned previously (Sunley, 1983). In this development the vibrating wires were stretched across a steel annulus which could be glued into a cored hole and overcored as before. However, this is not yet commercially available.

Techniques for measuring pre-existing surface stresses are also available. One of these is the photoelastic stress cell, where a photoelastic plug is bonded into a cored hole and overcored as before, and the stresses before and after overcoring are measured photoelastically. However, reported experience with this device in the field has shown inconsistent results.

Another variation on this theme is the flat-jack technique (Kim and Franklin, 1987) which can be used to measure the stress in one direction near the surface of a masonry structure. This technique involves fixing two pins into the surface in line with the direction in which the stress is to be measured and accurately measuring the distance between them. A slot is cut into the structure halfway between the pins and perpendicular to the

direction of the stress and the distance between the pins is re-measured. A flat hydraulic jack is inserted into the slot and pressurized until the pins return to their original spacing. The hydraulic pressure required to achieve this is equal to the pre-existing stress.

Several potential sources of error must be guarded against with this technique. The masonry must be sound, the slot should be grouted to ensure intimate contact with the jack, and the jack should be calibrated because edge effects can cause the pressure exerted to be less than the internal pressure.

Although this technique only measures stress in one direction, an array of flat jacks at different angles can be used to find the principal stresses. This will only be possible in fairly constant stress fields, because these arrays take up quite a large area.

(d) Displacement measurement

The long-term measurement of displacement usually employs well-known general techniques, and little purpose would be served in repeating their details here. Instead notes of caution will be sounded where appropriate.

Surveying must be the most common technique used. The accuracy obtained will depend to some extent on the accuracy of the theodolite or electronic distance measurement (EDM) device used and the skill of the operator, but more particularly on the stability of the datum and having a means of proving it. There are some large blocks of concrete near a bridge over the River Ouse in Yorkshire which could reasonably have been expected to remain where they were put. Unfortunately nature is not noted for being reasonable and they have moved in a most unreasonable manner!

Photogrammetry is a promising method as it allows many points on an inaccessible structure to be recorded at once and detailed measurements to be made later in the laboratory. It is based on the examination of stereoscopic pairs of photographs, taken over a period of time, of a structure with markers attached. The positions of these markers

FIGURE 6.7 Illuminated tunnel profile showing severe distortion. (*Courtesy of Photarc Surveys Ltd.*)

at different times can be measured in the laboratory and corrected for camera lens distortion to give the structural displacements. Again, the stability of the datum is of vital importance (Kalaugher, 1987).

Both the above techniques are for measuring absolute displacements. It is sometimes desirable to measure the displacement of one part of a structure relative to another – across a crack for example. A gap gauge or tell-tale can be used for this. As these devices are described in detail in Chapter 5, Section 5.6, no further comment is given here.

(e) Profile measurement

The techniques described here have been developed specifically for measuring the profiles of railway tunnels, but the idea has wider applications.

The deformation of tunnel profiles, which may take place over a long period of time, can be measured manually from markers set into the lining. The equipment used for this can be a calibrated lightweight telescopic rod or a distometer. The Distometer uses a length of Invar wire attached across part of the profile between two pins and a measuring device. A separate length of wire is normally required for each dimension to be measured. It has been found to be capable of measuring distances consistently to ± 0.05 mm. However, these manual methods are slow and labour intensive.

A photographic technique has been developed which consists of a light source mounted on one trolley which is rigidly connected to another trolley carrying a photogrammetric camera. The light source may be a beam that rotates and describes a line around the tunnel lining or a beam that shines through a circumferential slot to produce a line of light around the lining. The trolley is positioned at a marker and a photograph is taken of the line of light which defines the tunnel profile. An example is shown in Fig. 6.7.

This equipment can be used to photograph profiles at pre-marked positions along the tunnel, and the results can be digitized and stored in a computer database. The consistency to which a profile can be measured has been found to be ±5 mm.

A laser system has also been developed. The laser and a light receiver are mounted on a trolley and are rotated so that the laser light describes a path around the tunnel profile. The time taken for the light to travel to the tunnel lining and back to the receiver is converted into distance, from which a profile can be constructed. The consistency to which a profile can be measured using this system has similarly been found to be ±5 mm.

The digitized profiles from both systems can be used to produce a profile comparison by computer, which will indicate any points on the profile where changes have occurred outside the specified threshold levels.

6.4.3 Transient loading

In order to measure the effects of transient loading the devices used must be capable of giving a continuous output, which in practice means an analogue output. The data are usually more useful in digital form for later analysis and this can be obtained on-line or later in the laboratory. In either case, the digitization must be sufficiently frequent to allow accurate reproduction of the signal. When the analogue-to-digital conversion is to be performed on-line, the digitization rate chosen must be based on a reasonable estimate of the frequency of the event to be recorded. It is usual to allow a minimum of ten digital measurements to be made for each cycle of the analogue signal.

The following devices have been found to be suitable for measuring the effects of transient loading such as strain, displacement and acceleration.

(a) Electrical resistance strain gauges

The electrical resistance strain gauge is basically a flat coil of copper wire made as a printed circuit, which is firmly stuck to the structure. When the structure is strained, the cross-section of the copper changes and therefore the resistance of the gauge changes. This change in resistance is proportional to the strain and is measured using a Wheatstone bridge (Perry and Lissner, 1962).

The gauge length required depends on the material of the structure being tested and the strain gradient induced. Strain gauges are readily available with gauge lengths ranging from less than 1 mm, for high strain gradients in homogeneous materials, to 100 mm, for small strain gradients in concrete. Therefore the longer gauge lengths are required for direct attachment to masonry structures.

It is often not possible to attach electrical resistance strain gauges directly to masonry because the surface may be too uneven, cracked or wet. In these circumstances strains have been successfully measured using gauges stuck to thin metal strips and encapsulated. The ends of the metal strips are then attached rigidly to the structure. The strips need to be designed so that they are not thin enough to buckle under compression and not thick enough to stiffen the structure locally and affect the measurements.

It is possible to eliminate thermal strains by wiring a temperature-compensating gauge into the opposite arm of the Wheatstone bridge to that containing the active gauge. This gauge is usually placed on a separate piece of the same material as the active gauge, so that it does not experience the strain due to loading but is subject to the same thermal strain. It is also possible to eliminate the bending or direct strain components by the way in which they are electrically connected to the Wheatstone bridge. If gauges on each side of a component subjected to bending and direct stress are connected into the same arm of the Wheatstone bridge, the net change in resistance due to the bending strain will be zero and only the direct strain will be measured.

The electrical resistance strain gauge is a versatile device well suited to measuring transient strains. However, it has one major disadvantage in that it tends to suffer from zero drift over long periods and hence is unsuitable for measuring long-term cumulative effects.

(b) Measurement of relative displacement

Here we are concerned with measuring the displacement of one part of a structure relative to another quite close to it by means of a device long enough to bridge the two positions. Two devices have been found to be suitable for this purpose: the linearly variable differential transformer (LVDT) and the linear potentiometer.

These two types of instrument work on different principles, but both have a cylindrical body with a rod able to move in and out of it. The electrical output varies as the rod moves, and this output is calibrated to give the distance moved by the rod. Therefore the relative displacement can be measured by attaching one end of the body to one part of the structure and the end of the rod to another part. The LVDT can be obtained with a measurement range of 10 mm and the linear potentiometer with a range of up to 200 mm. The accuracy for both instruments is ±0.01 mm.

Another device – the capacitance probe – is capable of measuring relative displacements and does not require a physical connection between the positions to be measured. It works on the principle that the capacitance between two metal surfaces changes with the distance between them. However, it has a limited range of operation.

(c) Measurement of absolute displacement

Absolute displacements due to transient loadings are difficult to measure because of their small size and the distance between the position to be measured and a stable datum. However, the advent of lasers has provided the essential link.

A set of equipment which has been used successfully on many occasions is the Jenzer Oculus laser system (Jenzer, 1977). This is composed of a standard theodolite and tripod with a

FIGURE 6.8 Laser datum station.

laser fixed to one of the legs. The light from the laser is passed through a fibre optics light guide to a laser eyepiece on the theodolite telescope, through which it can be positioned and focused. This then forms a self-contained datum (Fig. 6.8). The laser beam is focused on a detector attached to the position where deflection is to be measured. The detector (Fig. 6.9) has a ground glass screen on which the spot of light from the laser falls. A lens system focuses this spot of light onto a 50 mm square photosensitive plate. The electrical output from each side of this plate is proportional to the position of the spot of light on it. Therefore as the structure, and consequently the detector, moves relative to the laser beam, the dynamic displacement in two perpendicular directions can be recorded. The detector is calibrated by adjusting its position relative to its attachment using micrometer adjusters in the two directions.

FIGURE 6.9 Laser detector.

The accuracy with which the displacements can be measured depends on the distance between the laser and the detector and the conditions of the atmosphere through which the beam passes. Under good conditions, e.g. at night, an accuracy of ±0.1 mm can be achieved over 100 m. In less ideal conditions, e.g. in sunlight, the accuracy could be reduced to ±0.5 mm over 100 m. The stability of the datum can also affect the accuracy, because, although long-term stability is not required, the tripod must be positioned where it will be unaffected by ground-borne vibrations.

(d) Accelerometers

An accelerometer produces a continuous electrical record of acceleration in a given direction. It is a small device which is easy to use and calibrate, and consequently it has often been used for investigating masonry structures. The trouble is that structures fail because of excessive stress or strain, not excessive acceleration, which is very much a second-order effect.

Accelerometers have been used to make comparative measurements of the effects of different loading events or the effect of a loading event on different structures. This usage may be able to identify that one loading event causes twice the acceleration of another, but as the acceleration cannot be related to the strength of the structure, not much will have been learned.

Accelerometers can be used to obtain displacement by double integration of the acceleration signal. This enables displacements due to transient loading to be measured in locations which are otherwise not easily accessible. However, the displacements obtained in this way are not very accurate at low frequencies. What is meant by low frequency depends upon the characteristics of the particular accelerometer, but problems with accuracy have been experienced when the frequency of oscillation was below 20 Hz.

(e) Acoustic emission

As a structure is subjected to an increasing load the strain energy stored within it increases. Microcracks develop during the loading process, and these are accompanied by a small release of strain energy in the form of elastic stress waves which travel through the structure at the speed of sound for that material. These are referred to as acoustic emissions.

The elastic stress waves can be detected by piezoelectric crystal accelerometers attached to the structure. However, not all acoustic emissions will be due to the applied loading; some background noise will be generated by other causes.

Therefore it is necessary to establish the level of background noise and to set a threshold level so that only acoustic emissions above this level are recorded. To quantify the level of acoustic emission it is usual to count the number of times that the threshold level is exceeded for a given loading event.

Acoustic emission has been used for testing masonry structures and an example is given by Hendry and Royles (1985). They demonstrated a correlation between acoustic emission counts and structural displacement, and suggested that the technique could be used for finding the start of non-linear behaviour non-destructively.

6.4.4 The environment

The environment can cause loads on structures whose effects can be measured using the techniques already described, particularly those under the heading of long-term loading. Temperature is another environmental effect and its measurement will be discussed here.

Other environmental effects are more complex and can only be studied in a controllable way by simulation in the laboratory. Given that money is no problem, any environment can be simulated using special experimental rigs. However, this is not usually the case, and so two simulations which use standard equipment will be dealt with, namely weathering and freeze–thaw.

The weather and load history environment that the structure has endured can affect the strength of the structural fabric and so strength testing of masonry materials is also mentioned.

(a) Thermocouples

The measurement of temperature has already been mentioned under the heading of thermography. This method provides a means of quickly scanning the temperatures over a surface.

Thermocouples provide a means of accurately measuring localized temperatures on the surface and can be inserted inside a masonry structure. A thermocouple utilizes the thermoelectric phenomenon that an electric current flows in a closed circuit consisting of two different metals when the two junctions are at different temperatures. A wide variety of different metals can be used for the couple depending on the temperatures that are to be measured. A copper–constantan couple is suitable for temperatures up to 400°C, and iron–constantan is suitable for temperatures up to 900°C. In the interests of standardization, the chromel–alumel couple has become an industry standard and this is suitable for temperatures up to 1000°C.

To put this technique to practical use involves welding the two wires together at one end and locating this where the temperature is to be measured. The other ends of the wires are connected to a commercially available measuring instrument which measures the temperature at this junction electronically. A direct temperature measurement is then made using the current flowing and the temperature at the measuring instrument.

A thermocouple can be used for continuously recording temperatures or it can be disconnected and reconnected for making periodic measurements.

(b) Simulation

Freeze–thaw is a common mode of deterioration for masonry structures. It occurs when the masonry is saturated or cracks are filled with water which then freezes. As ice occupies a greater volume than the water from which it is formed, expansive pressures are generated which are capable of extending cracks and spalling exposed surfaces. This process can be simulated in a commercially available insulated cabinet which can be heated and cooled. The volume is similar to a domestic chest freezer and so the size of the samples is limited. Also, the heating – cooling cycle takes about 12 h and so the process is slow.

Weathering is a general term for all environmentally imposed conditions, and so the only true simulation is the real environment. If it is necessary to assess the durability of repair materials some acceleration of the natural processes is desirable. With this in mind the Weatherometer was designed and made available commercially. The Weatherometer is basically a large drum with a rotating platform inside it. Specimens can be placed on this platform and as they rotate they are subjected to alternate spraying and drying. How this process relates to the real environment is problematical, but it can be useful for studying the comparative durability of different repair materials.

(c) Strength testing

Testing in the laboratory is virtually the only way of estimating the strength of old masonry material, which varies widely. British Standards give a guide to the methods of testing, but they only refer to new bricks and concrete. Whole bricks removed from structures can be tested by the methods described in BS 3921, but their usually irregular shapes will mean that they should be capped to conform to the surfaces of the testing machine. To estimate the strength of brick and mortar composites it is necessary to remove a whole section of brickwork from the structure, as it is difficult to reproduce old mortar. If it is sufficiently important to do this, BS 5628: Part 2 gives guidance for the testing of such specimens.

The stone blocks in a masonry structure are far too large to be removed and tested in a laboratory machine and so cores are usually taken. A guide to the testing of concrete cores is given in BS 6089, and the parts discussing the method of testing and size of specimen are helpful when testing stone cores. Any stone cores extracted that contain cracks should be discarded because they will give unrealistically low strengths.

A portable tool for measuring the flexural strength of masonry *in situ* has recently been developed by the Building Research Establishment. Known as the BRENCH, it comprises a long lever (about 800 mm long) with a set of adjustable jaws at one end, and it weighs only 10 kg. A hole must first be made above the brick or block to be tested and those on each side are isolated by cutting down the adjoining perpends.

The adjustable jaws are clamped to the unit under test and the operator presses down on the end of the lever until the unit breaks free. The load at which this occurs is displayed and can be converted to stress by a calibration chart. Thicknesses of masonry up to 215 mm brick headers can be tackled.

6.5 Summary

It is hoped that the reader, having borne with this chapter, much of which will be common knowledge to the experienced engineer, will be convinced of the need for a systematic approach to the investigation of defects in masonry structures. The main points to be considered can be summarized as follows.

1. Identifying the cause of a defect can help the engineer to choose long-lasting, as opposed to cosmetic, remedial action.
2. The history of a structure may provide clues to the cause of a defect. If this is not available, it may be an idea to start one for the benefit of our successors.
3. Defects can be caused by a structural or material deficiency and this consideration will affect the investigations required.
4. The construction details of a masonry structure determine the way that it supports the applied loading and may also be the root cause of defects. As this information is not usually available from drawings or obvious from the outside, it needs to be found by more thorough investigation.
5. The loading experienced by a structure can cause defects, and different investigation methods will be required depending on whether the loading is transient or applied over a longer period.
6. Defects will be caused by the total forces existing in the structure and it is necessary to consider the importance of the effects of pre-existing forces in addition to those measured during an investigation.

7. The environment is another factor to consider when trying to establish the cause of defects. Its effects are not easy to quantify, but some examples have been given.

8. Indirect and non-destructive techniques can be convenient to use, but until confidence in their accuracy is built up through experience, direct and visual checks should be employed.

6.6 References and further reading

Clifton, J. R. (1985) Non-destructive evaluation. *Proc. Symp. on Rehabilitation, Renovation and Preservation of Concrete and Masonry Structures,* (ed. G. Sabnis), p.19–29, American Concrete Institute, Detroit, MI.

de Vekey, R. C. (1988) Non-destructive test methods for masonry structures. *Proc. 8th Inter. Brick and Block Masonry Conference.* (ed. J. W. de Courcy) Sept., Dublin. Vol 3, p.1673–81. Elsevier Applied Science Publishers, London.

Dunnicliffe, J. and Green, G. E. (1988) *Geotechnical Instrumentation for monitoring field performance,* Wiley Interscience, New York.

Hendry, A. W. and Royles, R. (1985) Acoustic emission observations on a stone masonry bridge loaded to failure. *Proc. 2nd Int. Conf. on Structural Faults and Repair,* Engineering Technics Press.

Hobbs, B. and Wright, S. J. (1986) An assessment of ultrasonic testing for structural masonry. *Proc. 1st Intern. Masonry Conf.* (ed. H. W. H. West), December, London. Pp. 42–5. British Masonry Society.

Holt, F. B. and Eales, J. W. (1987) Non-destructive evaluation of pavements. *Concrete Intern.,* Vol. 19, June p.41–5. American Concrete Institute, Detroit, MI.

Jenzer, R. (1977) New method to detect dynamic movement between track and train. *ZEV Glasers Annalen,* Vol. 8/9, p. 391–4. (See also Technical Data Sheet for Jenzer Oculus Type OCL.2, Jenzer A. G. Instrumentation, Geroldswil, Switzerland.

Kalaugher, P. G. (1987) Structural Inspection using photographic colour transparencies: planning a monitoring exercise. *Proc. Int. Conf. on Structural Faults and Repair.* July. Engineering Technics Press.

Kim, K. and Franklin, J. A. (1987) Suggested methods of rock stress determination. *Int. J. Rock Mechanics and Mining Sciences.* 24 (1), p. 53–73.

Komeyli-Birjandi, F., Forde, M. C. and Whittington, H. W. (1984) Sonic investigation of shear-failed reinforced brick masonry. *Masonry Intern.* 3, November, p.33–40.

Komeyli-Birjandi, F., Forde, M. C. and Batchelor, A. J. (1987) Sonic Analysis of Masonry Bridges. *Proc. Int. Conf. on Structural Faults and Repair,* July, Engineering Technics Press.

Morice, P. B. and Base, G. D. (1953) The design and use of a demountable mechanical strain gauge for concrete structures. *Mag. of Concrete Research,* 5 (13), p. 37–42.

Noland, J. L., Atkinson, R. H. and Kingsley, G. R. (1987) Non-destructive methods for evaluating masonry structures. *Proc. Int. Conf. on Structural Faults and Repair, July,* Engineering Technics Press.

Noland, J. L., Kingsley, G. R. and Atkinson, R. H. (1988) Utilization of Non-destructive Techniques into the evaluation of masonry. *Proc. 8th Int. Brick and Block Masonry Conference.* (ed. J. W. de Courcy), September, Dublin. Vol. 3. p. 1693–703. Elsevier Applied Science Publishers, London.

Perry, C. C. and Lissner, H. R. (1962) *The Strain Gauge Primer,* McGraw-Hill, New York.

Pristone, G. and Roccati, R. (1988) Testing of large undisturbed samples of old masonry. *Proc. 8th Int. Brick and Block Masonry Conference* (ed. J. W. de Courcy), September, Dublin. Vol. 3, p. 1704–12. Elsevier Applied Science Publishers, London.

Sunley, V. K. (1983) The Rail Force Transducer and its possible use in other structures. *Colloq. on Instrumentation of Structures,* July, International Association for Bridge and Structural Engineering, Zurich.

Tyler, R. J. (1974) Measuring bending moments. *Tunnels and Tunnelling.* September, 6, p. 56–61.

Wang, Q. and Wang, X. (1988) The evaluation of compressive strength of brick masonry *in situ. Proc. 8th Int. Brick and Block Masonry Conference.* (ed. J. W. de Courcy), September, Dublin. Vol. 4, p. 1725–31. Elsevier Applied Science Publishers, London.

Chapter seven

Assessment of the load carrying capacity of arch bridges

M.A. Crisfield and
J. Page

7.1 Introduction

In this chapter we review both theoretical and experimental work on the analysis of masonry arch bridges. We start with the very early work and lead on to the assessment method at present in common use in the United Kingdom, the so-called MEXE (Military Engineering Experimental Establishment) method. Its mode of use is described and its limitations are discussed. Because of these limitations, work aimed at either replacing or updating the MEXE method is in progress at the Transport and Road Research Laboratory (TRRL) and elsewhere. We briefly review this

research and then concentrate on the work being undertaken by the TRRL. On the experimental side, this work includes both full-scale and model tests to destruction; on the theoretical side, it involves both 'mechanism' and non-linear finite-element analyses.

Figure 7.1(a) shows a side elevation of a typical masonry arch bridge and Fig. 7.1(b) illustrates two possible cross-sections. For the majority of bridges the fill between the outer spandrel walls is some form of earth or granular material which may or may not involve more substantial backing near the abutments (Fig. 7.1(a)). For some bridges, mainly of longer span, internal spandrel walls are provided. The work described in this chapter is mainly related to bridges without such internal walls.

7.2 Previous work

7.2.1 Early work

Howe (1897), who gave a fascinating history of arch bridges, believed that they were first used by the Chinese as early as 2100 BC. He also noted that a four-course brick arch, dating from about 1540 BC, was found in 'Campbell's tomb' in

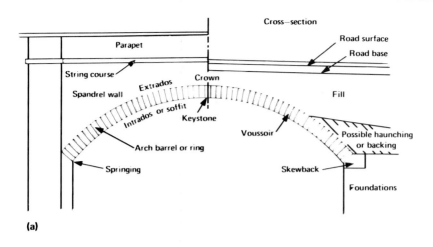

(a)

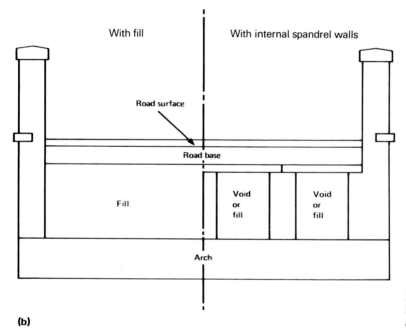

(b)

FIGURE 7.1 Typical masonry arch bridge construction: (a) elevation; (b) cross-sections.

Egypt. Despite this early history, the same author noted that 'the theory of the masonry arch has been and is now quite unsatisfactory from a practical point of view since we are unable to discover the directions and magnitudes of the forces caused by the spandrel filling'. We shall return to this point in Sections 7.2.2 and 7.5.1.

Reviews of the early literature relating to the analysis and design of arches have been given by Pippard and Chitty (1951), Heyman (1969, 1972, 1976, 1980, 1982), Withey (1982) and Tellet (1983). The brief summary given here relies heavily on the work of Heyman. Some of the earliest work (de la Hire, 1695, 1712; Couplet, 1729–30) involved both theoretical analysis and experiments and was related to the lines of thrust

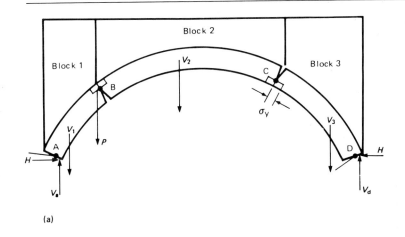

(a)

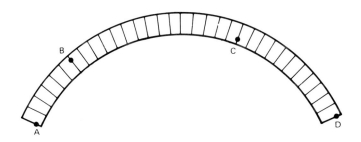

(b)

FIGURE 7.2 The mechanism method: (a) mechanism with equilibrating forces; (b) arch divided into elements.

and collapse mechanisms. Gregory (1697) established that the correct shape for an arch carrying its own weight was a catenary and observed that 'when an arch of any other shape is supported, it is because within the thickness some form of catenaria is included'. Heyman (1982) noted that this statement can be interpreted as defining the lower bound theorem of plasticity (Section 7.2.3). Couplet's concepts were later extended by Coulomb (1773).

Navier (1833) introduced the middle third rule which was applied to masonry arches by Rankine (1898) such that the line of thrust was constrained to lie within the middle third of the arch in order to avoid any tensile stresses. Further graphical work on the line of thrust was presented by

Barlow (1846) and Fuller (1875). Castigliano (1879) developed elastic methods via the theorems of minimum strain energy. However, for masonry arches he effectively applied an iterative non-linear no-tension analysis that is closely related to today's non-linear finite-element analyses (Section 7.2.5). Any parts of the arch that were found to be in tension were assumed to be cut out and a new analysis was performed on the resulting modified arch.

7.2.2 Rankine's comments relating to the fill or backing

Much of the previous work on arches has related to the arch ring alone. However, as pointed out by

83

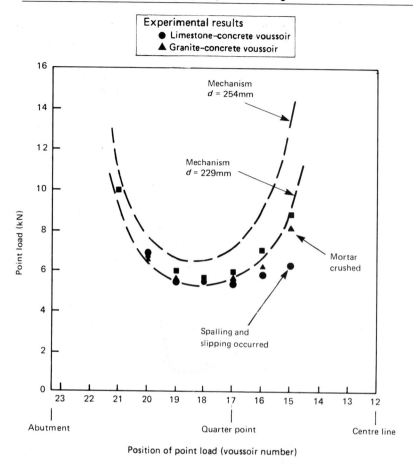

FIGURE 7.3 Collapse loads against load position for the arches of Pippard and Chitty (1951).

Howe (1897), Alexander and Thompson (1900, 1916) and Rankine (1898), many arches have considerable backing in their construction (Fig. 7.1(a)). This backing may consist of properly laid bricks or stones, in some cases bonded with mortar. Rankine writes 'The backing of an arch consists of block-in-course, coursed rubble, or random rubble and sometimes of concrete.' Strong backing is particularly likely for semi-circular or elliptical arches which would probably otherwise have been unstable during construction. Rankine also writes: 'many semi-elliptical arches may be treated as approximately hydrostatic arches (arches built of a figure suited to fluid pressure – that is, pressure of equal intensity in all directions)' and 'To give the greatest possible security to a hydrostatic arch, especially if the

span is great compared with the rise, the backing ought to be built of solid rubble masonry up to the level of the crown of the extrados.' Only that portion of the arch above the 'joints of rupture' was treated as the real arch and analysed with the aid of the middle third rule. Further, 'it is below these joints that conjugate (horizontal) pressure from without is required to sustain the arch, and that consequently the backing must be built with squared side-joints.' Unfortunately the procedure for computing these 'joints of rupture' is a little obscure. However, for 'most circular arches, the angles of rupture (which relate to a line normal to the arch ring at the joint of rupture) lie between 44° and 55° to the horizontal' (Rankine, 1898).

7.2.3 Work by Pippard, Heyman and others

Pippard and co-workers (Pippard *et al.*, 1936; Pippard, 1948, 1952; Pippard and Chitty, 1951; Pippard and Baker, 1957) made significant contributions involving elastic methods, collapse mechanism procedures and experimental work. In addition, Pippard (1948, 1952) made a major contribution to the existing assessment method (see Section 7.3). This was based on an elastic analysis involving a limiting compressive stress of 1.39 N/mm² (13 ton/ft²) and, in the initial stages of development, a middle half rule. This was to replace the more traditional middle third rule because both experiment and theory clearly indicated that there was a very considerable reserve of strength following first cracking. In addition to his elastic work, Pippard and his colleagues developed a tabular 'mechanism procedure' that was designed to compute the instability of the voussoir arch. This technique can be considered to be related to the mechanism procedures of Couplet and Coulomb and involved the collapse mechanism shown in Fig. 7.2(a). The assumptions are similar to those of plasticity theory (Heyman, 1982):

1. no tension
2. infinite compressive strength
3. infinite value for the modulus of elasticity, E
4. no sliding between voussoirs

Having introduced the four hinges A, B, C and D, we can now find the collapse load P (Fig. 7.2(a)) which have contributions from both the weights V_1, V_1 and V_3 of the three blocks (Fig. 7.2(a)) which have contributions from both the arch ring and the associated fill. Pippard assumed that hinge B was always under the load while hinge D was (as in Fig. 7.2(a)) at the far springing. The positions of hinge C and, to a lesser extent, hinge A, had to be obtained by trial and error using a rather tedious tabular method of computation. In relation to plasticity theory, this procedure can be considered an upper bound

method. TRRL's mechanism programme automates the procedure by dividing the arch ring into elements (Fig. 7.2(b)).

Pippard and co-workers conducted a series of experiments involving concrete voussoirs, some with cement mortar and others with weaker lime mortar. The fill was simulated by hanging weights from the voussoirs. The experimental results relating to the weaker mortar are plotted in Fig. 7.3 in conjunction with two sets of mechanism solutions (Crisfield, 1985a). Better agreement is obtained when the ring thickness is reduced from its true thickness of 254 mm to 0.9 times this value (229 mm). This agreement was obtained because 'in a number of experiments, the tension crack only extended to within 38 mm (85%) of the true depth' (Pippard and Baker, 1957). Pippard and co-workers also applied a repeated loading involving 10^5 cycles. They found that, although the load to produce first cracking was significantly reduced, the final collapse value was hardly altered.

Pippard did not use the terminology 'plastic method' in relation to his mechanism analyses. In contrast, Heyman (1969, 1980, 1982) related his techniques directly to the plasticity theorems of limit analysis. Thus an upper bound method would involve an assumed hinge configuration and the lowest upper bound should give a thrustline between hinges that is everywhere within the arch. In this case the safe or lower bound theorem would be satisfied because, 'If a thrust-line can be found, for the complete arch, which is in equilibrium with the external loading (including self weight), and which lies everywhere within the masonry of the arch ring, then the arch is safe' (Heyman, 1982).

For a quick approximate solution, Heyman (1980, 1982) assumed that the worst loading was at the quarter-point (see Fig. 7.3) with hinges at the load point, the springings and the centre of the arch. Simplifying assumptions were also made in relation to the geometry so that a simple single equation or tabular equivalent could be used directly to define the required depth of arch for a given loading. This procedure therefore involved the 'geometric factor of safety' (Heyman, 1969, 1980, 1982) which related the required depth to the actual depth. In relation to the simplified geometry, the thrust line did lie within the arch and hence the solution satisfied the safe theorem. For more accurate computations, Heyman proposed a semi-graphical procedure involving an initial estimate of hinge positions followed by new calculations if the thrust line lay outside the arch. The thrust line has also formed the basis of a semi-graphical appraisal method by Walklate and Mann (1983) in which the line of thrust was limited to the middle third.

7.2.4 Other mechanism methods and work relating to the structural contribution of the fill

Both Heyman and Pippard ignored any structural contribution from the fill. This assumption is in some conflict with the work of Rankine, Thompson and Alexander, and Howe (Section 7.2.2). Chettoe and Henderson (1957) conducted a series of tests on real arch bridges and measured deflections and abutment spread. As a result of these tests they advocated an elastic method of analysis in which no allowance should be made for any contribution from the spandrel walls. In addition, they found that there was no direct evidence of composite action with the fill. There was some debate on this point in the discussion of Chettoe and Henderson's paper, and they conceded that the point was 'not proven'. However, they argued that there 'seems to be some justification for considering the effective span to be reduced (as Rankine) where the vault is soundly haunched and backed, so that it can be assumed that part of the stiffly haunched portion of the vault is, in fact, part of the abutment'.

Davey (1953) also conducted a series of tests on real bridges but, in contrast with the work of Chettoe and Henderson, these tests were taken to collapse. Unfortunately they all involved loads at the crown. However, in relation to the contribu-

tion of the fill, Davey tested one bridge with loading and deflections at the quarter point and found that it was 2.5 times stiffer in the presence of the fill than in its absence.

Other work on the mechanism analysis of arches has been described by Crisfield (1985a, b, 1987) and Crisfield and Packham (1987) who directly introduce an approximate procedure for allowing for the lateral resistance of the fill by using its 'passive resistance'. The hinge near the springing remote from the load can also be raised with a view to representing 'rigid backing'. Similar procedures have been adopted by Harvey and Smith (1985), although the precise manner in which they allow for the 'passive resistance' is not entirely clear. Other work on the mechanism method has been described by Delbecq (1982) and Davies (1985). The former included an allowance for compressive yielding in the arch ring.

7.2.5 The finite-element method

A considerable amount of finite-element research has been devoted to the analysis of arches and shells. The first application of finite elements to masonry (or, more specifically, brickwork) arches appears to be that of Sawko and Towler (1981) and Towler (1981, 1985) who compared their theoretical solutions with their own experimental results on a series of brickwork arch models. This work showed the potential of the finite-element method not only in computing the collapse load but also in providing data on the extent of cracking due to abnormal loading. In both this work and some of the subsequent analyses by Crisfield (1984) no allowance was made for interaction with the fill. Crisfield (1985a, b) later showed that, in these circumstances, the finite-element method should give lower collapse loads than the mechanism method but that Towler's results did not always achieve this end. Work with the mechanism method on deep arches indicated that absurdly low collapse loads could be produced if no allowance was made for the lateral resistance of the fill or backing (Crisfield, 1985b,

1987). Hence, non-linear springs were introduced into the finite-element model (Crisfield 1985a, b, 1986) with a view to simulating the lateral resistance of the fill. More recent work by both Crisfield (1985b, 1986, 1987) and Towler (1985) has attempted to model the fill directly. In Crisfield's case, this was achieved with the aid of the Mohr–Coulomb yield criterion.

7.2.6 Experimental work

The contributions of Chettoe and Henderson and of Davey, which involved full-scale tests, have already been discussed in the previous sections as has model work by Pippard and Towler. The full-scale tests undertaken as part of TRRL's programme of research will be described in Section 7.4, and model work will be covered in Sections 7.5.2(h) and 7.7.

7.3 The current assessment method

7.3.1 The method

In the current method the load-carrying capacity of the arch barrel alone is assessed. This capacity will be influenced by the strength of the spandrel walls, wing walls, fill, foundations etc., but these are not directly taken into account. The reader is referred to Department of Transport *Departmental Standard BD 21/84* (1984) and *Advice Note BA 16/84* (1984) for complete advice on its application; a summary is given below.

A 'provisional axle load', based on the geometry of the arch, is first calculated. Five modifying factors are then applied to provide an allowable axle load for a two-axle bogie. The permitted axle load for a single axle or for a three-axle bogie can be calculated from this and can then be modified for the case where the longitudinal road surface profile of the bridge causes one axle of a bogie to lift off either partially or fully.

The geometric data required are as follows: the span L (m); the rise r_c (m) of the arch barrel at the crown; the rise r_q (m) of the arch barrel at the quarter-points; the thickness d (m) of the arch barrel adjacent to the keystone; the depth h (m) of fill (including road surfacing) at the arch crown. A provisional axle load (PAL) is then calculated either from a nomogram or from the following expression:

$$PAL = 740(d+h)^2/L^{1.3} \text{ tonnes} \qquad (7.1)$$

The limits of applicability of the formula are 1.5 m and 18 m for span L, and 0.25 m and 1.8 m for $d+h$. The following factors are then applied to the calculated value of PAL.

1. **Span:rise factor** F_{sr} It is assumed that flat arches are not as strong as deep arches. A span:rise ratio of 4 or less is given a factor of 1.0, and this is reduced for span:rise ratios greater than 4.

2. **Profile factor** F_p The ideal arch profile is assumed to be parabolic and for this shape the rise at the quarter-points is given by $r_q = 0.75r_c$. The profile factor for $r_q/r_c \leq 0.75$ is taken to be unity, and is less than unity for $r_q/r_c > 0.75$.

3. **Material factor** F_m The material factor is determined from the material type and thickness of the arch barrel, and the material type and thickness at the arch crown of the fill.

4. **Joint factor** F_j The strength and stability of the arch barrel depend to a large extent on the size and condition of the joints. The joint factor is determined from the width and depth of the joints, and the quality of the mortar.

5. **Condition factor** F_c The estimation of the preceding factors is based on quantitative information obtainable from a close inspection of the structure, but the condition factor depends much more on an assessment of the importance of the various cracks and deformations that may be present and how far they can be counterbalanced by indications of good material and workmanship. The value of F_c varies between zero and unity.

The five modifying factors are then applied to the value of PAL to give a modified axle load as follows:

$$\text{modified axle load} = F_{sr}F_{p}F_{m}F_{j}F_{c} \, \text{PAL} \qquad (7.2)$$

This represents the allowable axle loading for a double-axle bogie with no 'lift-off'. Permitted axle loads for single axles and for tri-axle trailers can then be calculated for situations where axle lift-off may or may not occur. If the axle loads calculated in this way fall below the weight limits permitted by the Construction and Use Regulations, then the gross vehicle weight restrictions to be applied to the bridge are defined.

7.3.2 History and theory

In 1933 the owners of bridges in Great Britain were empowered to restrict the maximum weight of vehicles permitted to cross any given bridge, and did so on the basis of methods of strength assessment then thought appropriate. However, experience showed that vehicles very much in excess of the permitted weight could be carried without apparent ill effects. It seemed that the assessment methods used were unduly conservative. Therefore in 1936 the Building Research Station (BRS) began an investigation on behalf of the Ministry of Transport into the behaviour under load of various types of bridge, in particular masonry arch bridges. In the years between 1936 and the outbreak of war many loading tests were carried out on selected bridges, and three of them were tested to destruction. In addition BRS commissioned Professor Pippard to undertake a model and theoretical study of the behaviour of arch bridges under load.

In 1942 the then Ministry of Supply wanted to be able to classify bridges according to their capacity to carry military loads. A method based on the work of Pippard *et al.* (1936) and Davey (1953) was evolved and by the 1950s this had become the so-called MEXE method. It was subsequently adapted for civil use and appears in

its most recent form in Department of Transport *Departmental Standard BD 21/84* (1984) and *Advice Note BA 16/84* (1984).

Pippard used an elastic analysis and is known to have made the following assumptions in his calculations of load-carrying capacity (Pippard and Chitty, 1951; Heyman, 1982; Larnach, 1987).

1. The arch was parabolic;
2. The arch was two pinned;
3. Its span:rise ratio was 4.0;
4. The arch section increased from the crown to the abutments because of an assumption made to simplify the analysis;
5. Load was applied as a point load at midspan. An arch rib is weakest under the action of a point load at about quarter-span, but Pippard argued that, allowing for transverse dispersion of the load, a greater width of arch would be available to carry the load at quarter-span than at the crown;
6. Fill density was the same as masonry density (a value of 140 lb/ft³ (2240 kg/m³) was assumed);
7. The fill had no structural strength;
8. A limiting compressive stress of 13 ton/ft² (1.39 N/mm²) was applied;
9. In early calculations, a middle-third rule was adopted in order to prohibit tensile stresses. A less restrictive middle-half rule was later substituted but even this was eventually abandoned. However, Pippard noted that 'the maximum tensile stress produced in no case exceeded 100 lb/sq.in (0.69 N/mm²). This contention has recently been disputed by Larnach (1987).

Pippard himself allows that the assumptions are 'sweeping' and many further assumptions must be involved in the conversion of the basic calculations into the assessment method, for instance in the definitions of the various modifying factors. Heyman notes that a curiosity of the method is that the arch ring thickness d does not

TABLE 7.1 Description of bridges tested

	Bridgemill	Bargower	Preston	Prestwood	Torksey	Shinafoot
Span (square) (mm)	18300	10000	4950	6550	4900	6160
Span (skew) (mm)	–	10360	5180	–	–	–
Angle of skew (deg)	0	16	17	0	0	0
Rise at midspan (mm)	2850	5180	1636	1428	1155	1185
Arch thickness at crown (mm)	711	558	360	220	343	540
Arch shape	Parabolic	Segmental	Elliptical	Originally segmental	Segmental	Segmental
Arch material	Sandstone	Sandstone	Sandstone	Brick	Brick	Stone
Arch density (kN/m³)	2.1	2.7	2.3	2.0	2.1	2.5
Fill density (kN/m³)	2.2	2.1	2.2	2.0	2.0	2.0
Spandrel wall thickness (mm)	–	1400	610	380	380	365
Parapet thickness (mm)	–	400	370	–	380	365
Spandrel/parapet material	Sandstone	Sandstone	Brick	Brick	Brick	Stone
Total width (mm)	8300	8680	5700	3800	7805	7030
Fill depth at crown (mm)	203	1200	380	165	246	215

enter directly into the MEXE calculations (except in the calculation of F_m). Instead, the total thickness $d+h$ of ring and fill at the crown is used. Larnach has shown that this can lead to violations of the compressive permissible stress criterion (point 8). Larnach also notes that Pippard's calculations are based on combined dead and live loading but shows that it is possible for the limiting compressive stress to be exceeded for dead load alone. In addition, he found it difficult to relate the final tables to the original calculations. Nevertheless, the method is easy to use and has served well, but is now considered to be conservative, particularly for longer spans. It also has an additional shortcoming in that spans are limited to 18 m and distorted arches cannot be assessed. Consequently, the TRRL has embarked on a research programme to revise or replace it.

The current research is described in the following sections and the possibilities for a revised assessment method are briefly discussed in the final section. It should be noted that *BD 21/84* permits the use of limit state methods of analysis for masonry arch bridges.

7.4 Experimental work

The following factors affect the traffic load-carrying capacity of arch bridges: span, rise, width, arch shape including distortions, arch thickness, depth of fill, arch material including defects, fill material including surfacing, quality of mortar, thickness of spandrel walls, degree of bond between arch and spandrel walls, strength and stiffness of foundations, and applied load including its position, form and distribution through the fill and surfacing.

It is clear that no series of full-scale tests of a realistic size can hope to explore the effects of all these factors, and so the test programme is concentrating on variations in span, span:rise ratio and arch material. The bridges used in the tests become available to TRRL because of road realignment schemes etc. At the time of writing, six full-scale tests have been completed. A description of the bridges is given in Table 7.1 (NB Arch and fill densities are given in units suitable for use by the TRRL mechanism and finite-element programs (see Sections 7.5 and 7.6). Some of the values are assumed). References to detailed

(a)

(b)

(c)

(d)

(e)

(f)

descriptions of the bridges and the tests are also given in Table 7.1. Photographs of the bridges are shown in Fig. 7.4.

7.4.1 The load tests

Tests on Bridgemill and Bargower bridges were conducted by Professor Hendry of Edinburgh University, under contract to TRRL, and the remainder were performed by Page and Grainger of TRRL. Load was applied along a transverse line at road surface level at a position above the arch calculated (by the TRRL mechanism method – see Section 7.5) to require the lowest load to produce collapse. This position was at a quarter-span point for Bridgemill, Prestwood, Torksey and Shinafoot and at a third-span point for Bargower and Preston. In practice the line was made wide enough to avoid a premature failure of the fill. This width was 750 mm for all the bridges other than Prestwood; for Prestwood it was 300 mm because of the low collapse load that was expected. The load was applied to the full width of the bridge between parapets (in the case of Prestwood Bridge it was applied to the full width of the bridge because the parapets were missing).

Load was applied using hydraulic jacks reacting, via a loading frame, against ground anchors embedded in the ground beneath the bridge and passing through holes drilled through the bridge deck. Load cells were installed in the loading frame to measure the load being applied. In the case of Prestwood Bridge, where a small collapse load was expected, load was applied via hydraulic jacks reacting against kentledge supported on a frame above the bridge deck which rested on plinths on the road surface clear of the arch.

Displacements during loading were measured by surveying because this technique requires little equipment to be attached to the bridge. Up to 20 targets were attached to the bridges, mostly on the soffit of the arch. Two other techniques were tried at Preston Bridge: high resolution moiré photography and photogrammetry. The acoustic emission technique (see Chapter 6, Section 6.4.3 (e)) has been used during several of the tests to try to detect the development of cracking during loading. The arch soffits and spandrel walls were sprayed with white emulsion paint to make it easier to see the development of cracks during loading. Still cine and video photographs were taken as the tests proceeded.

The test procedure is to apply a predetermined load and then to cut off the oil supply to the jacks until the various measurements are completed. A further increment of load is then applied and the process is repeated; the aim is to apply about 20 increments before collapse occurs. Towards the end of the test it is likely that the applied load will decrease as the displacement increases; at this stage predetermined increments of displacement rather than load are applied. The maximum loads that were applied to the bridges before collapse are listed in Table 7.2.

Curves of applied load against vertical displacement of the arch ring beneath the load line are shown in Fig. 7.5. In all cases the curve is non-linear from the very start and there is no obvious point at which non-linearity can be said to have commenced. Prestwood was unusual in that no deflection was detected until a load of about 80 kN or a third of the collapse load had been applied. The load at which the first visible signs of damage occurred was high: about half the maximum load at Bridgemill, three-quarters the maximum load at Prestwood and the maximum load at Torksey and Shinafoot.

FIGURE 7.4 (*facing page*) The six bridges: (a) Bargower Bridge; (b) Bridgemill Bridge; (c) Preston Bridge; (d) Prestwood Bridge; (e) Torksey Bridge; (f) Shinafoot Bridge.

TABLE 7.2 Maximum applied loads and comparison with MEXE analysis

	Bridgemill	Bargower	Preston	Prestwood	Torksey	Shinafoot
Maximum applied load (kN)	3100	5600	2100	228	1080	2500
MEXE analysis Provisional axle load (t)	14.13	109.5	47.8	9.5	32.5	39.9
F_{sr}	0.715	1.0	1.0	0.92	1.0	0.832
F_p	0.98	0.69	0.77	0.92	0.92	0.97
F_m	1.32	0.95	1.09	0.7	0.87	1.27
F_j	0.9	0.81	0.81	0.73	0.73	0.8
F_c	0.9	0.7	0.8	0.5	0.7	0.9
Modified axle load (t)	10.6	40.7	26	2.06	13.3	29.4
Single-axle L_s (t)	18.5	65.5	30.4	2.78 (1.78)[a]	14.9	38.3
$2L_s$ (i.e. two single axles side by side) (kN)	363	1285	596	17.5[b]	292	751
Maximum applied load/$2L_s$	8.5	4.4	3.5	13.1[b]	3.6	3.4

[a]The calculated value has been modified using a mechanism analysis to make allowance for the distorted arch shape. The modified value is slightly different from that given by Page (1987).
[b]The bridge was not wide enough for two axles side by side.

7.4.2 Collapse modes

(a) Bridgemill Bridge (Hendry *et al.*, 1985)

Collapse was not achieved (because of inadequate stroke to the jacks). However, it is believed that little further increase in applied load would have been obtained. At the maximum load, a 'hinge' was visible beneath the load line and the acoustic emission measurements indicated the formation of hinges close to the two springings and between the load line and the further springing.

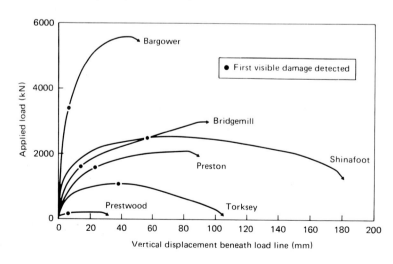

FIGURE 7.5 Load–displacement curves for the TRRL bridge tests.

FIGURE 7.6 Preston Bridge shortly before collapse.

(b) Bargower Bridge (Hendry *et al.*, 1986)

The longitudinal cracks, about 1.0–1.5 m away from the faces of the arch, developed as the load increased, indicating that the central more flexible section of the arch was splitting from the outer sections which were stiffened by the spandrel walls. There were signs of hinges developing beneath the load line and close to the arch crown but failure occurred due to compression in the arch ring causing horizontal splitting of voussoirs which then progressively dropped out of the ring.

(c) Preston Bridge (Page, 1987)

As at Bargower, the central section of the arch attempted to split from the outer edges stiffened by the spandrel walls. Hinges were seen to be developing beneath the load line, close to the adjacent springing and roughly halfway between the loadline hinge and the further springing. However, collapse occurred as the result of a crushing failure of the arch voussoirs beneath the load line. Figure 7.6 shows the condition of the arch shortly before collapse.

(d) Prestwood Bridge (Page, 1987)

A classical mechanism failure occurred, with a hinge developing beneath the load line, two close to the springings and one between the load line hinge and the further springing. Negligible material damage occurred other than at the hinges. Figure 7.7 shows the bridge shortly before collapse.

FIGURE 7.7 Prestwood Bridge shortly before collapse.

(e) Torksey Bridge (Page and Grainger, 1987a)

Three hinges developed, one beneath the load line, one close to the nearest springing and one close to the arch crown. Failure then occurred by a 'snap-through' of these hinges. The spandrel walls remained standing.

(f) Shinafoot Bridge (Page and Grainger, 1987b)

A four-hinged mechanism failure occurred, although the hinges were not as well defined as at Prestwood because of the random nature of the masonry. Both spandrel walls had previously collapsed.

7.4.3 Comparisons with the MEXE analysis

The MEXE analysis provides a permissible axle load and is therefore not directly comparable with a collapse load. An indication of the factor of safety provided by MEXE can be obtained by calculating the permissible load on two single axles side by side and comparing this with the collapse load. This comparison is provided in Table 7.2 where, following MEXE, units of tonnes have been adopted (1 tonne = 9.80665 kN). It should be remembered that the legal weight limit for a single axle is at present 10.5 tonnes in the United Kingdom.

7.5 The modified mechanism method of the Transport and Road Research Laboratory

The basic mechanism method has already been discussed in Section 7.2.3. The main modifications that have been introduced in the TRRL mechanism programme are as follows.

1. An allowance is made for compressive yielding at the hinges so that the centres of rotation A–D lie inside the arch (Fig. 7.8). Strictly, the introduction of a compressive yield stress implies that the material of the arch ring exhibits infinite ductility. Clearly, this is not true and hence a reduced compressive strength σ_y should be adopted;
2. An approximate allowance is made for the lateral resistance of the fill which relates to an input angle of friction. Details will be given in Section 7.5.1;
3. The procedure is fully computerized so that the worst hinge configuration and worst loading position are automatically obtained. However, it is possible to specify various hinges or all of them. The programme also computes the lowest collapse load and plots the hinge positions and thrust line.

In relation to Fig. 7.8 a virtual work relationship involves

$$P\Delta p + V_1\Delta_1 = V_2\Delta_2 + V_3\Delta_3 + H_f\Delta_h +$$
$$\tfrac{1}{2}d_a^2\sigma_y + \tfrac{1}{2}d_b^2\,\sigma_y\,(1 + \Delta\theta_2) + \tfrac{1}{2}d_c^2\theta_y$$
$$(\Delta\theta_2 + \Delta\theta_3)\quad \tfrac{1}{2}d_d^2\sigma_y\Delta\theta_3$$

where Δ_1, Δ_2 and Δ_3 are the virtual displacements undertaken by the centroids of the dead weights of blocks 1, 2 and 3 respectively (the latter include the weight of both the arch and the associated fill (see Fig. 7.1)), $\Delta\theta_1$, $\Delta\theta_2$ and $\Delta\theta_3$ are virtual rotations, d_a, d_b, d_c and d_d are the depths of the compressive yield blocks, and H_f is the total horizontal force assumed to be provided by the passive resistance of the fill. Details will be given in Section 7.5.2. The relationships between Δ_1, Δ_2, Δ_3, Δ_h, $\Delta\theta_1$, $\Delta\theta_2$ and $\Delta\theta_3$ are found from basic kinematics, and the depths d_a, d_b, d_c and d_d of the stress blocks are found from statics in conjunction with an iterative technique (Crisfield and Packham, 1987). In concept, equation (7.1) is applied for the complete range of possible hinge positions until the minimum collapse load P_{min} is obtained. To this end, the arch barrel is divided into a set of elements (Fig. 7.2(b)) which should strictly coincide with the stone voussoirs if hinges are only to occur at the mortar joints. However, this would clearly be impractical for brickwork arches, and current practice is to provide a large number (say 50) of such elements. For a given set of hinge positions, the weights V_1, V_2, and V_3 of the blocks are simply computed by summing the

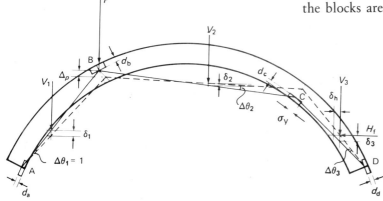

FIGURE 7.8 Mechanism with virtual deformations.

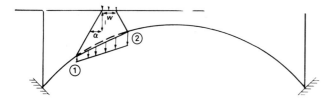

FIGURE 7.9 Assumed distribution through the fill.

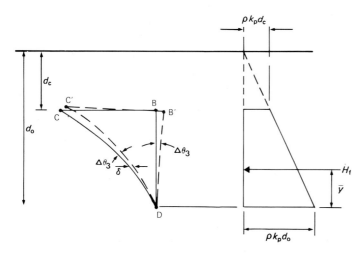

FIGURE 7.10 Assumed 'passive action' from the fill.

weights of the included elements in conjunction with their associated vertical strips of fill.

Allowance is made for distribution through the fill and surfacing so that the initial load length (w in Fig. 7.9) is increased by the time that the distributed load reaches either the ring extrados or its centre line. The user can specify either of these options and must also input the distribution factor α (see Fig. 7.9). In computing the loads that are applied to the arch, the pressures at points 1 and 2 (Fig. 7.9) are assumed to be inversely proportional to their depths. The pressure distribution between points 1 and 2 is then assumed to vary in a linear manner and the $P\Delta p$ term in equation (7.3) is replaced by $\Sigma P\Delta p$ where the P are the loads on the ring elements resulting from the distributed pessure and the Δp are the equivalent virtual displacements. Details are given by Crisfield and Packham (1987).

The programme changes the hinge positions in a search for the lowest load. To this end a logic circuit is used to minimize the number of trial configurations. This circuit uses the gradient of the computed collapse load in relation to the hinge position.

7.5.1 Approximate allowance for the lateral resistance of the fill

It will be shown that the mechanism method gives absurdly low collapse loads for deep arches unless some allowance is made for the lateral resistance of the fill or backing. It has already been mentioned that a possible procedure for simulating the effects of sound backing involves raising the hinge near the springing that is remote from the load. This can easily be achieved by a change in the data relating to the permitted range of nodes for hinge D (Figs. 7.2 and 7.8). However, while this procedure may possibly be valid when the user is certain that very substantial backing or haunching is available, this knowledge will often be unavailable. Consequently, it is assumed that the portion CD of the arch in Fig. 7.8 meets a lateral resistance H_f from the fill as it rotates. The virtual work contribution of $H_f\Delta_f$ is therefore included in equation (7.3). It is further assumed that this work is equivalent to the work that would be obtained when a rigid block DCB (Fig. 7.10) with a frictionless surface DB rotates into the soil. In these circumstances, the concept of 'passive pressure' can be invoked and the force H_f is given by

$$H_f = \tfrac{1}{2}\rho k_p (d_o^2 - d_c^2) \tag{7.4}$$

where k_p is the passive resistance factor which is given by

$$k_p = \frac{1 + \sin \varphi}{1 - \sin \varphi} \tag{7.5}$$

Here φ is the angle of friction of the fill. The force H_f acts at a distance $\bar{y}$ above D (Fig. 7.10), where

$$y = d_0 - \tfrac{2}{3}\frac{d_0^3 - d_c^3}{d_0^2 - d_c^2} \tag{7.6}$$

and $\Delta_f = \bar{y}\Delta\theta_3$. It can be shown (Crisfield and Packham, 1987) that this procedure leads to an overestimation in comparison with the work that would be done by a smooth surface CD. At the cost of considerable extra complexity, it might be possible to incorporate the latter action. However, there are many other strengthening factors, such as the possible contributions from the spandrel walls, that have been neglected. Consequently, it is reasonable to use the proposed simple procedure, at least as a research tool, to assess the experimental and other theoretical results.

TABLE 7.3 Results from the mechanism program, destructive tests and the MEXE assessment method

	Bridgemill	Bargower	Preston	Prestwood	Torksey	Shinafoot
Span/rise	6.44	2.0	3.17	4.59	4.28	5.20
Span/thickness	25.7	18.6	14.4	29.8	14.2	8–15.8
Approx. shape	Segmental	Semicircular	Semi-elliptical	Distorted	Segmental	Segmental
Load position[a]	1/4	1/3	1/3	1/4	1/4	1/4
Figure	7.11	7.12	7.13	7.14	7.15	7.16
Experimental collapse load (kN)	3100	5600	2100	228	1060	2525
Program upper bound (kN)	5495	13873	1873[b]–9216[c] 1259[d]	168	2668	22400
Program lower bound (kN)	1816	0.1	154	42	705	965
Factored (3) MEXE (kN)[e]	1089(8.5)	3854(4.4)	1790(3.5)	53(12.9)[f]	877(3.6)	2253(3.4)
Program best estimate (kN)	2545–3640	5751–7726	1248[b]–1761[c] 1686[d]	177	1415	2330
Fill factor k_p	3.0	3.0	4.0–3.0	3.0	3.0	3.0
σ_y	4.0–10.0	4.0–10.0	4.0	4.0	4.0	4.0

[a] As approximate ratio of internal span.
[b] Thrust line clearly outside the arch.
[c] Position of hinge D limited to account for the effect of backing.
[d] Solutions with automatically reduced k_p.
[e] The values in parentheses give the achieved safety factor in relation to the MEXE assessment.
[f] Modified MEXE values (Section 7.5.2) because the arch is distorted.

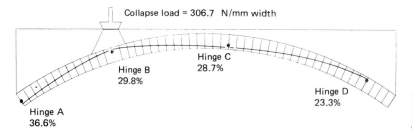

FIGURE 7.11 Mechanism solution for Bridgemill Bridge (σy = 4 N/mm²; kp = 3.0).

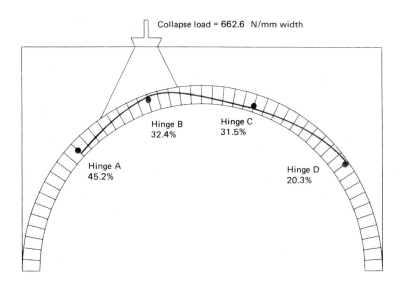

FIGURE 7.12 Mechanism solution for Bargower Bridge (σy = 4 N/mm²; kp = 3.0).

Recent experimental work by Melbourne (in press) has indicated a very appreciable build-up of lateral pressure due to traffic loading. Given this starting point, we can anticipate that, as a result of an excessive overload at B in Fig. 7.8, the arch would experience a significant reduction in lateral pressure on the left-hand portion AB as well as some build-up on the right-hand portion CD (say, as H_f in Fig. 7.8). The net effect would still involve a considerable lateral resistance against sway to the right which, in the limit as the deflections become large, could approach those assumed in the present 'ultimate limit' calculation.

7.5.2 Application of the mechanism programme

The programme has been used to compute the collapse strength of a range of bridges tested in TRRL's recent research programme (see Section 7.4). Before detailing these solutions, it should be emphasized that there are many factors that can significantly influence the experimental results. In particular, the direct contribution of the spandrel walls can either be large or negligible. In practice, a tool such as the current mechanism program would be used to compute realistic but safe lower bounds. It is rather a different matter to predict the results from an experiment. The material and other factors used in the following comparisons should not be taken as a guideline for future

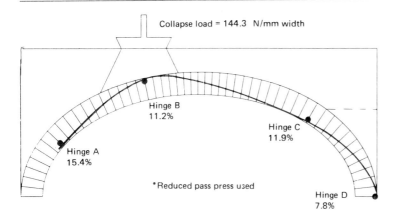

Collapse load = 144.3 N/mm width

Hinge B
11.2%

Hinge C
11.9%

Hinge A
15.4%

*Reduced pass press used

Hinge D
7.8%

FIGURE 7.13 Mechanism solutions for Preston Bridge (σy = 4 N/mm²; kp (reduced) = 2.13).

assessment purposes but rather as a first attempt to investigate the possibilities of the computer program.

The dimensions and densities of the full-scale bridges have already been given in Table 7.1. Table 7.3 summarizes the mechanism results for which some of the computed hinge positions and thrust lines are shown in Figs. 7.11–7.16. Table 7.3 also includes the MEXE solutions from Table 7.2 after they have been multiplied by an assumed 'safety factor' of 3.0. In this table and the following discussion, the terms 'upper bound' and 'lower bound' are used rather loosely in an engineering sense and do not relate to the plasticity theorems (see Section 7.2.3 and Harvey and Smith, 1985).

For the upper bound analyses, the values of the passive fill factor k_p, the compressive strength σ_y in the arch and the load-distribution factor α used in the upper bound analyses were 4.6 ($\varphi = 40°$), 1000 N/mm² (effectively infinite) and 0.5 respectively. For the lower bound solutions, these values were 0.01 (approximately zero), 4 N/mm² and zero respectively. For the upper bound solutions, the loads were distributed to the centreline of the archring. For most of the best-estimate solutions, σ_y was set to 4 N/mm² and k_p to 3.0 ($\varphi = 30°$) but other values were also considered (see foot of Table 7.3). For these solutions, the distribution factor α (Fig. 7.9) was set to 0.5 and the loads were distributed to the extrados.

(a) Bridgemill Bridge

The compressive yield strength used in the analysis must be reduced below that of the parent stone in order to allow for the mortar, relate to the stone-work and compensate for the limited ductility. For ashlar stone, with a compressive strength of 44 N/mm² (as measured for Bridgemill) *BD 21/84* gives a characteristic strength of about 7.5 N/mm². Consequently, the first best-estimate solution involved σ_y = 10 N/mm². However, a better agreement with the experimental collapse load was obtained when σ_y was reduced to 4 N/mm² (Fig. 7.11). Such a reduction could be justified because Bridgemill arch was very shallow and the mechanism method does not allow for geometric effects relating to snap-through. Hence, a reduced compressive strength can be used to allow for these effects artificially (Crisfield and Packham, 1987).

(b) Bargower Bridge

The upper and lower bound results for the deep semicircular Bargower Bridge (Fig. 7.12) show a very wide range. This is typical of solutions for deep arches for which the strengths depend strongly on the fill or backing and, in terms of the mechanism programme, on the value adopted for k_p. If a compressive strength σ_y of 10 N/mm² is assumed for the stonework, in comparison with a recorded value of 33 N/mm² for the parent stone, the best-estimate solution is 7726 kN which is

38% higher than the experimental collapse load. However, the load on this bridge was cycled several times before the maximum load was reached. Consequently, the final failure, which was associated with compressive material damage, may possibly have been affected by the previous cycling. This possibility is reflected in an alternative best-estimate mechanism solution (see Fig. 7.12 and Table 7.3) which, with σ_y set to 4 N/mm², gives a collapse load (5751 kN) that is very close to the experimental value.

(c) Preston Bridge

Considerable backing was found near the springings of the semi-elliptical Preston Bridge (to the level of the broken line in Fig. 7.13). As a first attempt to allow for this backing, the best-estimate solution involved a high k_p factor of 4.0 (equivalent to $\varphi = 37°$). However, this solution gave an unacceptable thrustline that lay outside the arch and involved an outward rather than an inward reaction at the springing. This difficulty was overcome within the computer program by automatically reducing the passive pressure factor k_p to make this reaction zero (Fig. 7.13). The resulting thrustline lay within the archring and involved hinge positions that were in reasonable coincidence with those obtained from a finite-element analysis.

An alternative procedure to account for the backing involved raising the hinge position of the effective right-hand springing. A solution that was 20% too low was obtained when the right-hand hinge D was constrained not to form below hinge 45 (halfway up the backing (Fig. 7.13). A

similar constraint on the right-hand hinge was required in order to produce an upperbound solution that lay above the experimental value. The difficulties with the thrustlines and negative reactions seem to be limited to elliptical arches. No such problems were encountered when the k_p value was set to zero. However, the resulting 'lower bound' collapse load was very low (154 kN).

(d) Prestwood Bridge

The arch barrel for Prestwood Bridge (Fig. 7.14) was severely distorted so that the MEXE method was not strictly applicable (see also Section 7.3). Consequently, the MEXE value given for this bridge in Table 7.3 was obtained by reducing the full MEXE value by a factor of 0.64 which was the reduction ratio which the mechanism program solutions indicated would result from the distortion when the bridge was loaded at the internal quarter-point. The mechanism results given in the table relate to the measured distorted shape (Fig. 7.14). The results relating to a perfect segmental shape were about 56% stronger.

Prestwood was the only bridge for which the experimental collapse load did not lie between the computed upper and lower bounds. It was also the only bridge tested with the line load spanning across, rather than inside, the spandrel walls. This, coupled with narrowness of the bridge, would have led to a more significant contribution from the spandrel walls and could have accounted for the extra (about 36%) strength.

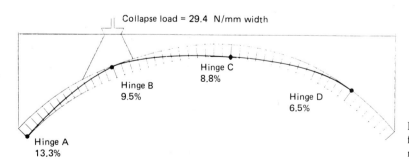

Collapse load = 29.4 N/mm width

Hinge C
8.8%

Hinge B
9.5%

Hinge D
6.5%

Hinge A
13.3%

FIGURE 7.14 Mechanism solutions for Prestwood Bridge (σ_y = 4 N/mm²; k_p = 3.0).

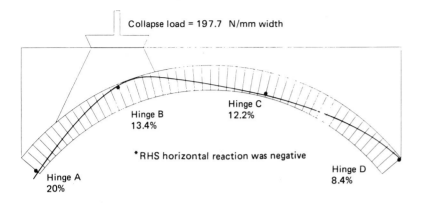

FIGURE 7.15 Mechanism solution for Torksey Bridge (σ_y = 4 N/mm²; k_p = 3.0).

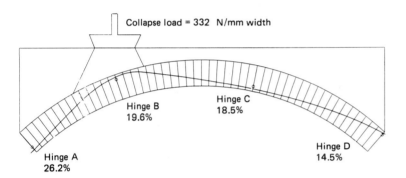

FIGURE 7.16 Mechanism solution for Shinafoot Bridge (σ_y = 4 N/mm²; k_p = 3.0).

(e) Torksey Bridge

The best-estimate mechanism solution for Torksey Bridge (Fig. 7.15) was about 33% too high. Experimentally, however, there was evidence of some form of snap-through behaviour before the formation of all four hinges. For this bridge, the spandrel walls were completely detached from the arch barrel so that no additional stiffening effect would be expected.

(f) Shinafoot Bridge

Shinafoot Bridge was made from rubble masonry and had a very variable ring thickness. The present version of the mechanism program will not handle variable thickness rings (although this deficiency is being corrected). Consequently, for the upper bound solution in Table 7.3 the thickness was set to the mean measured value at the crown (540 mm), while for the lower bound solution it was set to the minimum measured thickness of 390 mm. A value of 470 mm (halfway between the two previous values) was adopted for the best-estimate solution (Table 7.3 and Fig. 7.16). The resulting collapse load was very close to the experimental value (see Table 7.3).

(g) The MEXE solutions

Only for Preston Bridge did the MEXE assessment method, when combined with a factor of safety of 3.0, give a collapse load that lay closer to the experimental value than the mechanism solutions. For both this bridge and the Torksey and Shinafoot bridges, the MEXE method indicated safety factors of about 3.5. The inferred safety factors for the other bridges were significantly higher (see Table 7.3).

(h) Solutions for a model bridge

A part of a research programme sponsored jointly by the Science and Engineering Research Council

101

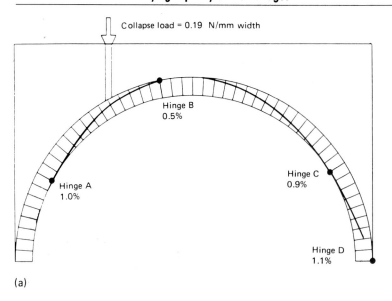

Collapse load = 0.19 N/mm width

Hinge B
0.5%

Hinge C
0.9%

Hinge A
1.0%

Hinge D
1.1%

(a)

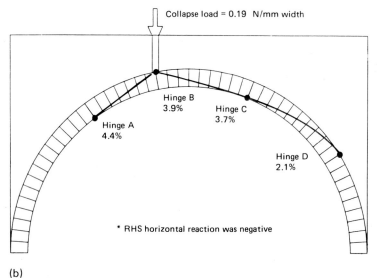

Collapse load = 0.19 N/mm width

Hinge B
3.9%

Hinge C
3.7%

Hinge A
4.4%

Hinge D
2.1%

* RHS horizontal reaction was negative

(b)

FIGURE 7.17 Mechanism solutions for the Bolton model: (a) without passive pressure; (b) with passive pressure.

(SERC) and the TRRL, a model bridge with a span of 1 m was tested at Bolton Institute of Higher Education. It was made from model bricks and was 500 mm wide with 10 mm-down limestone aggregate for the fill. The rise was 500 mm, the arch thickness was 53 mm and the depth of fill at the crown was 100 mm. The compressive strength of the brickwork was estimated to be 4.0 N/mm². From the analysis viewpoint, the interesting features of this model are firstly that the load was applied directly to the arch and secondly that there were no spandrel walls although the fill was constrained from moving laterally.

The experimental collapse load was 4.4 kN but this related to a tensile strength in the mortar and the final load, at which a plateau was recorded, was about 3.5 kN. For comparison purposes, the

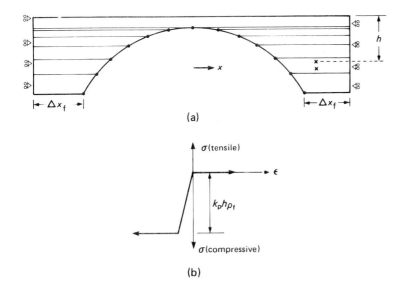

(a)

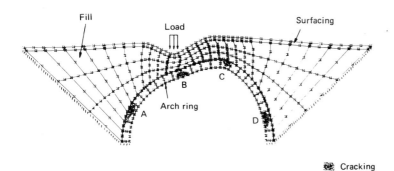

(b)

FIGURE 7.18 Idealization for lateral fill effects with simple finite-element (one-dimensional) model: (a) arch with fill elements; (b) assumed horizontal stress–strain relationships in fill.

FIGURE 7.19 Computed deformations (exaggerated scale) and crack zones for Bargower Bridge.

latter should be treated as the collapse load because collapse would probably have occurred at this loading if the mortar had been unable to sustain tension.

If this arch is analysed with allowance for the weight of the fill but with no allowance for any horizontal resistance, the computed collapse load (see Fig. 7.17) is only 0.096 kN or about 3% of the experimental load! However, when allowance is made for horizontal resistance by introducing a passive pressure factor k_p of 5.83, the computed collapse load is increased to 3.34 kN which is very close to the experimental value. The computed hinge positions also compared reasonably with the experimental positions (Crisfield and Packham, 1987). This agreement was obtained when the

angle of friction φ for the fill was set to 45° (to give k_p = 5.83). The draft report (Walker and Melbourne, 1987) quotes a very high value of 55° which was obtained from a shear-box test. Although there was no direct contribution to the strength from any spandrel walls, it is possible that the strength of the model was influenced by friction between the fill and the retaining side-walls.

7.6 The non-linear finite-element methods of the Transport and Road Research Laboratory

In this section we shall briefly discuss the TRRL non-linear finite-element methods (for further details see Crisfield, 1984, 1985 a,b, 1986). We

103

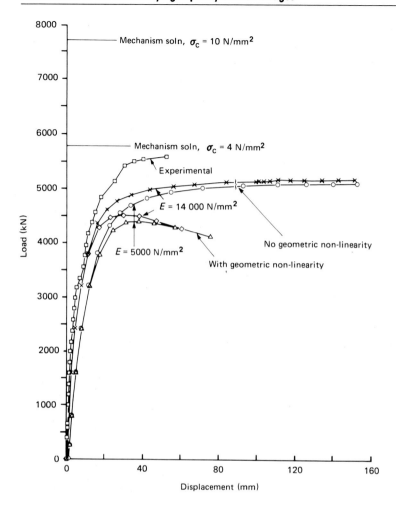

FIGURE 7.20 Relationship between load and deflection under load on ring for Bargower Bridge.

shall also describe their application to the analysis of two of the full-scale bridges and compare the findings with those from the mechanism method.

Two levels of idealization are adopted. Both are essentially two dimensional and involve a unit width of bridge. In the more sophisticated 'continuum' model both the ring and fill are idealized using two-dimensional elements, while in the simpler, much cheaper, model the fill is idealized via special horizontally acting elements (Fig. 7.18) that are effectively non-linear springs (Crisfield, 1985a, b). The latter have no tensile strength and reach a maximum compressive strength governed by the 'passive resistance' of the fill (Fig. 7.18(b)).

These elements add no additional degrees of freedom to those required for the effectively one-dimensional ring element.

The element divisions within the ring are not necessarily chosen to coincide with the voussoirs. Instead, the material properties given to the ring are meant to model the combined stone–mortar joint or brick–mortar joint assembly. Eight-noded isoparametric elements are adopted for the continuum model with conditions of plane stress assumed for the ring within which plane sections are assumed to remain plane. This is achieved by adopting 'hierarchical shape functions' (Crisfield, 1986) and setting the mid-depth variables to zero.

In addition, the material properties are resolved in radial and tangential directions and only the latter are degraded to account for tensile cracking or compressive yielding. A no-tension criterion for cracking is generally adopted.

In order to obtain a reasonable representation of this cracking response, it was necessary, for both levels of modelling, to introduce a special numerical integration scheme within the ring whereby, for each element, two Gaussian integration points were used in the circumferential direction and eight in the radial direction. This was needed in order to be able to reproduce classical mechanism solutions for arches without fill but with equivalent vertical loading. For such analyses, both the compressive yield strength and the modulus of elasticity E were set to very high values and reasonable agreement was obtained with the experimental solutions (Crisfield, 1985a). As both E and the compressive strengths are reduced below these high values, the finite-element solutions must reduce. In the case of E, this reduction will only occur if the effects of geometric non-linearity are included so that, in the limit, a snap-through type of failure can be modelled. The choice of E (and the compressive strength σ_y) is a little difficult because the finite-element model is meant to represent the combined voussoir – joint assembly. Preliminary studies indicate that an appropriate value of E may be about a third of that of the parent stone.

For the continuum model, the fill is treated as a set of two-dimensional elements under plane strain. The latter condition is assumed to be imposed by the lateral restraint of the spandrel walls. The fill is modelled as a near-cohesionless frictional material ($c = 0.0001$ N/mm²) obeying the Mohr–Coulomb yield criterion with an associated flow rule. A low value of E is adopted and free slip is introduced between the arch and fill via a special slip element. In the examples presented here, unless otherwise stated, the E value of the fill was set to 40 N/mm² and the angle of friction φ was set to 30°. The surfacing is generally modelled with the aid of a weak con-crete model, although in some circumstances no such surfacing is introduced and the structure is loaded (numerically) directly onto the ring.

7.6.1 Bargower Bridge

Figure 7.19 shows the deformed mesh and cracking zones for Bargower Bridge, and Fig. 7.20 compares the computed and measured load–deflection relationships on the arch ring under the load. These results were obtained using the more sophisticated continuum model. In line with the previous mechanism analysis, the compressive yield strength of the ring was set to 10 N/mm². A concrete model with $E = 5000$ N/mm², a compressive strength of 5 N/mm² and a tensile strength of 0.5 N/mm² was adopted for the surfacing.

Figure 7.20 shows two sets of solutions: in the first, the full E value (14000 N/mm²) was adopted; in the second, this E value was reduced to 5000 N/mm² in an attempt to account for the E value of the stone–mortar assembly. The first solution gave better agreement with the experimental results, but this could be fortuitous because the load–deflection relationship could also be stiffened by increasing the E value of the fill. Nonetheless, if the effects of geometric non-linearity are neglected, the computed collapse load is effectively independent of the E value. The inclusion of geometric non-linearity (in the ring only) does lower the load but, as anticipated, not as significantly as for the much shallower Bridgemill arch. The finite-element method indicated that first cracking (tensile stresses above zero) would occur on the intrados under a load of less than 800 kN.

It can be seen from Fig. 7.20 that, with $\sigma_y = 10$ N/mm² (as in the finite-element analysis), the mechanism solution is 50% higher than the collapse load given by the finite-element method. This difference appears to be related to the approximations inherent in the use of 'passive resistance' in the mechanism procedure (see Section 7.5.1). Figures 7.12 and 7.19 can be used to

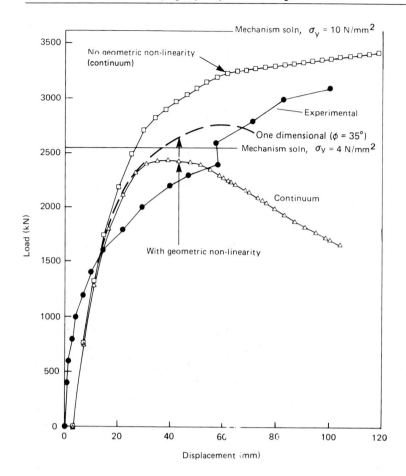

FIGURE 7.21 Relationship between load and deflection under load on ring for Bridgemill Bridge.

compare the hinge positions given by the two methods. They are in reasonable agreement although the right-hand hinge D is rather higher for the mechanism procedure. It can be seen from Fig. 7.19 (and even more clearly from a similar plot with further exaggeration of the deformations) that the significant deformations occur at hinge B and (less so) at hinge C. Relatively little deformation is computed at the remaining two hinges. This observation is consistent with the experimental results for both this and other bridges in which it is often difficult to detect hinges A and D.

7.6.2 Bridgemill Bridge

In line with the previous mechanism analysis, σ_y was set to 10 N/mm². The E value of the ring was set to 5000 N/mm² which is about a third of the value of the parent stone (15000 N/mm²). In order to ease the numerical modelling, the load was applied directly to the arch ring and the surfacing was treated as ordinary fill.

The numerical and experimental load–deflection relationships are compared in Fig. 7.21 which also gives mechanism solutions for σ_y values of 4 and 10 N/mm². The one-dimensional model gives a solution that is in reasonable agreement with the more sophisticated continuum model, particularly when allowance is made for the higher φ value (35° compared with 30°) (Crisfield, 1985b). However, the precise form of the load–deflection relationship for the former does depend on the lengths of the horizontal spring-like elements that

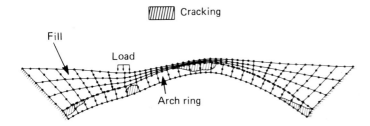

FIGURE 7.22 Computed deformations (exaggerated scale) and crack zones for Bridgemill Bridge.

are used for the fill.

The mechanism solution with σ_y = 10 N/mm^2 is in reasonable agreement with the finite-element solution obtained without considering geometric non-linearity (which also used σ_y = 10 N/mm^2). The mechanism solution obtained with σ_y = 4 N/mm^2 agrees reasonably with the finite-element solution obtained with geometric non-linearity, thus showing that it may be possible to compensate for these effects by artificially reducing the yield stress in a mechanism solution. The computed cracking zones (Fig. 7.22) are also in reasonable agreement with the mechanism hinges (Fig. 7.11). The better agreement between the theoretical and experimental solutions for the analysis without geometric non-linearity is probably fortuitous. There are many factors not included in these two-dimensional analyses, particularly the contribution of the spandrel wall.

The finite-element method indicated that cracking (tensile stresses above zero) would have occurred on the extrados at the springings under dead weight alone. Live loading then increased the cracking at the springing near the load (hinge A in Figs. 7.2, 7.3 and 7.8), while the cracks at the far springing closed (hinge D, Fig. 7.2). Cracking (tensile stresses above zero) was recorded on the intrados below the load when this load was less than 180 kN.

7.7 Some comments on the use of scaled models

It is well known that, if the same materials are used in models and prototypes, some artificial loading should be introduced to the former in order to compensate for dead-load effects. Often, however, such extra loading is omitted and, as a consequence, the frictional forces play a far less significant role in relation to small-scale models than they do in relation to the prototype. Conversely, a small tensile strength will lead, in comparison with the prototype, to misleadingly high collapse strengths. These effects have been investigated by analysing both a model and prototype for which experimental results are available (Hendry *et al.*, 1986).

It was mentioned in Section 7.5.2(h) that the

TABLE 7.4 Results from mechanism program model and prototype for Bargower Bridge

| | Collapse load (kN) | | Ratio r | s^3/r |
	Full scale	Model		
Experiment	5600	90	62.2	2.2
Mech: $\sigma_y \rightarrow \infty$	9686	60.75	159.4	1.0
Mech: σ_y=10 N/mm^2	7726	58.24	132.6	1.2
Mech: σ_y=4 N/mm^2	5751	54.54	105.4	1.51
Mech: σ_y=4 N/mm^2	5751	195.52	29.4	$5.44 \approx s$
$\rho_{mod}=s\rho_{pr}$				

For mechanism solutions α=0.5.

Bolton model bridge (Fig. 7.17) produced a maximum load of 4.4 kN which dropped to a plateau of 3.5 kN. Hendry *et al.* (1986) tested a 5.42:1 scale model ($s = 5.42$) of Bargower Bridge which also failed due to tensile failure (in an abutment block) and also gave a load–deflection response that fell sharply beyond the maximum load. It is interesting to compare the experimental and mechanism solutions for both the model and the prototype (see Table 7.4).

When the compressive strength of the ring is taken as infinite, the mechanism program indicates that the full-scale bridge should be 160 times stronger than the model, while the experimental ratio is only 62. The figure of 160 is almost exactly s^3, where s is the scaling parameter. This is the ratio that would be expected for a no-tensile arch with the resistance largely afforded by friction in the fill. It would therefore appear that the relative strength of the model has been significantly increased in relation to the strength of the prototype by the tensile strength of the mortar.

These findings were reinforced by finite-element analysis which gave almost identical load–deflection relationships for the prototype and the model once the loads in the latter were divided by s^3 and the deflections increased by s^2. A very small change in the tensile strength of the ring (from 0.001 N/mm² to 0.02 N/mm² as a tensile yield stress with no subsequent softening) had almost no effect on the prototype but led to a 10% increase in the strength of the model. The numerical analyses also showed that the provision of a small cohesion in the fill significantly affected the model results but had little effect on the prototype. Table 7.4 shows that, when the compressive yield stress σ_y in the model is reduced from infinity to 4.0 N/mm², the ratio of the strength of the prototype to the strength of the model, according to the mechanism procedure, is reduced from 159.4 to 105.4. Hence the reduced compressive strength has a more damaging effect on the prototype than on the model. These findings were confirmed by studies of the thrust

lines for the mechanism method and the stress states at the Gauss points for the finite-element analyses.

In summary, if scaled models are to be used to represent the behaviour of prototypes, both the compressive and tensile strengths must be reduced. In particular, very weak mortar should be used in the models. If not, the observed collapse modes and loads may both be unrepresentative of the prototype. Theoretically, the only alternative to providing lower strengths for the models is to provide higher densities. If the densities of both arch and fill are increased by a factor of s, and the same compressive and tensile (zero) strength are used, the mechanism program will give the same collapse mode for model and prototype with the latter giving a strength that is s^2 higher (see Table 7.4 for $\sigma_y = 4$ N/mm²).

7.8 Discussion and conclusions

7.8.1 The MEXE method

Doubts relating to the theoretical basis of the MEXE method were expressed in Section 7.3.2. However, comparisons with the results from a range of full-scale tests (Tables 7.2 and 7.3) have indicated no lack of safety. In contrast, for at least two of the bridges, the inferred safety factors (Table 7.3) were very high. Therefore the search for either an enhanced or an alternative assessment method can be justified on economic grounds.

7.8.2 The modified mechanism method

For five out of the six full-scale bridges, the mechanism method could be used to compute upper and lower bound solutions that bracketed the experimental collapse loads. Unfortunately, for a deep semicircular arch and a semi-elliptical arch, the range between these bounds was very wide. With one exception, the best-estimate solutions lay closer to the experimental results than did those obtained from the factored (by 3.0)

MEXE method. This exception involved a semi-elliptical arch for which the mechanism program had difficulty in producing sensible solutions unless the effective right-hand springing was moved up into the backing. In contrast with the MEXE method, the mechanism method can be used to estimate the strength of distorted arches. The program is easy to use and only requires about a minute of computer time (on the Cyber 720 at TRRL). It could easily be modified to run on microcomputers and could form the basis of a new assessment method (see Section 7.8.4).

Many factors, including the bonding with the spandrel walls, for example, can influence the strength of a masonry arch and it could be fortuitous that a computer program gives reasonable agreement with the experimental results. For at least three of the tests, the strength did appear to be influenced by a contribution from the spandrels and yet the mechanism method makes no such allowance.

Numerical comparisons with the full-scale tests have clearly indicated that, for deep arches, some allowance must be made for the lateral resistance of the fill or backing if sensible solutions are to be obtained. This finding was reinforced by work on a deep semicircular model bridge (Walker and Melbourne, 1987) which was tested without spandrel walls. When no allowance was made for any horizontal resistance in the fill, the computed collapse strength was only 3% of the experimental value, but reasonable agreement was obtained when allowance was made for a passive resistance. (However, it is possible that the experimental strength was influenced by friction between the fill and the retaining side-walls).

Two methods of treating the lateral resistance are available to users of the mechanism program. In the first, the hinge remote from the load is constrained not to lie below some point above the springing but below the top of any assumed firm backing. In the second, allowance is automatically made for the lateral resistance by invoking the passive resistance of the fill which depends on its angle of friction. The latter approach has the advantage of being less dependent on direct knowledge regarding the backing.

7.8.3 The finite-element methods

After a number of judgements have been made regarding the properties (in particular the effective E values), it has been shown that the finite-element method can give load–deflection relationships that are reasonably correlated with the experimental results. In addition, the programs compute the levels of cracking as the loading is increased. The latter computations indicate that, unless allowance is made for 'at rest' lateral pressure from the fill, the criterion of first cracking or first reaching tension in the ring cannot usefully be used for assessment purposes because such cracking is computed at very low load levels.

For deep arches, it is essential to model the fill as well as the arch ring. If this is achieved with the aid of plane strain modelling, including fill and surfacing elements, the resulting computer program is both too expensive and too complex for use in routine assessments. However, it can provide valuable research data. The cost of the analysis can be very significantly reduced by adding horizontally acting non-linear spring elements to the arch ring.

In the limit, when the E value of the ring is made very high and the effects of geometric non-linearity are excluded, the finite-element method should give collapse loads that coincide with those of the mechanism method. In the absence of fill, but in the presence of equivalent vertical loads, this coincidence has been shown to occur. In the presence of lateral fill effects there is reasonable correlation, but, because of the assumptions made in the passive resistance approach for the mechanism method, it is not complete.

The finite-element procedures overcome one important drawback of the mechanism method – its inability to consider snap-through effects whereby for shallow arches failure may occur, partly as a result of geometric non-linearity, before the formation of all of the hinges. This effect could be worsened by moving abutments. Only one of the full-scale arches that was tested was particularly flat and this appeared to have very firm abutments.

7.8.4 Future assessment methods

If the MEXE method is to be replaced, a number of options suggest themselves. These include proof loading, the mechanism method, the finite-element method or a revised MEXE procedure. Each will be considered in turn, and a possible strategy will be proposed.

Davey (1953) and others have advocated the use of proof load tests to assess structural adequacy. Davey argued that the structural complexity and variability of arch bridges was too great to allow their strength to be calculated with any reasonable degree of accuracy, and hence load tests should be used. It can equally be argued that load tests are meaningless without structural calculations to interpret them, and if calculations are not accurate enough for strength calculations then, equally, they are not likely to be accurate enough for the interpretation of load tests. In addition TRRL ultimate load tests have indicated that it may be difficult to obtain useful information from the early loading stages such as would be observed in a load test. In particular, as discussed in Section 7.4.1, there is no obvious load at which non-linearity can be said to commence. Also, the first observation of cracking can occur at a very late stage – in some circumstances almost simultaneously with collapse (Fig. 7.5). Hence a load test can give no real guide to the reserve of strength and is not advocated as the primary means of assessing arch bridges. Rather, it should be used when a calculation has left questions unanswered which such a load test might resolve.

It has already been indicated that the continuum finite-element method is likely to be too expensive and complex to use for a routine assessment procedure. We are therefore left with the modified mechanism method, the simple one-

dimensional finite-element method and a revised MEXE procedure. The first two of these options will now be considered together.

The object of the research is to provide a method of assessing the traffic loads that arch bridges may be permitted to carry. The work so far has concentrated on the calculation and measurement of collapse loads under a simple load regime applied at a fixed point. To use the mechanism or the simple finite-element program which calculates a collapse load, factors must be derived to take account of the difference between line loads and axle loads and bogie configurations, the effect of road surface irregularities and of the hump of many arch bridges, and the many repetitions of load that are applied, and to provide an acceptable margin of safety against collapse. The latter may include a geometrical (Heyman, 1982) as well as a load factor.

Hence if either the mechanism or the simple finite-element method were to be used as an assessment tool, factors such as those currently incorporated in the MEXE method would need to be introduced to account for effects such as the material, joint and condition factors (see Section 7.3.1). In addition, effective widths would need to be included to account for the transverse distribution. For the mechanism method, work would be required (and is currently in hand) to allow the treatment of multiple wheel loads.

So far we have concentrated on the arch ring, but we have also mentioned the importance of the fill or backing as well as the rigidity of the foundations. The spandrel walls play an important part in containing the fill but they can also make a direct contribution to the strength if a good structural connection exists between the arch and the walls. In such circumstances, it may be possible to take their contribution to the load-carrying capacity into account. However, problems with spandrel walls are very common: they develop local bulges or they move outwards over a large area, either sliding over the arch extrados (as at Prestwood) or taking the strip of arch beneath the wall with them (as at Torksey and Bargower).

The cause is thought to be the outward pressure generated in the fill by traffic, particularly where it can run close to the edge of the bridge, or the effect of freeze–thaw cycles or perhaps a combination of the two. When these problems are present the wall will no longer be contributing to the stiffness of the arch. Hence for a standard assessment procedure it is expected that no direct contribution should be considered although, in special circumstances, it may be possible to make some allowance.

Having considered the relative merits of the modified mechanism method and simple finite-element technique, we are currently of the view that, largely on account of its simplicity, the former will be more appropriate for routine assessment purposes although the latter could play a role in special circumstances. However, much further work is required before the program can be used as an assessment tool. In particular, rules must be provided relating to the properties (compressive yield strength σ_y and k_p value) or hinge positions in the presence of backing. Because the mechanism method cannot directly handle snap-through effects, reduction factors may be required for shallow arches or those deemed to have weak foundations. The latter could be indirectly introduced via a reduced compressive strength. The finite-element method could help in the generation of these reduction factors.

Finally, there could be scope for a 'refined MEXE method'. This could take the form of nomograms or charts as at present, but it could be modified to account for the results from the full-scale and model tests as well as the theoretical results from the mechanism and finite-element methods.

Further experimental work is required to investigate the effects of multiple wheel loads, transverse distribution from wheel loads and skew arches. The latter effects have not been considered in the tests that have so far been undertaken. Further work is also required in relation to serviceability under repeated loading, maintenance, repair and some scientific quantification of the effect of defects. This should relate to the spandrel walls as well as the arch ring. The non-linear finite-element method may have an important role to play in such work.

7.9 Acknowledgements

The work described in this chapter forms that part of the programme of the Transport and Road Research Laboratory undertaken up to the end of 1987. It is published by permission of the Director. The work was carried out in both the Bridges Division and the Structural Analysis Unit of the Structures Department of the TRRL. Thanks are due to Dr J. C. Cheung, Structural Analysis Unit, TRRL, and Mr J. W. Grainger, Bridges Division, TRRL, who made important contributions. The work of Professor Hendry of Edinburgh University, who carried out the first two full-scale tests in TRRL's programme, is acknowledged as is the enthusiastic support of Mr J. Steel of the Scottish Development Department and the guidance of Dr G. P. Tilly of TRRL. Crown Copyright. Any views expressed in this chapter are not necessarily those of the Department of Transport. Extracts from the text may be reproduced, except for commercial purposes, provided the source is acknowledged.

7.10 References and further reading

Alexander, T. and Thompson, A. W. (1900) *The Scientific Design of Masonry Arches with Numerous Examples*, Dublin University Press, Dublin.

Alexander, T. and Thompson, A. W. (1916) The scientific design of masonry arches. *Elementary Applied Mechanics*, Chapter 12, Macmillan, London.

Barlow, W. H. (1846) On the existence (practically) of the line of equal horizontal thrust in arches, and the mode of determining it by geometrical construction. *Min. Proc. Inst. Civ. Eng.*, 5 162.

Castigliano, C. A. P. (1879) *Théorie de l'Equilibre des Systèmes Elastiques et ses Applications*, Augusto Federico Negro, Turin; Translated by E. S. Andrews in *Elastic Stresses in Structures*, Scott, Greenwood, London, 1919; reprinted, with an introduction by G. A. E. Oravas, as *The Theory of Equilibrium of Elastic Systems and its Applications*, Dover Publications, New York, 1966.

Chettoe, C. S. and Henderson, W. (1957) Masonry arch bridges: a study *Proc. Inst. Civ. Eng.*, 723–55.

Coulomb, C. A. (1773) Essai sur une application des règles de maximis et minimis à quelques problèmes de statique, relatifs à l'architecture. *Mémoires de Mathématique et de Physique, présentés à l'Académie Royale des Sciences par Divers Savants* 7, 343.

Couplet, P. (1729/30) De la poussée des voûtes. *Histoire de l'Académie Royale des Sciences,* pp.79, 117.

Crisfield, M. A. (1984) *A Finite Element Computer Program for the Analysis of Masonry Arches.* Rep. LR1115, Transport and Road Research Laboratory, Crowthorne, Berks.

Crisfield, M. A. (1985a) *Finite Element and Mechanism Methods for the Analysis of Masonry and Brickwork Arches.* Res. Rep. 19, Transport and Road Research Laboratory, Crowthorne, Berks.

Crisfield, M. A. (1985b) Computer methods for the analysis of masonry arches. *Proc. 2nd. Int. Conf. on Civil and Structural Engineering Computing,* Vol. 2, pp.213–20, Civil-Comp. Press, Edinburgh.

Crisfield, M. A. (1986) *Finite Elements and Solution Procedures for Structural Analysis,* Vol. 1, *Linear Analysis,* Pineridge Press, Swansea.

Crisfield, M. A. (1987) Numerical methods for the nonlinear analysis of bridges. *Proc. 3rd Int. Conf. on Civil and Structural Engineering Computing,* Vol. 2, pp.66–74, Civil-Comp. Press, Edinburgh.

Crisfield, M. A. (in press) Plasticity computations using the Mohr–Coulomb yield criterion. *Eng. Comput.*

Crisfield, M. A. and Packham, A. J. (1987) *A Mechanism Program for Computing the Strength of Masonry Arches.* Res. Rep. 124, Transport and Road Research Laboratory, Crowthorne, Berks.

Crisfield, M. A. and Wills, J. (1986) Nonlinear analysis of concrete and masonry structures. In *Finite Element Methods for Nonlinear Problems* (eds. P. G. Bergan *et al.*), pp.639–52, Springer.

Davey, N. (1953) *Tests on Road Bridges.* National Building Studies Res. Pap. 16, HMSO, London.

Davies, S. R. (1985) The assessment of load carrying capacity of masonry arch bridges. *Proc. 2nd. Int. Conf. on Civil and Structural Engineering Computing,* Vol. 2, pp.203–6, Civil-Comp. Press, Edinburgh.

Delbecq, J. M. (1982) *Masonry Bridge-Stability Evaluation.* Report, Structures Department, SETRA, Gagneux.

Department of Transport (1984) *Departmental Standard BD 21/84, The Assessment of Highway Bridges and Structures.*

Department of Transport (1984) *Advice Note BA 16/84, The Assessment of Highway Bridges and Structures.*

Franciosi, C. (1986) Limit behaviour of masonry arches in the presence of finite displacements. *Int. J. Mech. Sci.,* 28 (7), 463–71.

Fuller, G. (1875) Curve of equilibrium for a rigid arch under vertical forces. *Min. Proc. Inst. Civ. Eng.* 40 143.

Gregory, D. (1697) Catenaria. *Philos. Trans.,* **231**, 367.

Harvey, W. J. and Smith, F. W. (1985) Assessment and design of masonry arch structures using a microcomputer. *Proc. 2nd. Int. Conf. on Civil and Structural Engineering Computing,* Vol. 2 pp.207–12, Civil-Comp. Press, Edinburgh.

Hendry, A. W., Davies, S. R. and Royles, R. (1985) *Test on Stone Masonry Arch at Bridgemill–Girvan.* Contract. Rep. 7, Transport and Road Research Laboratory, Crowthorne, Berks.

Hendry, A. W., Davies, S. R., Royles, R., Ponniah, D. A., Forde, M. C. and Komeyli-Birjandi, F. (1986) *Test on Masonry Arch Bridge at Bargower,* Contract. Rep. 26, Transport and Road Research Laboratory, Crowthorne, Berks.

Heyman, J. (1969) The safety of masonry arches. *Int. J. Mech. Sci.,* **11**, 363–85.

Heyman, J. (1972) *Coulomb's Memoir on Statics: An Essay in the History of Civil Engineering.* Cambridge University Press, Cambridge.

Heyman, J. (1976) Couplet's engineering memoirs, 1726–33. In *History of Technology* (eds. A. R. Hall and N. Smith), Mansell, London.

Heyman, J. (1980) The estimation of the strength of masonry arches. *Proc. Inst. Civ. Eng.,* Part 2, 69 921–37.

Heyman, J. (1982) *The Masonry Arch,* Ellis Horwood, Chichester.

de la Hire, P. (1695) *Traité de Mécanique,* Paris.

de la Hire, P. (1712) Sur la construction de voûtes dans les édifices. *Mémoires de l'Académie Royale de Sciences,* p.69.

Howe, M. A. (1897) *A Treatise on Arches,* Wiley, New York, and Chapman and Hall, London.

Larnach, W. J. (1987) *Report on an Investigation into the Basis of the "MEXE" Method for the Assessment of the Load Carrying Capacity of Masonry Arch Bridges.* Unpublished report commissioned by the Royal Armament Research and Development Establishment.

Melbourne, C. (in press) A new arch construction technique. *Struct. Eng. Rev.*

Navier, C. L. M. H. (1833) *Resumé des Leçons Donnés à l'Ecole des Ponts et Chaussées sur l'Application de la Mécanique à l'Etablissement des Constructions et des Machines,* 2nd edn, Paris.

Page, J. (1987) *Load Tests to Collapse on Two Arch-bridges at Preston, Shropshire, and Prestwood, Staffordshire.* Res. Rep. 110, Transport and Road Research, Laboratory, Crowthorne, Berks.

Page, J. and Grainger, J. W. (1987a) *Load Test to Collapse on a Brick Arch Bridge at Torksey.* Work. Pap. WP/B/134/87, Transport and Road Research Laboratory, Crowthorne, Berks (unpublished).

Page, J. and Grainger, J. W. (1987b) *Load Test to Collapse on a Rubble Masonry Bridge at Shinafoot, Tayside.* Work. Pap. WP/B/136/87, Transport and Road Research Laboratory, Crowthorne Berks. (unpublished).

Pippard, A. J. S. (1948) The approximate estimation of safe

loads on masonry bridges. *Civil Engineer in War,* Vol. 1, pp.365–72, Institution of Civil Engineers, London.

Pippard, A. J. S. (1952) *Studies in Elastic Structures,* Arnold, London.

Pippard, A. J. S. and Baker, J. F. (1957) *The Analysis of Engineering Structures,* Arnold, London.

Pippard, A. J. S. and Chitty, L. (1951) *A Study of the Voussoir Arch.* National Building Studies, Res. Pap. 11, HMSO, London.

Pippard, A. J. S., Tranter, E. and Chitty, L. (1936) The mechanics of the voussoir arch. *J. Inst. Civ. Eng.,* **4** 281.

Rankine, W. J. M. (1898) *A Manual for Civil Engineering,* Griffin, London.

Sawko, F. and Towler, K. (1981) Structural behaviour of brickwork arches. *Proc. 7th Int.Conf. on Loadbearing Brickwork* (ed. H. W. West), British Ceramic Society, London.

Tellet, J. (1983) A review of the literature on brickwork arches. *Proc. 8th Int. Symp. on Loadbearing Brickwork, Tech. Sec. 2 – Structures,* Building Materials Section, British Ceramic Society, London.

Towler, K. (1981) *The Structural Behaviour of Brickwork Arches.* Ph.D. Thesis, University of Liverpool.

Towler, K. (1985) Applications of non-linear finite element codes to masonry arches. *Proc. 2nd. Int. Conf. on Civil and Structural Engineering Computing,* Vol. 2, pp.197–202, Civil-Comp. Press, Edinburgh.

Towler, K. and Sawko, F. (1982) Limit state behaviour of brickwork arches, *Proc. 6th Int. Brick Masonry Conf., Rome, May 1982,* pp.422–9.

Walker, P. J. and Melbourne, C. (1987) *Draft Report to TRRL on the Load Tests to Collapse of Three Model Brickwork Masonry Arches,* Bolton Institute of Higher Education (unpublished).

Walklate, R. P. and Mann, J. W. (1983) A method for determining the permissible loading of brick and masonry arches. *Proc. Inst. Civ. Eng., Part 2,* **75** 585–97.

Withey, K. (1982) *Assessment of the Masonry Arch Bridge.* Work. Pap. WP/B/28/82 Transport and Road Research Laboratory, Crowthorne, Berks. (unpublished).

PART THREE

Deterioration: causes, characteristics and counter measures

Chapter eight

Deterioration of materials

8.1 Bricks and brickwork - by J. Morton

8.1.1 Introduction

That brickwork, properly designed and constructed, is one of our most durable materials has never been in doubt. From the remains of the City of Babylon or the brick masonry forming much of the Forum in Rome, brickwork is clearly seen to be capable of passing the test of time. In Britain, the structures built for the canals of the eighteenth century and the railways of the nineteenth century demonstrate the ability of well-conceived, well-constructed and well-maintained brick masonry to provide what, in modern terms, is an extremely long life span.

What factors influence whether a brick structure will be successful or not? What are the properties of bricks and brickwork which affect durability? How do these properties interact with the need for a carefully considered programme of maintenance and repair?

8.1.2 Saturation and freezing

The most severe test to which masonry can be subjected is to undergo freeze–thaw cycles while in a state of total saturation. Saturation on its own, without freezing, provides few problems; similarly, the continual freezing and thawing of brickwork which is wet but not saturated causes no difficulties. It is the combination of saturation with cyclical freezing and thawing which constitutes the harshest test.

In many ways the temperate climate in Britain is not particularly helpful; our winters produce freeze–thaw cycles with great frequency – even several in a single night. In contrast, in the more severe winters of other countries in the same latitudes (e.g. much of Canada and Central Asia) there may be less than three weeks between the 'no frost' season and the permanently frozen period; in such conditions, brickwork may undergo only 10–20 freeze–thaw cycles per year.

If freeze–thaw action on saturated masonry is so critical, what can be done to ensure optimum life? Clearly, without completely insulating a structure, nothing can be done to alter its climatic environment. The answer lies more in matching the known performance of the type of brick used to the expected conditions and, especially, in protecting against total saturation. The precept 'Don't allow brickwork to become saturated' must be engraved on the thinking.

8.1.3 Frost resistance

Until recently, bricks were specified in the relevant British Standard as either 'special' or 'ordinary' in quality. While ordinary quality bricks perform perfectly satisfactorily in the walls of normal buildings, the majority are not able to withstand the extreme exposure conditions of total saturation accompanied by freeze–thaw cycling. Special quality bricks, however, are able to withstand such conditions.

TABLE 8.1 Guidance notes on the selection of bricks

Situation	Clay bricks[a]	Calcium silicate bricks[b]
Near ground level (particularly one course below and two courses above)[c]		
Well-drained site or poorly drained site not subject to freezing	FL, FN, ML, MN	Classes 3–7
Poorly drained site subject to freezing	FL, FN	Classes 3–7
Free-standing walls[d]		
With effective coping	FN, MN, FL, ML	Classes 3–7
With flush capping	FN, FL	Classes 3–7
Unrendered parapets and chimneys[d]		
With effective coping	FL, FN, ML, MN	Classes 3–7
With flush capping	FN, FL	Classes 3–7
Earth-retaining walls[d]		
Waterproofed retaining face and effective coping	FL, FN, ML, MN	Classes 3–7
Waterproofed retaining face	FN, FL	Classes 4–7
No waterproofing of retaining face[e]	FL	Classes 4–7
Brick-on-edge cappings	FL, FN	Classes 4–7
Facing brickwork to concrete retaining walls		
Low risk of saturation	FN, MN, FL, ML	Classes 3–7
High risk of saturation	FN, FL	Classes 4–7
Drainage and sewerage manholes		
Within 150 mm of ground level	FN, FL	Classes 3–7
More than 150 mm below ground level	FL, FN, ML, MN	Classes 3–7

[a] Suitable designation of bricks (to BS 3921, Table 3).
[b] Suitable brick classes (to BS 187, Table 2).
[c] Care should be taken to ensure that paved surrounds do not channel water into the brickwork.
[d] Excluding coping (which should be to BS 5642: Part 2: 1983) or capping.
[e] Some staining may occur if waterproofing is omitted.
For recommendations as to the minimum mortar designations in various situations see Table 4.2.

In 1985 the British Standard for clay bricks (BS 3921) was revised and, in terms of their frost resistance, the 'ordinary' and 'special' qualities for bricks were withdrawn and replaced by the terms Designation F and Designation M. As the letters indicate, Designation F is a frost resistant brick while a Designation M brick is moderately frost resistant.

8.1.4 Sulphate content of bricks

With the advent of modern cementitious mortars containing ordinary Portland cement (OPC) to BS 12, the sulphate content of the brick becomes important in relation to sulphate attack on the mortar. All bricks made from earth clays contain sulphates, although the amount present is normally quite small. Nevertheless, these sulphates can migrate in aqueous solution if the brickwork is saturated and percolate down into the mortar beds, with results which are described in Section 8.3.2 and evidenced by horizontal cracking within the bed joint itself.

It follows that, where saturation is an inevitable consequence of the microclimate, consideration should be given to using bricks which are low in salt content. In BS 3921:1985 these are characterized by the Designation L for low sulphate content, as compared with designation N for normal sulphate content. In positions of extreme exposure Designation FL bricks should be preferred to ensure frost resistance and low salt content.

8.1.5 The selection of bricks and mortars for durability

Guidance is given in BS 5628:Part 3, Table 13, and in BDA *Design Note No. 7* on the choice of bricks and mortars most appropriate for particular situations as regards durability. Table 8.1 has been derived from that Standard and gives recommendations for bricks for new work or for the renewal of major elements of brick structures.

The subject of mortars is more fully covered in Section 8.3, from which it is apparent that they also may be vulnerable to frost and sulphate attack. Broadly speaking, the denser and stronger the mortar, the less is its propensity to frost attack. The same applies to sulphate resistance, since salts in aqueous solution are less able to percolate through its microstructure. For recommendations as to the minimum mortar designations in various situations see Table 4.2.

When undertaking patch repairs, such as replacing 2 or 3 m² of brickwork within a much larger area, there is always the difficult choice between trying to match the existing soft brickwork and using harder bricks and mortar in accordance with modern practice for new work. This is often an area of debate and judgement; some of the following information may be relevant when making a decision.

8.1.6 Movement

A characteristic of brick as a material is its long-term expansion, by a very small amount, when the ceramic material has been removed from the kiln. This can be thought of as irreversible movement and is often referred to as 'moisture expansion'. It is important to recognize this phenomenon and to place it in context with the repair and/or maintenance required. Movement, perhaps more than any other design consideration, is a matter of experience and judgement. However, certain guidelines can prove useful.

When undertaking repair work of a significant nature, such as completely replacing parapet walls to a bridge, the construction can take modern form and modern design guidance can be used. Movement joints in clay brickwork should be incorporated at 10–12 m centres and the width of the joint should be assessed using the guidance given in BS 5628:Part 3, allowing 1 mm of movement per metre run (see BDA *Design Note No. 10*). It is important to recognize that these joints will close with time; in other words, they are predetermined gaps into which the brickwork can expand. It is essential that any filler put into such gaps to form a backing to the sealant should be sufficiently compressible. Foamed cellular polyethylene or polyurethane are useful here because of their high compressibility; highly incompressible fillers **must** be avoided in expansion joints in clay masonry.

In smaller patch repairs some attempt should be made to assess the effect of brick movement on both the surrounding structure and the newly constructed brickwork. This is again a matter of judgement to which certain basic principles can be applied. The amount of expected movement is proportional to the maximum dimension. However, any strains – and hence stresses – developed within the new masonry as a result of its slight expansion within confined restraints are the same for any length. Forces developed within confined brickwork are independent of size and should not be a cause for concern. 'Locking in' by the slight expansion should be considered as a benefit. Equally beneficial when positioning patch repairs to large areas of brickwork built with soft mortars is the ability of the older masonry to 'move' and 'give' since the mortars are generally 'softer' in the elastic sense.

Finally, to complete the picture of movement,

it must be remembered that all masonry will expand or shrink with changes in temperature. This will affect the new and old brickwork equally, with the only difference in performance being the greater ability of the old masonry to 'give' because of its softer mortar. Like most other materials, masonry is subject to creep, which can be beneficial inasmuch as the high stresses generated will dissipate with time by factors of up to a half. Thus, when large areas of repair are present within a more major structure, the stresses generated by movement may be contained by the dead weight from the masonry above. The build-up of stresses will not be as high as otherwise predicted owing to the stress relief which creep affords. The Brick Development Association is unaware of any failure of brickwork in compression which has been the direct result of **uniform** compressive forces generated by movement expansion.

8.1.7 Sizes

Depending upon the nature of the repair, it is often essential to use bricks of a similar size to those already present and difficulties can ensue if the latter are significantly at variance with the current British standard size of 215 mm x 65 mm x 102.5 mm. Although some brick manufacturers stock small quantities of standard special **shaped** bricks, it should never be assumed that special **sizes** can be purchased from stock nor, indeed, that they will be available as readily or as quickly as would standard-sized bricks. Liaison with the brick manufacturers must therefore be established well in advance.

In certain types of repair work, it is also worth remembering that standard bricks vary in size. This is the direct consequence of the manufacturing process. Whilst such variations are small and not significant in large areas of walling, they become more significant in repairs to small sections of brickwork, such as piers where sharp arises are required at a separation of two or four bricks. In such situations – but normally only in such situations – bricks may need to be specially

selected for size, either from those delivered to the site or by the manufacturer. In the latter event, initial contact and discussion with the manufacturer is essential – again, well in advance of the required delivery date.

8.1.8 Colour matching

In repair work which requires brick replacement, some attempt should be made to replace bricks with those of a similar colour and texture. In Britain bricks are available in a variety of different colours and types, thus offering a wide range from which to seek a match.

Providing a suitable match is therefore not usually a major difficulty. What can become an acute embarrassment, however, is where bricks are selected, ordered and delivered to the site and the repair is effected prior to the brickwork being cleaned. The existing surface is often a totally different colour after cleaning. If major cleaning is contemplated, a sufficiently large area should be similarly cleaned before a colour match is made and the replacement bricks are ordered.

8.1.9 Reinforced and prestressed brickwork

Perhaps not surprisingly, the techniques for reinforcing and prestressing concrete are now being applied to masonry. Masonry is, after all, a superb compressive material. It is now feasible to design and construct a reinforced bending member using steel reinforcement to carry the tension generated by the bending and the brickwork to carry the compressive stresses. Similarly, post-tensioned masonry can be used. By applying compressive stress through a post-tensioned bar, the magnitude of the tensile stress developed in a bending member can be reduced or eliminated altogether. Although the greatest potential of these techniques is in new build, they can offer economic alternatives in certain types of repair and maintenance work. The use of grouted cavity construction for the replacement of bridge parapet walls is one such application, provided that a suitable anchorage detail can be incorporated at the base. The full potential of such systems has not yet been fully realized, since in many ways they are still at the development stage.

8.1.10 Summary

Of the measures described in this chapter which can be taken to optimize the durability of brickwork, the overriding thinking required when considering repair and maintenance of existing structures is the elimination of total saturation by the use of details which minimize the amount of wetting, even in positions of extreme exposure.

8.2 Stonework - by B.A. Richardson

8.2.1 Nature of deterioration

Masonry deterioration can be attributed to direct chemical attack or to physical disruption, usually due to crystallization damage through the presence of salts or through freezing, but sometimes due to thermal or hygroscopic movement. Physical and particularly chemical deterioration can be strongly influenced by biological activity. These forms of deterioration are not confined to natural stone but extend equally to clay brick and tile, as well as cement-based materials such as mortar, render and concrete.

Simple leaching is usually insignificant, except in the case of the more soluble magnesian carbonate limestones, but rainfall normally consists of carbonic acid, formed by the absorption of atmospheric carbon dioxide by rainfall, and this weak acid attacks the carbonates in natural stone to form the more soluble bicarbonates. In polluted urban and industrial areas the much stronger sulphur and nitrogen acids may be formed in rainfall, greatly increasing the rate of carbonate erosion and even causing slow deterioration of silica. The rate of deterioration depends on the nature of the stone in terms of crystalline form and size as well as pore volume and size. The

relationship governing the rate of leaching appears to be complex but actually obeys the well-known natural law of mass action; the rate of leaching is greatest for stones with a high total porosity coupled with small crystallite size as this involves a large volume of water or acid attacking a small volume of stone. Thus an impervious carbonate such as a marble will suffer only surface etching whereas a stone with a high total porosity comprising pores of small diameter (a microporous stone) will tend to retain moisture owing to limited ventilation in the pores and powerful capillarity. When water retention occurs in this way it permits protracted absorption of atmospheric pollutants and thus prolonged acid attack of the stone components, so that deterioration tends to be more severe.

Stone deterioration does not depend upon acid leaching alone. In fact, acid leaching results in the formation of soluble salts, sulphates and nitrates if sulphur or nitrogen acid gases are present as atmospheric pollutants. As these salts accumulate through evaporation of water, they absorb water of crystallization; the number of water molecules absorbed per salt molecule depends on the nature of the salt and thus the nature of the atmospheric acid and the stone with which it reacts. Calcium sulphate has two molecules of water of crystallization, whereas calcium nitrate has four, magnesium sulphate has seven and magnesium nitrate six. The growth of salt crystals depends upon the number of water molecules absorbed during crystallization, and as this crystal growth stresses and damages the stone it is clear that a magnesian limestone is generally likely to be less resistant than a calcium limestone to crystallization damage induced by atmospheric pollution.

The natural durability of a stone thus depends on a number of factors. One factor is the ability of the stone to resist leaching which will increase its porosity and reduce its cohesiveness. Resistance to leaching depends upon the relationship between total porosity and crystallite size; leaching is more severe for microporous stones which tend to hold the moisture and thus absorb atmospheric pollutant gases over a longer period. Salts generated by the action of rainfall that has absorbed pollutant gases tend to crystallize in the evaporation zone close to the stone surface. Crystal growth invariably occurs as a result of the absorption of water of crystallization. The degree of expansion depends on the nature of the salt involved which in turn depends on the stone and the pollutant gases, whilst resistance to crystallization stress induced by crystal growth depends on the porous properties of the stone; a macroporous stone is more resistant to crystallization stress than a microporous stone, apparently because the latter tends to hold water for a protracted period. This relationship also holds for crystallization stress resulting from freezing, so that the pore-size distribution within a stone, i.e. the proportions of microporosity and macroporosity or small and large pores, will generally define durability in the most reliable way within a particular group of stones of similar structure, although resistance to crystallization also depends on the cohesiveness of the stone so that stones of very low total porosity tend to be more durable in any case.

The soluble salts resulting from acid rainfall attack will depend on the nature of the stone, so that a stone of given durability must be predominantly macroporous if it is a magnesian limestone, perhaps less macroporous in the case of a calcium limestone and even less macroporous in the case of a sandstone. These macroporosities are related to the salts that are likely to be formed. However, salts may be derived from other sources, the most obvious being water flow from one stone to another. Thus if a calcium limestone is exposed to washings from a magnesian limestone, it must be particularly macroporous if it is to be durable, just as if it were a magnesian limestone itself. Whilst washings from magnesian limestones onto calcium limestones frequently cause salt crystallization damage, perhaps the best-known example of this effect occurs when washings from calcium limestone walls are carried onto sandstone paving when severe deterioration of the

sandstone can occur, even if a sandstone of normally excellent durability is involved. Salts can also be introduced from other sources, particularly dampness rising from footings or emerging from retained soil. In construction, a decision has to be made as to whether it is more realistic to employ a naturally durable material or to introduce special precautions against dampness spreading from these sources; damp-proof courses are a normal precaution in buildings.

The chemical and physical deterioration that has been described can take various forms. In rural areas leaching may be principally due to carbonic acid formed by the absorption of carbon dioxide into the rainfall. This carbonic acid will have no effect on siliceous stones, but it must be appreciated that a sandstone with a carbonaceous matrix must be considered as a carbonaceous stone as it will be affected in the same way as limestones by leaching in the form of bicarbonate. During dry periods, the bicarbonate will decompose to redeposit the carbonate so that leaching tends to involved redistribution of carbonate and its concentration at the evaporation surface. As a result, the surfaces tend to become progressively denser, whilst the stone at the limit of rain absorption will tend to become progressively weaker. Naturally this acid will attack the smaller crystallites in particular and, as these often comprise the cementing matrix in both limestones and sandstones, the stone may become friable in the area affected by leaching. Spalling damage often results through this surface densification accompanied by weakening beneath the densified surface.

Where acid pollutants are present the damage will be caused more rapidly. Calcium sulphate may contribute to surface densification, but crystallization in the weakened area beneath this densified surface will tend to induce spalling. With some predominantly macroporous stones the surface densification is often absent and damage is then limited to very slow surface erosion, apparently resulting from stressing of the surface pores during crystallization of the salts.

Thus St Paul's Cathedral in London is constructed from predominantly macroporous Portland limestone, and the damage on horizontal surfaces of copings and cornices catching rainfall tends to be progressive erosion occurring at a rate of about 10 mm per 100 years. Apparently, calcium sulphate is unable to form a surface skin on these horizontal areas exposed to rainfall, but a dense surface patina of calcium sulphate occurs on adjacent vertical areas which eventually forms a darkened but apparently durable surface, presumably because it effectively prevents further rainwater absorption. However, where rainwater penetration can occur from another direction, spalling of the vertical surface may occur through crystallization beneath the densified layer, and very severe spalling of this type is, in fact, common on stones that are less macroporous and less cohesive, such as Bath limestones. This spalling tends to occur to a consistent depth and is thus often described as contour scaling. On stones that are relatively microporous with a high total porosity, spalling of a stone to a thickness of perhaps 10 mm may occur, whereas on stones with a much lower total porosity there may be simply a tendency for a very thin veneer to peel away at intervals.

Reference has been made to the fact that limestones composed of magnesium carbonate tend to be much less durable than those composed of calcium carbonate as magnesium salts are more soluble and their expansion on crystallization is greater. However, it must be emphasized that the physical properties of the stone, particularly the total porosity, the degree of macroporosity and the crystallite sizes, may be more important than chemical composition. In particular it should be appreciated that many traditional building limestones contain magnesium in the form of a double carbonate with calcium known as dolomite and, where stones consist entirely of this mineral in its common rhombohedral crystallite form, they tend to be particularly durable. For example, the Huddlestone used on York Minster is a stone of this type and has proved to be extremely durable. In contrast, attempts at conservation during the last 100 years have often involved replacement of this magnesian limestone with calcareous limestones which have often deteriorated extremely rapidly through absorption of washings from the adjacent magnesian limestone, despite their excellent durability in normal circumstances.

There is an enormous range of natural stone of different types but, even within a single quarry, stones vary from bed to bed and also within a bed. The durability reputation of a particular stone has often been established through service, yet this particular stone was removed from the quarry long ago and current production may be entirely different. Reputation cannot be considered a reliable method for stone selection, although it indicates the properties of stone that may be available from particular geological deposits. Samples of stone must be checked to ensure that they are likely to be reasonably durable, and this is best achieved by checking the porosity characteristics. It will be apparent from previous comments that total porosity is largely irrelevent, except that a high total porosity usually means a low cohesive strength and comparatively rapid leaching losses. Pore-size distribution is far more important as it enables the water-retaining properties of the stone to be assessed; a stone that retains water for an excessive period is susceptible to both leaching and crystallization damage, particularly when exposed to urban atmospheric pollution.

The earliest method used for this type of assessment was known as the determination of the saturation coefficient. This is a relatively simple process in which a stone is permitted to absorb water naturally and the amount of absorption is compared with the porous volume determined by vacuum impregnation. Despite the simplicity of the method, the saturation coefficient gives an excellent indication of comparative durability when applied to stones of a single geological and structural type, particularly stones derived from a single bed which require routine checking. Comparative durability assessments of stones of a wider type are more difficult and must involve determinations of pore-size distribution, i.e. the proportions of various pore sizes that occur. Usually this is achieved most simply by defining a critical pore diameter such as 0.005 mm; pores exceeding this size are considered to be macropores. The normal method involves saturating the stone with water and then applying a pressure or vacuum which will remove the water from the macropores alone. In the classical method samples of saturated stone are placed in contact with a suitable capillary bed and a vacuum of 600 cm of water is drawn, representing a pressure sufficient to evacuate the macropores alone, i.e. the pores with diameters in excess of 0.005 mm. However, the preparation of the samples is difficult and consequently the test is too expensive for routine use. An alternative method, known as the Penarth pressure method, was developed some years ago in the author's laboratory. In it the small sample of stone is replaced by a hole drilled in a block of stone. A probe is sealed into the hole and water is injected to saturate the stone; the eventual water flow rate is an indication of total porosity. Air is then introduced at a pressure which will remove water from the macropores alone; the air flow rate indicates the macroporosity. In this way it is possible to determine macroporosity as a proportion of total porosity, thus indicating the probable durability of the stone. This method has the particular advantage that stone can be tested in service on a building or other structure and the results compared with stone being considered for repairs, so that a stone type that has weathered badly can be avoided and repair stone can be matched with a type that has proved durable.

Pore-size distribution indicates only the probable durability of stones of average porosity because resistance to leaching and crystallization also depends on both the total porosity and the cohesiveness of the stone. Thus a stone of low porosity is likely to be much more durable than a stone of very high porosity, and these factors must be taken into account when assessing pore-size

distribution results. Whilst saturation coefficient and pore-size distribution tests are usually used for routine durability assessments, it is considered that a laboratory crystallization test is likely to give more reliable results when stones of different types need to be compared. Crystallization of sodium sulphate during a wetting and drying sequence and the consequent weight loss through spalling or erosion from the stone surface is the method that is generally used, but freeze–thaw sequences are preferred in some countries. Obviously these methods assess only resistance to crystallization; acid tests are used to assess direct resistance to polluted atmospheres and various strength tests are also necessary. These tests are not peculiar to natural stone but are equally appropriate to brick and cement-based materials.

Whilst the method of assessing stone durability involving the drilling of a hole can be used on existing structures to establish the basic porosity characteristics in order to help with matching of replacement stone or to explain deterioration, it is important to appreciate that apparently severe deterioration may have occurred over a period of many centuries, indicating a stone that is, in fact, very reliable in service. Alternatively, the most severely deteriorated stones may be relatively recent replacements. It is therefore essential to establish the history of the masonry before considering any remedial or replacement work.

Unfortunately, stone deterioration is not confined to the relatively simple chemical and physical characteristics of the stone and the surrounding atmosphere that have already been described. For example, the Portland limestone of St Paul's Cathedral in London is exposed to an atmosphere polluted with sulphur dioxide. This generates rainfall containing sulphurous acid which should react with the calcium carbonate of the stone to form calcium sulphite. However, chemical analysis of the stone shows that it contains only calcium sulphate. Whilst it is possible that this has been produced through chemical oxidation, it is a strange coincidence that the occurrence of calcium sulphate is always associated with the presence of high concentrations of sulphating bacteria, particularly *Thiobacillus* spp. The atmospheric sulphur dioxide is derived mainly from burning oil. The heavy oil used by power stations and for heating large buildings contains large quantities of sulphur and, whilst the Clean Air Acts have efficiently restricted particulate emissions, the present requirements limit only sulphur dioxide concentrations, without controlling total emissions. As a result, operators blow extra air up flues if permitted sulphur dioxide concentrations are likely to be exceeded, so that sulphur dioxide pollution is now a particularly serious problem in urban and industrial areas.

In industrial areas oxides of nitrogen also occur in the atmosphere, usually from burning coal, but an exactly similar oxidation sequence is observed as with sulphur dioxide; whilst nitrites would be expected to occur on stone, only nitrates are found and nitrating bacteria of the *Nitrobacter* spp. are invariably present.

In rural areas there is a further source of nitrogen compounds which can lead to this form of deterioration. Ammonia is often present in significant concentrations, particularly from the breakdown of animal urine, and where ammonia is absorbed into porous structural materials *Nitrosomonas* bacteria are invariably present which convert ammonia to nitrites, after which *Nitrobacter* spp. convert nitrites to nitrates, as previously described for industrial pollution.

Bacteria are, of course, active only on damp surfaces, but the same conditions also support algae. Whenever suitable conditions of dampness, warmth and light occur, algae will develop extremely rapidly, frequently within an hour or two of rainfall, becoming apparent as a bright green, but occasionally dark green, brown or pink, coloration. Drying results in the death of the algae which then form deposits of dirt or humus on the stone surface. Deposits may also arise from other sources, such as wind-blown dirt, bird droppings and leaves from overhanging trees. These accumulating deposits permit higher plants to develop, particularly mosses but also liverworts, grasses and eventually trees. At the same time, deposits of dead algae and other organic matter permit the development of fungi.

Lichenized fungi or lichens may occur. These consist of a fungus, normally an *Ascomycete,* and an alga in symbiotic relationship, with the alga usually being present within the fungus. The fungal hyphae penetrate into the stone by exploring cracks and crannies and also by generating organic acids such as oxalic acid. In limestones these acids enable the lichen to dissolve stone material which is usually redeposited as calcium oxalate in or close to the surface of the stone. In some cases calcium oxalate will accumulate within the lichen itself, eventually killing it, although a lichen 'fossil', consisting of the calcium oxalate deposit, will remain. If the calcium oxalate is deposited immediately beneath the growth but within the stone, as in the case of some crustose lichens, the past presence of lichen growth can be detected on the stone by patches of dense calcium oxalate deposits, even if the growth is entirely removed by brushing, scraping or the use of biocides. Acids generated by lichen, and indeed by mosses also, not only damage carbonaceous stones such as limestones or sandstones with a carbonating matrix but will also attack silica and cause etching damage on granite and even glass surfaces, while the run-off can cause very severe deterioration of metal gutters, even of lead.

Lichens can be classified into three main types (foliose, fructicose and crustose) depending on the form of their thalli or surface growth. Foliose lichens have thalli in the form of leaves or scales which are attached to the surfaces by threads, whereas fructicose lichens have branching thalli which are attached to the surface at their base. The crustose type has a thallus that forms a crust in close contact with the stone surface; if the crust is removed by scrubbing or scraping, the thallus will regrow over a period of perhaps a year to its original dimensions. In many cases, the older or central part of the thallus becomes relatively inactive, perhaps as a result of accumulations of calcium oxalate, and will fall away to disclose a clean stone surface which will then be recolonized by the growth. In most cases crustose lichen activity is associated with densification of the immediate surface of the stone owing to deposition of calcium oxalate, but severe weakening of the surface may occur at a greater depth and, if crystallization also occurs through the presence of soluble salts or through freezing, the densified surface may spall away. In polluted atmospheres many species of lichen growth are suppressed or controlled, but some resistant species may be able to develop; pollution from, for example, fertilizer factories may result in accelerated lichen activity which may, in turn, lead to spalling or contour scaling at frequent intervals and so to progressive loss of the stone surface.

Whilst bacteria, algae, lichens, mosses and higher plants represent the most severe growth problems on masonry surfaces, slime fungi sometimes occur and must therefore be mentioned. These fungi do not possess a distinct structure but simply form a film of slime on the stone surface; it frequently incorporates trapped algae which give it a green, dark green or black colour. Slime fungi can occur on heavily shaded areas and in tunnels or wherever continuous dampness exists coupled with a suitable source of nutrient.

One of the most serious chemical deterioration problems in masonry structures is sulphate attack, in which sulphuric acid or sulphates react with the tricalcium aluminate in any cement which may be present in mortar, render or concrete, forming calcium sulphoaluminate. The change results in considerable expansion, as each molecule of this complex absorbs 31 molecules of water of crystallization; eventually the calcium sulphoaluminate decomposes to sulphate, leading to collapse of the expansion and loss of cohesiveness (see also Section 8.3.2). Such sulphate attack can occur under a variety of circumstances in wet or saturated conditions.

Clearly, the deterioration of stone and stone work, whether it is chemical, physical or biological, is always associated with the presence of water. Structures must therefore be designed, as far as possible, to avoid unnecessary accumulations of water in masonry surfaces; alternatively, where such accumulations are unavoidable, care must be taken to use only stone of adequate natural durability. Whilst other methods of controlling moisture, such as the use of water repellents, may be useful in some circumstances, it must be appreciated that water repellents can exaggerate deterioration if they encourage crystallization to occur beneath the treated surface. For this reason, water-repellent treatments should never be applied to stone in contact with soil moisture, such as soil-retaining masonry. It is also advisable to avoid their use as a means of controlling algal and other growth on stone; the use of suitable biocides will give more reliable results without the dangers associated with water-repellent treatments.

8.2.2 Cleaning techniques

Cleaning is not always considered desirable as it is sometimes suggested that some buildings and structures were designed to be black! This suggestion has been made particularly forcibly in relation to public buildings in Edinburgh and Glasgow as these are often constructed of sandstones which darken uniformly. In fact, many of the important buildings in Scotland were actually constructed before coal came into general use as a fuel and thus before buildings commenced to darken so seriously in this way, and there is really no proper justification for the suggestion that the designers anticipated that the buildings would become uniformly black. All architects wish their buildings to have the most attractive final appearance as soon as construction is completed, so that if they are designing a black building it will certainly be constructed from black materials rather than from white or fawn stone which will progressively blacken over a period of a century or more. Thus there can never be any valid aesthetic opposition to cleaning. Indeed, dirty buildings appear neglected and this is probably the greatest stimulus to cleaning buildings in the United Kingdom. Whilst masonry is generally cleaned for aesthetic reasons, it is also necessary to consider whether the cleaning process may decrease or increase the rate of deterioration.

The causes of dirt accumulation are not always clear. Dirt is usually attributed to soot deposits but, whilst this was true in the past and carbon particles can often be found trapped in the surface of old masonry, soot has not been a general cause of dirt deposits since the introduction of the Clean Air Acts, yet buildings still become dirty. Detailed investigation often shows that biological growth is responsible for the dirt. For example, even uniform black dirt can be attributed to the fruit bodies of minute crustose lichens or fungi such as *Aureobasidium pullulans*. In many cases simple control of these growths by the application of a biocidal treatment will result in progressive natural cleaning, and it is always worthwhile treating a small area with a biocide in order to check whether it induces the required effect without the necessity for any more drastic remedial work. In many cases the biocidal treatment will not be completely effective and the residual dirt can be attributed to non-biological factors which will require different cleaning treatments.

In a rural atmosphere a limestone surface will often change little in appearance with time but it may become slightly dirty and this dirt is sometimes difficult to remove. The calcium carbonate of which the limestone is composed is slightly soluble in the carbonic acid normally formed as rainfall absorbs carbon dioxide from the atmosphere. This slight dissolution of the stone is followed during a dry period by reprecipitation, particularly at the surface where there is a danger that the formation of the carbonate precipitate will trap dirt particles which have become attached to the surface during wind and rain or which may be derived from the dead algae and other biological growth. Where an area is exposed to direct rainfall the solution action tends to be particularly powerful, so that slight erosion of the stone surface is normal and the stone always has a

clean appearance. In areas sheltered from direct rainfall, such as under string courses, cornices and sills, this washing action will not occur and dirt will often accumulate. As limestone is slightly soluble in water the easiest method of removing this dirt is water soaking followed by light brushing or the application of a high pressure water jet. The soaking period required to achieve the necessary softening will vary from several hours to several days, but it is important to appreciate that only a gentle trickle of water is necessary, sufficient to keep the surface damp without resulting in excessive water penetration to the interior or inconvenience to passers-by. If brushes are used for cleaning they should be stiff bristle or nylon; alternatively, if the deposits are particularly hard, non-ferrous or stainless steel wire brushes can be used.

In urban atmospheres the condition of limestone will be rather different as atmospheric sulphur acids dissolved in the rainfall will result in the formation of a patina or surface deposit of calcium sulphate. This deposit can be particularly hard, and it is sometimes said that it protects the stone and that its removal will be harmful. In fact, there is a danger when a dense surface of this type occurs that soluble salts will accumulate beneath it, eventually crystallizing and causing the formation of massive blisters, flakes or spalls. The patina is therefore best removed at intervals before such damage can occur and the exposed surface of the stone washed to remove salt deposits. This is one of the most significant arguments in favour of regular cleaning of limestone for, if spalling occurs, damage may be quite deep and stone replacement may be necessary.

Dirt may often become trapped within these calcium sulphate deposits. The easiest method of freeing the dirt and removing the patina is again a water spray, followed by cleaning with light brushing or high pressure water jets, possibly with the introduction of polymer or grit to the latter to increase the abrasive or cutting action.

In the case of heavy calcium sulphate deposits preliminary soaking is essential, often for prolonged periods, even if high pressure water jets are to be used, in order to soften the encrusting deposits which are otherwise rather harder than the stone beneath. In some situations steam cleaning may be an advantage for, although it offers little technical advantage over prolonged water soaking and is rather more expensive, it is capable of softening more persistent encrustations without the use of enormous quantities of water. Some years ago steam cleaning was used in conjunction with caustic soda or soda ash but this results in the formation of harmful soluble salts which later crystallize to cause damage to the cleaned surface. Chemical cleaning methods should never be employed in normal circumstances.

A carbonaceous sandstone can usually be cleaned using the water-soaking, high pressure jets or brushing techniques normally employed for limestones, but non-carbonaceous sandstones and grit stones will not respond to these methods and it is necessary to employ much more severe cleaning techniques such as dry or wet grit-blasting or the use of hydrofluoric acid. Granite can also be cleared with dilute hydrofluoric acid, but solutions of ammonium bifluoride can be used instead; they decompose on the stone to liberate hydrofluoric acid and act in precisely the same way. Great care must be taken in selecting an appropriate cleaning method. For example, hydrofluoric acid must not be used under any circumstances on a limestone or carbonaceous sandstone, and ammonium bifluoride or any other compound likely to form salts within a stone must never be used on any porous stone. Grit-blasting techniques must be carefully controlled in order to avoid unnecessary damage to relatively soft stones, and it is important to remember that dry grit-blasting in particular can generate silica dust, both from the grit in some systems and from silica within the surface, introducing a danger of silicosis to the operators, so that proper protection is essential. Indeed, the safest technique for cleaning sandstones is to use high pressure water jets if they are capable of achieving the desired results and wet grit-blasting, i.e. high pressure water jets

incorporating cutting aids such as polymers or grit, whenever a more powerful cutting action is necessary. The use of power tools with wire brushes or grinding discs should always be avoided as they can cause severe and unsightly damage which is very difficult to repair.

Concrete, cast stone, rendering and brickwork can be cleaned as if they are natural stones. Thus high pressure water jets should be used whenever possible. If these do not achieve an adequate result, water soaking should be considered, preferably followed by high pressure water jet cleaning, although stiff bristle, nylon or non-ferrous wire brushes can be used for localized difficult areas. If these methods are inadequate then carefully controlled wet grit-blasting should be employed, but this should always be entrusted to a specialist company employing operatives who are fully experienced in selecting nozzles, pressures and grit to achieve adequate cleaning without unnecessary damage. This is, in fact, a summary of the approach to cleaning that should be adopted for all surfaces, whether of natural stone or other materials, and all other cleaning methods should be rejected, particularly any which employ acids, alkalis or salts.

It has already been explained that much of the dirt on masonry is actually biological growth which can perhaps be best controlled by simple biocidal treatments. Water is always involved in cleaning techniques and inevitably encourges biological growth; biocidal treatments following cleaning are therefore recommended in all cases.

8.2.3 Biocidal treatments

It must be accepted that moss, lichen and algal growth often imparts a very attractive mellow appearance, but it is equally certain that it conceals the true appearance of the structure and may cause damage. In fact, damage is unlikely on some relatively inert brick, tile and stone surfaces, but severe structural deterioration can occur in circumstances such as those described in Section 8.2.1. Whether growth needs to be encouraged or controlled depends largely on the circumstances. It might be considered desirable to encourage growth on the surface of a new extension or repair to an old building, whereas control might be essential when the growth or its acid washings are causing damage. Deliberate attempts to encourage growth are not usually successful. Perhaps a more sensible alternative, particularly if an attempt is being made to match new surfaces with old, is to remove the growth from the old surfaces, which will then look clean and well maintained and often far more attractive than the growth-encrusted surfaces.

In some circumstances growth control is essential; the problem is then to decide upon the most economic method for doing so. Although lichen growth can survive in very dry conditions, algal growth in particular is entirely dependent upon a high moisture content and, as algal growth represents the start of the humus chain that later permits mosses and higher plants to become established, it would seem that the control of moisture content might represent an important means of controlling growth. Whilst this is essentially true, it is equally important to appreciate that algal growth is sometimes evidence of dampness defects that require attention. For example, heavy algal growth may indicate leaks or blockages in a downpipe, and repairs to these leaks and blockages will control the problem. Often algal growth occurs along the base of a wall, indicating rising dampness, but control of the algal growth will simply change the appearance without alleviating the dampness which may result in other deterioration such as crystallization spalling through the presence of salts or through freezing. Water-repellent treatments can be used to reduce the moisture content of surfaces and can thus achieve excellent control of algal growth, yet there are circumstances in which these treatments can induce severe deterioration, such as when a surface is able to absorb water from another source, e.g. rising dampness. Whilst water-repellent treatments may inhibit growth if applied to new clean surfaces, water repellents should never be applied to surfaces on which growth is already present, and it is therefore necessary to remove growth before applying such a treatment or, indeed, a masonry paint; biocidal treatments are frequently used as pretreatments in these cases.

Growths of moss, lichen and algae can be removed by dry or wet scrubbing, but this is very tedious and must be considered unrealistic where extensive areas are to be cleaned. In contrast, a biocidal treatment achieves an almost magical result: it is simply sprayed onto the contaminated stone surface, causing the growth to die and become brittle so that it is removed naturally through the action of wind and rain. This effect is less marked on structures that have been permitted to become encrusted with heavy lichen and moss growth over many years, as only limited penetration of the biocide can usually be achieved from an initial spray. In these circumstances, the initial application is usually sufficient to cause the death of the growth, with perhaps some brushing after a period of about two months in order to remove any remaining growth. The clean surface is then treated again with biocide to ensure a uniform and generous inhibitory treatment.

An extensive range of biocides is available but many are unsuitable for masonry for various reasons. Whilst they must obviously be toxic to the growth, they must be completely harmless to the operatives applying them. They must also cause no damage to the stone to which they are applied, either by direct action on the stone substance or by leaving deposits which, in conjunction with other factors, may result in damage, e.g. by increasing water retention, so that frost damage becomes more likely, or by chemical reactions which give unsightly discoloration. The activity of many toxic systems such as calcium hypochlorite and sodium hypochlorite, sold as chlorinated lime or bleaching powder and bleach respectively, is only transient.

Other treatments such as sodium pentachlorophenate (also known as sodium pentachlorophenoxide) will leave a deposit within the stone surface and achieve some degree of inhibition. However, the deposit is slightly volatile and leachable and the inhibitory activity decreases as the deposit becomes depleted. A mixture of sodium or potassium methylsiliconate and a sodium or potassium phenate, i.e. a salt of a phenol such as pentachlorophenol or *ortho*-phenylphenol, is an efficient eradicant and tends to prevent subsequent rainwater absorption, causing the growth to become dry and brittle so that it is readily removed by light brushing or even becomes detached without further treatment during periods of bad weather. Unfortunately, such treatments suffer from the very serious disadvantge that they are likely to cause severe staining on light-coloured surfaces. Furthermore, the introduction of alkali metal salts into stone must always be suspect, as there is a risk of the development of soluble salts which may cause significant crystallization damage to the stone surface. Copper, zinc and magnesium salts, as well as fluorides and silicofluorides, must be rejected because of the danger of damaging salt formation in the treated stone.

Several outstanding biocidal treatments have been developed in recent years and are now very widely used. Amongst the inorganic salts, the borates have proved of particular interest. Sodium borates, such as borax, react with atmospheric carbonic acid to give a deposit of sodium carbonate and boric acid, the latter being almost insoluble in water. There is, of course, a danger that the atmospheric sulphur acids will form sulphites and sulphates which can cause crystallization damage, but this is reduced with the proprietary borate Polybor, which is much more soluble in water than borax but contains less sodium whilst having a higher boric acid equivalent. It is therefore easier to use than borax, as well as being more active and less likely to cause damage through the formation of salts. Polybor is comparatively inexpensive and is probably the best general purpose toxic treatment available today, but its limitations should be appreciated; whilst its inhibitory action will per-

sist for several years, its eradicant action is not very powerful. On reasonably porous stone a 4% solution of Polybor will give freedom from algae for about two and a half years and freedom from moss and lichens for a substantially longer period.

On structures where moss and lichen are more important than algae, the effective life of Polybor treatment will be much greater, perhaps 10 – 20 years, depending on local conditions. Concentrations need to be varied according to the porosity of the stone – perhaps 2% on a very porous stone, increasing to perhaps 10% on marble and granite.

In recent years tributyltin oxide, solubilized using a quaternary ammonium compound, has been used to an increasing extent. Quaternary ammonium compounds ('quats') are particularly suitable as solubilizers as they possess biological activity of their own, apparently by rupturing the cell walls of algae and lichen as well as by a true toxic action. Generally, quats are powerful eradicants, but they also possess an affinity for a variety of substrates and are therefore capable of producing a persistent inhibitory action. Quats have long been used for pharmaceutical purposes against both bacterial and fungal infections. Over the years the costs of the more popular compounds have reduced progressively, and they can now be considered for industrial purposes including biocidal masonry treatments. Good water solubility and powerful biocidal action appear to be associated particularly with compounds containing an alkyl chain of about 14 carbon atoms together with an aryl group, as in benzalkonium chloride and several proprietary compounds such as Gloquat C. These quats can be used in concentrations as low as 1% (i.e. 2% of the 50% concentrate usually supplied) to give an effective life against algae of about two and a half years. Mixtures of tributyltin oxide and quat, such as Murosol and Thaltox (now sold only under the Murosol label), are to be preferred where lichen and moss are present or where a more persistent inhibitory action is desired.

In summary, it can be said that many of the

proprietary formulations that are marketed for masonry treatment, including bleaching powder and bleach (calcium hypochlorite and sodium hypochlorite), sodium pentachlorophenate, sodium *ortho*-phenylphenate, mercurials and generally all metal or alkali metal salts, cannot be recommended. The best low cost general treatment is probably provided by the proprietary borate Polybor or the quaternary ammonium compounds which are safe, easy to use and relatively inexpensive. Tributyltin oxide solubilized with a quaternary ammonium compound will give a much more powerful eradicant and more persistent inhibitory action. These formulations are much more reliable where growth consists of lichen rather than algae, but the inhibitory action is eventually lost from the immediate surface of the stone, permitting algal growth to develop; this can readily be controlled by applying a quaternary ammonium treatment alone when the growth has developed to a stage when it is considered unsightly.

Water repellents can be used to inhibit growth on clean surfaces but it will be appreciated that they should be applied only to new stone or to stone that has already been cleaned using a biocide. They should never be applied to surfaces which are liable to internal wetting from rising dampness or other sources (see Section 8.2.5), as the treated stone surface may spall owing to crystallization of salts or through freezing. In these situations it is best to employ a biocidal treatment alone, preferably one that possesses a degree of persistence which will enable it to prevent the re-establishment of biological growth for a prolonged period. Cleaning involving water should be followed by a simple algicidal treatment, preferably using a quaternary ammonium formulation, in order to prevent algal growth which occurs very rapidly after such cleaning.

8.2.4 Consolidant treatments

Stone preservatives have been intended to prevent deterioration of the stone without changing its appearance, and there has been, since about 1840, a long succession of proposals to that end. Even so, no stone preservative used as a surface treatment has yet met with any significant measure of success. Some have been followed, sooner or later, by scaling of the treated surfaces and have done more harm than good. The situation now is much the same as it was 100 years ago. This is not for want of apparently promising methods, but because of the fundamental difficulty that no surface treatment penetrates far enough to give the protection looked for.

This comment appeared in *Building Research Station Digest (First Series)* No. 128, which was first published in November 1959, and surprisingly, represents a situation that still prevails today; whilst the need for deep penetration has long been appreciated, it has not yet been reliably achieved through a method that is realistic for application to masonry.

As water is involved in all masonry deterioration processes, whether they are chemical, physical or biological, the most effective conservation technique might be the use of water repellents in order to prevent water penetration. It will be appreciated, however, that for this to be successful it would be necessary to ensure adequate sealing of all surfaces through which water could enter the masonry and, in the case of the structures covered in this book, these would include all faces in contact with the soil (including the foundations), the extrados of arches and the tops of bridge piers, none of which are readily accessible in existing structures. When such structures were built, precautions against rising dampness were rare and, if waterproofing was provided, it is unlikely to have remained wholly effective. The insertion of a damp-proof course and the (re)waterproofing of buried surfaces at the current time may prove impracticable. However, without them, the treatment of only the readily accessible faces with water repellents would be pointless (see also Section 8.2.5).

In the case of buildings and free-standing walls,

a different situation may apply. Various forms of consolidant have been used to improve the cohesiveness of stone by the replacement of cementing matrix that has been lost over the years as a result of weathering; examples include baryta and lime water which deposit carbonate and various silicates which deposit silica. However, all currently available preservative or consolidation systems which involve depositing material within the interstices of the stone suffer from the same disadvantages. Whilst the best treatments are clearly those that result in insoluble deposits without introducing soluble salts, there is still the danger that they will modify the stone structure so that it becomes microporous and more sensitive to crystallization damage. In addition, all these treatments are capable of only limited penetration and, in reducing the porosity of the surface, they may not significantly reduce water absorption but may seriously obstruct evaporation. Water accumulating behind the treated zone in this way, or diffusing from other sources, may freeze or absorb atmospheric acids with consequent danger of salt crystallization, as the restricted evaporation will encourage slow crystal growth, and particularly severe spalling of the treated layer. Certainly, one method of reducing these dangers is always to follow a preservation or consolidation treatment with a water-repellent treatment to reduce rainwater absorption; this is always advisable where there is a risk that the treatment may lead to restricted porosity of the stone surface. However, from a preservation point of view, a more sensible technique is to apply a water repellent to a considerable depth, or from a consolidation point of view to apply a water repellent which also possesses consolidating properties.

It has frequently been suggested that organic polymers should be used for the preservation and consolidation of masonry, but it must be appreciated that, if they are applied at reasonable concentrations in order to achieve adequate consolidation, they are likely to seal the stone surface completely, with a consequent danger of damage through interstitial condensation followed by crystallization behind the treated zone. If organic polymers are applied from dilute solutions in order to ensure that the porosity of the stone is preserved, they tend to lose much of their consolidating power and they also become particularly susceptible to loss by volatilization and oxidation. At the present time, these difficulties can only be overcome using silicone resins, but the normal water-repellent silicone resins consist of relatively large polymers which cannot penetrate deeply into stone.

Saturated polymer systems are essential if adequate life is to be obtained from treatments applied in dilute solutions in order to ensure that permeability of the surface is maintained, and silicones give much longer life than normal organic systems. As polymer solutions will not achieve adequate penetration, it is theoretically better to apply monomers and to form the polymers *in situ*. At the same time it will be appreciated that the purpose of the treatment, at least in most instances, is to achieve consolidation to an appreciable depth using a polymer system which also possesses water-repellent properties. These requirements limit the choice of monomer systems that are available, but considerable success has been achieved using trialkoxymethylsilane. This is the basic monomer used in the Brethane process developed by the Building Research Establishment.

It now seems certain that silane systems will be more widely adopted as masonry preservative and consolidant systems, although clearly further development work is essential before their effectiveness and reliability can be fully established.

These comments may appear unhelpful, but it is obviously foolish to submit buildings and other structures to treatments which are intended to be preservative or consolidants but which are actually ineffective and perhaps damaging.

8.2.5 Water repellents

In the preceding section we have already drawn attention to the futility of applying water-repellent treatment to exposed surfaces when water penetration can occur from another direction. Far from inhibiting deterioration, such treatment may accelerate it. Although silicone water-repellent systems to BS 3826 are designed to ensure sufficient permeability on normal substrates to permit trapped moisture to disperse, it should clearly be appreciated that dampness behind the treated area may lead to a concentration of salt crystals, causing the treated layer to spall away. Even if salts are absent, in the case of the structures covered by this book, the permeability of the repellent may not allow the moisture accumulation behind the treatment to disperse sufficiently quickly to avoid damage through freezing.

One method of reducing water penetration into walls is to increase their absorptive capacity, perhaps by the use of a thick porous rendering, a process which may also cover unsightly spalling. Unfortunately, it is not generally appreciated that the function of this rendering is to be porous and to hold rainfall so that it can subsequently disperse by evaporation, and often thin dense or waterproof renderings are employed. These are not usually sufficiently flexible to tolerate the seasonal thermal and moisture content changes in the structure, and eventually cracks develop. Water flowing down the wall can readily be absorbed into the cracks, yet the remaining rendering obstructs evaporation so that the final result is often progressively accumulating moisture. Attempts to reduce dampness by the use of cement slurry, paint, bitumen or plastic coatings should be avoided as they introduce severe condensation hazards; a permeable structure must never be sealed at the cold external surface as there is then a danger that humidity within the structure will result in condensation immediately beneath these external coatings, followed by severe frost spalling in cold weather.

8.3 Mortar - by R. C. de Vekey

8.3.1 Introduction

There are many mechanisms by which masonry, particularly in civil engineering structures, can

FIGURE 8.1 Loss of mortar by percolating water and frost causing bricks to drop out of a brick arch. (*Courtesy of R.C. de Vekey.*)

deteriorate. These mechanisms can be broken down broadly as follows:

1. chemical/biological attack due to water and water-borne acids, sulphates, pollution, chemicals released by growing plants
2. erosion by particles in flowing water and wind, by frost attack, by salt crystallization, by plant root action
3. stress-related effects due to movement of foundations, movement/consolidation/wash out of infill, vibration, overloading, moisture movement of bricks and blocks, thermal movement
4. staining due to efflorescence, lime, iron

Most chemical attack affects the mortar component of masonry because mortar is a less stable material than fired clay, stone and dense concrete. Erosion processes such as wind, water scour and frost attack both constituents but erode the softer of the two at a faster rate.

8.3.2 Chemical attack

(a) Water

Water percolating into masonry is always a potential source of damage. Absolutely pure water will have no direct chemical effect but some of the constituents of mortar are very slightly soluble and will leach away slowly. Rainwater contains dissolved carbon dioxide forming a very mild acid which dissolves calcium carbonate by production of the soluble bicarbonate. Thus lime mortars and weak OPC:lime mortars will eventually be destroyed by percolating rainwater because calcium carbonate is their main binding agent. Strong OPC mortars with well-graded sand are less susceptible, partly because the calcium silicate binder is less soluble but mainly because they are largely impermeable and so prevent percolation.

Typical characteristics of water leaching are loose sandy or friable mortar, loss of mortar from the outside of the joints, which gives a raked joint appearance, and in serious cases the loss of bricks

from the outer layer of masonry, particularly from tunnel and arch heads (Fig. 8.1). The process will sometimes be accompanied by staining due to reprecipitation of the dissolved materials.

Counter measures at the design stage are to channel water away or at least to allow it to escape via weep holes. In earth-retaining structures a porous back-fill and drain system is the usual method. Where contact with water is necessary, e.g. in channels, strong impermeable mortars should be used. Where erosion is already taking place mechanical repointing is probably the most effective remedy using a waterproofed or polymer-modified mortar if lower permeability is required (see Chapter 14).

(b) Sulphate attack

Sulphate attack is the next most common problem and is due to the reaction between sulphate ions in water solution and the tricalcium aluminate (C_3A) phase in set OPC mortars to form calcium sulphoaluminate or 'ettringite'. This process gives a net expansion and causes both local disruption of the mortar bed and stresses in the brickwork as a result of the expansion which can be of the order of several per cent. It must be stressed that it will only occur in wet or saturated conditions and where there is a source of water-soluble sulphate compound. It will never occur in dry or slightly damp masonry. The commoner sulphates are the sodium, potassium and magnesium salts, which are all freely soluble, and calcium sulphate, which is less soluble but will leach in persistently wet conditions. The sulphates may be present in groundwaters and can affect masonry below the damp proof course and masonry in contact with the ground such as retaining walls, bridges and tunnels. Sulphates are also present in some types of clay brick and will be transported to the mortar in wet conditions. Old types of solid brick with unoxidized centres (blackhearts) often have large amounts of soluble sulphates, as do some Scottish composition bricks, while semi-dry pressed bricks made from Oxford Clay (Flettons) have high levels of calcium sulphate. Sulphates do not attack pure lime mortars as there is no calcium aluminate present but may have some effect on hydraulic lime mortars. Sulphate-resisting Portland cement is deliberately formulated to have a low C_3A content but may be attacked in very extreme conditions. Sulphate attack is more likely in porous mortars, while rich dense impervious mortars are affected less despite their higher cement content.

Typical visible effects of such attack are expansion of the masonry where unrestrained and stress increases where restrained. Typically the mortar is affected more within the body of the wall than on the surface, and so small horizontal cracks are sometimes visible in the centre of each bed joint. In thick masonry vertical cracks often appear on the external elevations owing to the greater growth in the centre of the wall which remains wetter for longer than the outside as the outside can dry out by evaporation. This is very common where the water is leaking into the structure from faulty details or from ground contact. Horizontal cracking may also occur, but it is likely to be less obvious where there is a high vertical dead-load stress. Rendered masonry

exhibits a network of cracks termed 'map-cracking' (Fig. 8.2). If walls also have a face to an internal space which is kept dry, then efflorescence (growth of crystals) may occur as a result of transport of the sulphates to the surface.

There are three main methods of prevention at the design stage:

1. Keep the masonry dry;
2. Exclude sulphates;
3. Use mortars that are not affected by sulphates.

If sulphate attack is suspected in existing work, try to correct any faults that might be causing unintended wetting of the work. In serious cases it may be necessary to demolish and rebuild.

(c) Acid rain

There is no systematic evidence that rain acidified by sulphur dioxide from flue gases at the normal levels has a significant effect on mortar but in special cases such as proximity to a high level of gaseous pollution near industrial sites there may be a very significant increase in the deterioration rate of all forms of structure including masonry. Unlined chimneys are a particular case and can

FIGURE 8.2 Sulphate attack on rendered brickwork. (*Courtesy of R.C. de Vekey.*)

suffer severe acid/sulphate attack in the exposed parts where rain saturation or condensation occurs. This should be treated as a special case of sulphate attack.

(d) Acid groundwaters

In certain parts of the country, particularly where rain run-off is from peat moors or dense forest, the groundwater may become acidified by humic acid. This can be a serious hazard for mortars. Mortar affected by such attack will become soft and friable and will readily be eroded by frost, wind and water. Countermeasures are basically similar to those for water leaching.

(e) Waterborne pollutants

Masonry in regular contact with waterborne pollutants may be attacked, particularly in acid conditions. The range of possible materials is too great to detail but domestic, farm and industrial effluents, particularly if discharged at elevated temperatures, are likely to be harmful and to cause severe erosion of mortar. Advice on appropriate mortars is contained in BS 5628:Part 3 and in the Water Research Laboratory *Information and Guidance Note 4-10-01*.

(f) Plant life

Plant life is likely to have an effect on mortar similar to its effect on limestone of comparable porosity. Details are given in Section 8.2.

8.3.3 Erosion

(a) Frost

Frost is the principal eroding agent of masonry exposed to normal exterior conditions. Quite clearly it will not affect masonry buried more than a few feet and so will not affect deep footings or the insides of buried culverts. Its effect is due to the stresses created by the expansion of water on freezing in the pore system of materials and thus only occurs in water-saturated or near-saturated conditions in porous materials. Mortar is susceptible to frost damage, particularly in combination with any of the other forms of chemical deterioration. A particular case is frost attack on unhardened (green) mortar at the time of construction.

Typical effects are the spalling of small areas of the mortar to form a layer of detritus at the foot of the wall or just general softening and erosion of

FIGURE 8.3 Loss of mortar at the water-line by scour, followed by loss of bricks. (*Courtesy of R.C. de Vekey.*)

the mortar indistinguishable from chemical erosion.

Countermeasures are to protect work from rain and cold during construction or to use aerated mortars which are less affected. So-called accelerators and anti-frost additives for mortars are not effective and can cause corrosion damage to buried metal fixings. In very exposed work appropriate mortars should be used as recommended in Table 4.2. If possible, try to eliminate saturation of existing work, but if this is not possible then try mechanical repointing with mortar containing a waterproofer or polymer additive.

(b) Salt crystallization damage

Salt crystallization damage is an analogous process to frost attack and is due to expansive crystallization of hydrated salts in the pore structure. It is not common in the United King-

dom since it normally occurs in warm conditions where there is rapid drying of water causing the salts to crystallize out below the surface. In addition, there must be a source of water or water-containing salts. Its typical appearance is similar to that of frost damage but it will usually be associated with salt crystals (efflorescence). Treatment of affected work is very difficult and it is best dealt with by using appropriate materials and detailing at the design stage.

(c) Abrasion

Abrasion by particles in wind and water probably acts more in concert with other processes than alone (Fig. 8.3). The appearance will normally be of loss of surface and change of colour and texture with softer areas wearing faster than harder areas. At the design and construction stage the process can be controlled by use of tough impervious

FIGURE 8.4 Heave of brick parapet by plant growth. (*Courtesy of R.C. de Vekey.*)

mortars. In existing work the remedies suggested for water attack apply.

8.3.4 Stress-related effects

Mortar joints are only moderately good at resisting shear and tend to crack easily when subjected to bending or direct tension. Joint failure in compression is rare but step cracking or splitting of the mortar beds is common where movement or tension/shear forces occur. The most common source of direct tension failure of mortar beds is invasion by plant roots which then split the porous mortar as they grow (Fig. 8.4). There are no mortar remedies as such but repointing will often prolong the life of affected structures.

8.3.5 Staining

There are four common chemical staining processes affecting masonry:

1. leaching of coloured compounds, which in mortar are only likely to be iron impurities giving brown stains but could be copper from damp-proof courses giving green stains

2. lime staining due to leaching out of calcium compounds which form patches of white carbonates (recognizable by solubility in acid and evolution of carbon dioxide bubbles)

3. silica staining by leaching of soluble silica fines to give white surface deposits (insoluble in most acids) (Fig. 8.5)

4. efflorescence due to crystallization of soluble salts on the surface, again giving white areas (soluble in water).

Iron staining can be suppressed by repointing with a waterproofed mortar but is best avoided by using clean sands. Lime staining can be reduced by using mortars which do not contain lime or mortars formulated to reduce percolation. Silica staining is difficult to anticipate and avoid, but details which discourage water percolation will make it less likely. Silica stains are difficult to remove except with the very dangerous hydrofluoric acid. Efflorescence is usually temporary and is best either brushed off or left to be removed by weathering. Most stains except silica can be removed or reduced by careful water or acid treatment, but acid will attack poor mortars and so great care is necessary.

FIGURE 8.5 Probable silica staining. (*Courtesy of R.C. de Vekey.*)

8.4 References and further reading

8.4.1 Bricks and brickwork

Brick Development Association
Building Note 1, 1979, Bricks and Brickwork on Site.
Building Note 2, 1982, Cleaning of Brickwork.
Building Note 3, 1986, Bricklaying in Winter Conditions.
Design Note 7, 1985, Brickwork Durability.
Design Note 10, 1986, Designing for Movement in Brickwork.
Design Guide 2, Revised 1987, *Brickwork Retaining Walls.*
Practical Note 7, 1976, *Repointing of Brickwork.*
British Standards Institution
BS 3921:1985. *Specification for Clay Bricks.*
BS 5628 Part 3: 1985. *Code of Practice for the Use of Masonry. Materials and Components, Design and Workmanship.*
Building Research Establishment (1971) *Digest 89* (revised), *Sulphate Attack on Brickwork.*
Grimm, C. T. (1983), Durability of brick masonry: a review of the literature. *Proc. Symp. on Masonry: Research, Applications and Problems, Bal Harbour, December 1983, Spec. Tech. Publ. 871,* American Society for Testing and Materials, Philadelphia, PA.
Nash, W. G. (1986) *Brickwork Repair and Restoration* Attic Books, Builth Wells.

8.4.2 Stonework

British Standards Institution
BS 6270 Part 1: 1982. *Code of Practice for Cleaning and Surface Repair of Buildings. Natural Stone, Cast Stone and Clay and Calcium Silicate Brick Masonry.*
BS 6477: 1984. *Specification for Water Repellent Treatments for Masonry Surfaces.*
Building Research Establishment (1959) *Digest 128 (First Series), Stone Preservatives.*
Building Research Establishment (1972) *Digest 139, Control of Lichens, Moulds and Similar Growths.*
Building Research Establishment (1975) *Digest 177. Decay and Conservation of Stone Masonry.*
Building Research Establishment (1989) *Report 141 Durability tests for building stone.*
Clifton, J. R. (ed.) (1986) *Proc. Symp. on Cleaning Stone and Masonry, Louisville, April 1983, Spec. Tech. Publ. 935,* American Society for Testing and Materials, Philadelphia, PA.
Ireson, A. S. (1987) *Masonry Conservation and Restoration.* Attic Books. Builth Wells.
Richardson, B. A. (1980) *Remedial Treatment of Buildings,* Construction Press.
Schaffer, R. J. (1932) *The Weathering of Natural Building Stones.* Spec. Rep. 18, Department of Scientific and Industrial Research. Reissued by the Building Research Establishment, 1985.

8.4.3 Mortar

British Standards Institution:
BS 5628 Part 3: 1985 *Code of Practice for the Use of Masonry. Materials and Components, Design and Workmanship.*
Building Research Establishment (1971) *Digest 89* (revised), *Sulphate Attack on Brickwork.*
Building Research Establishment (1975). *Digest 177, Decay and Conservation of Stone Masonry.*
Building Research Establishment (1977) *Digest 200, Repairing Brickwork.*
Water Research Laboratory (1986). *Information and Guidance Note 4-10-01, Bricks and Mortar.*

Chapter nine

Defects originating in the ground

C.J.F.P. Jones

9.1 Introduction

Most of the masonry structures in the United Kingdom were built before the beginning of the twentieth century. The terrain they occupy varies from flat agricultural land to hills and valleys, providing foundations ranging from soft alluvial deposits to competent rock. Materials used in construction often reflect the terrain and the geology. Where stone could be found, this would be used; elsewhere, bricks were made from local clays. Serious distress in such structures is often associated with plastic yielding of the soil beneath them or that providing stabilizing thrust, accompanied by the collapse of those parts of the structure providing direct contact with the ground.

9.2 Settlement and ground strain

Many cases of structural distress are the result of localized differential settlement or ground strain.

The cause of the settlement is normally site dependent and is the result of geotechnical factors, material deterioration or outside forces such as flooding. Failure due to the latter may be sudden, and Leadbeater (1986) advises keeping a record of all bridges susceptible to scour and carrying out frequent inspections, particularly after flooding (see also Chapter 5, Section 5.2). Material deterioration may involve parts of the masonry or the top of timber support piling (see Chapter 10, Section 10.1). The resulting movement of different parts of a structure will lead to different defects, typically as shown in Figs. 9.1–9.5.

Thus differential settlement in bridge abutments may result in longitudinal cracking of an arch, which can be significant if the cracks are wider than 3 mm. Alternatively, subsidence of the sides of an abutment may cause diagonal cracking, starting near the sides of the arch at the springing line and spreading towards the centre of the barrel at the crown. Uniform movement of the abutments may result in lateral cracks within the arch. Many of these defects are not conducive to the application of empirical rules, although an indication of the consequence of different crack patterns resulting from settlement can be deduced from Table 9.1.

9.3 Consolidation and expansion

In some situations consolidation of the subsoil or the shrinkage of underlying clays may be

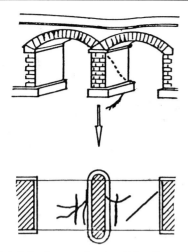

FIGURE 9.1 Settlement of a pier. (*After Cornet and Corte, 1981.*)

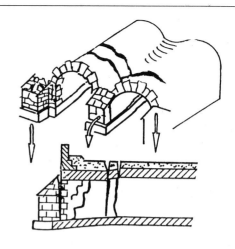

FIGURE 9.4 Settlement of one edge of a bridge. (*After Cornet and Corte, 1981.*)

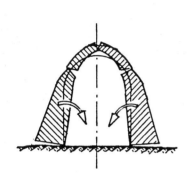

FIGURE 9.2 Forward rotation/movement of abutments. (*After Cornet and Corte, 1981.*)

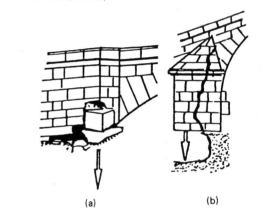

(a)　　　　　　　(b)

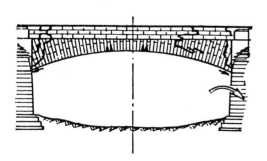

FIGURE 9.3 Rotation of abutment into the fill. (*After Cornet and Corte, 1981.*)

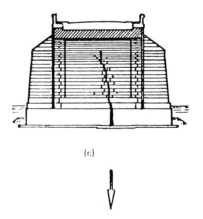

(c)

FIGURE 9.5 Effects of local settlement. (*After Cornet and Corte, 1981.*)

identified as the cause of differential settlement within a structure (Figs. 9.6). Similarly, the presence of expansive soils may cause localized cracking within an arch and cracks in spandrel and retaining walls. Cracks resulting from consolidation may occur early in the life of the structure and become stable.

Shrinkage and heave may occur at any time in the life of the structure if subsoil conditions are susceptible to a change in moisture content or in the level of the water-table. In many cases the cause may be identifiable and might typically be the result of a burst water main causing expansion or the growth of large trees close to the structure causing shrinkage. The development of differential movements due to shrinkage or expansion is often manifested by the appearance of clean new breaks in the masonry (Fig. 9.7).

9.4 Earth pressure

9.4.1 Behind abutments and retaining walls

In the majority of cases the lateral earth pressure behind abutments and retaining walls is constant

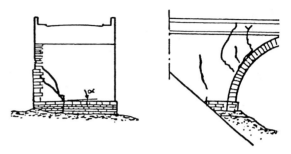

FIGURE 9.6 Effect of consolidation. (*After Cornet and Corte, 1981.*)

and stable, and at-rest soil conditions may be assumed. An increase in the back-fill stress may result in greater lateral loading and possible forward movement or tilting. This movement of an abutment may distort and crack the arch and spandrel walls it supports (Fig. 9.2). Alternatively, weakening of the supporting soil behind an abutment may lead to a reduction in lateral earth presure and outward movement (Fig. 9.3). A change in earth presure may be identified as the result of a specific geotechnical factor, such as a change in the local water-table or the presence of mining subsidence. The increases in traffic loading and in permissible vehicle weights in

TABLE 9.1 Defects affecting the stability and load-carrying capacity of an arch barrel

Defect (in arch barrel)	Possible cause	Condition factor F_c*
Longitudinal cracks	Differential settlement in abutments	
> 1 m spacing		0.6
< 1 m spacing		0.4
Lateral cracks/ deformation (dip in parapet)	Abutment movement Partial fracture of arch	0.6–0.8
Diagonal cracks	Subsidence at sides of abutment or pier	0.3–0.7
Cracks in spandrel walls at quarter-points	Flexibility of arch	0.8

After Department of Transport *Advice Note BA 16/84*.
*A condition factor of less than 0.4 requires repair or reconstruction of the bridge.

recent years may not only affect direct live loading on structures but also develop additional earth pressures.

9.4.2 Between spandrel walls

Arched structures are unique in that the load-bearing surface is frequently constructed on a soil infill supported between spandrel walls. The lateral pressure exerted by the infill on the spandrel walls has a significant influence on the behaviour of the structure. The critical nature of the fill is recognized in the current Bridge Assessment Code (Department of Transport Departmental Standard BD 21/84) by the use of partial

factors which vary with different fills (Table 9.2). Distress in spandrel walls is discussed in greater detail in Chapter 10, Section 10.2, and possible countermeasures are considered in Chapter 10, Section 10.4.8.

9.5 Uncontrolled water

The presence of water, in either the infill or the backing material, may have a deleterious effect. The resulting softening leads to plastic strain and increased soil pressures on spandrels and a potential reduction in the supporting thrust to an arch.

Failure of water pipes or drainage over a bridge

FIGURE 9.7 Effects of settlement due to subsoil shrinkage. (*Courtesy of British Rail.*)

TABLE 9.2 Fill factors

Filling	Fill factor
Concrete	1.0
Grouted material	0.9
Well-compacted material	0.7
Weak material (evidenced by tracking of the carriageway surface)	0.5

After Department of Transport *Departmental Standard BA 16/84*.

may cause the fill to become waterlogged. In addition, trenches excavated through the infill for public utility works can damage the waterproofing provided by road surfacing and lead to deterioration if reinstatement is not to the highest standard (Addy, 1985). Waterlogged fill has the potential to freeze, thus generating additional forces.

9.6 Mining subsidence

Mining of coal and other minerals results in earth movements in the vicinity of the excavated area. These movements, known collectively as subsidence, are three dimensional in nature, with any affected point having components of displacement along all three axes of a general Cartesian coordinate system. The displacements are imposed on any structure in the affected zone and may result in damage or even collapse unless adequate safeguards have been made or the necessary precautions taken. In the past, buildings and structures were sufficiently small and the mining methods sufficiently flexible that the effects could either be tolerated or be avoided by sterilizing the appropriate area for mining purposes.

Today, the demand for energy and the modern mining methods that have been developed to enhance output make sterilization of coal under a particular structure prohibitively expensive. At the same time, the effects of modern mining methods, in which settlements in excess of 1 m and ground strains of up to 1% are common, can cause serious disruption. Nearly all masonry and brick structures were built before the introduction of modern mining techniques and no structural precautions were taken during their construction to cater for large imposed differential ground movements.

Arched bridges are intimately connected to the ground through their substructure and their overall stability is dependent upon thrust acting through the abutments or piers. In areas of weak subsoil, piled foundations have been used which further connect the structure to the ground.

The consequence of the high degree of soil structure interaction exhibited by such structures is that any ground strains are imparted directly to the foundation and the arch. Accordingly, these structures are particularly vulnerable to mining movements. Although arches can withstand some differential settlements and spread of the springings, the strains associated with mining are often greatly in excess of the tolerance available in the structure. Subsidence damage is then severe, and complete reconstruction is frequently required.

9.6.1 Ground-movement patterns

The most common coal mining technique used in the past and still used today is the room and pillar method. In this method, coal is excavated in a series of intersecting adits, or rooms, leaving unmined coal, or pillars, to support the roof. The retention of pillars of unworked mineral influences the subsequent collapse of the working, and subsidence prediction may be complicated by the fact that later extraction of the pillars was practised but poorly documented. Surface movement resulting from old shallow mining is usually the result of a breakdown of the bridging strata between the coal pillars. Subsidence due to such void migration is unpredictable, although an assessment of the maximum subsidence may be possible given a knowledge of the mining method, the geometry of the workings and the geology.

Experience in the Yorkshire coal field has shown that little subsidence occurs in relatively level strata if the old room and pillar working is at a depth greater than 15 m below the top of the

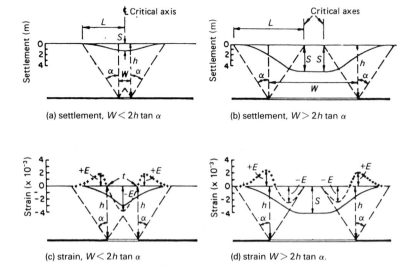

(a) settlement, $W < 2h \tan \alpha$

(b) settlement, $W > 2h \tan \alpha$

(c) strain, $W < 2h \tan \alpha$

(d) strain $W > 2h \tan \alpha$.

FIGURE 9.8 Cross-sectional views of a subsidence trough with ground settlement and strain patterns (S, maximum settlement; h, depth of seam from surface; α, angle of draw; W width of panel; L, distance between zero and maximum subsidence; $+E$, maximum tension; $-E$, maximum compression).

rock. This observation is consistent with the conclusions drawn by Piggott and Eynon (1977) from simple analytical models which account for strata collapse and bulking of fallen material. They recommended that precautions be taken if old mine workings are present at depths of less than ten times the extracted thickness below the top of the rock. Given an extracted thickness of 1.5–1.8 m, this results in a depth of 15–18 m.

Modern longwall mining entails the complete removal of a given thickness of coal. Excavation is performed by mechanical breaking tools that are drawn continuously across the working face so that no pillars are left in the mined-out area. The roof is allowed to cave in behind the face. The collapsed material, which is of greater bulk than the natural rock, extends some 6–10 m above the worked seam. Higher levels of strata sag into the trough which develops up to the surface and extends with the advancing face. The majority of measurable subsidence occurs within 12 months of working.

9.6.2 Deformation pattern

Observations have shown that surface deformation resulting from longwall subsidence takes the form of a shallow trough which covers an area larger than the projected surface area of excavated coal. Figure 9.8 shows cross-sectional views of the subsidence caused by longwall mining. Figures 9.8(a) and 9.8(b) show surface settlement, whereas Fig. 9.8(c) and 9.8(d) illustrate how tensile and compressive strains develop.

The limit of the subsidence trough is defined by the angle of draw, i.e. the line joining the edge of the excavated area to the point of zero effect at the surface. Experience in the United Kingdom shows that the angle of draw varies between $25°$ and $35°$, with the latter value being most representative of panels mined in the Pennsylvanian age shales and siltstones of Yorkshire and the Midlands (National Coal Board, 1975). In the case of a level seam with an angle of draw of $35°$, there is no subsidence at a distance 0.7 times the depth h of the seam outside the limit of extraction or 'area of influence'. It follows that the width:depth ratio of any seam influences surface subsidence. When the width:depth ratio is 1.4, maximum subsidence occurs at the centre of the trough. When the width:depth ratio is greater than 1.4, the subsidence trough develops a flat-bottomed profile. With a width:depth ratio of less than 1.4, unworked coal remains within the

FIGURE 9.9 Effects of compression strain on a retaining wall as a result of mining subsidence. (*Courtesy of C.J.F.P. Jones.*)

area of influence and the subsidence is not fully developed.

As the subsidence trough develops, the centre subsides vertically, while the remainder moves inwards towards the centre of the working, resulting in both vertical and horizontal movements. These differential horizontal movements cause strain in the ground, producing a zone of compression in the trough of the worked area and tension at the edges of the excavated area, as illustrated in Figs. 9.8(c) and 9.8(d).

Structural distortion will depend on both the imposed differential vertical and horizontal movements. As can be seen in Fig. 9.9, the maximum tensile and compressive strains along the subsidence trough develop at locations of maximum convex and concave curvature. This qualitative observation permits correlation of field measurements of horizontal strain with differential settlement, expressed as curvature. One such correlation developed by O'Rourke and Turner (1981) is shown in Fig. 9.10, where the maximum tensile strain ε_h is plotted relative to

the maximum convex curvature for subsidence profiles over longwall panels in both the United States and the United Kingdom. In all cases, maximum strains and curvatures occurred at the same locations.

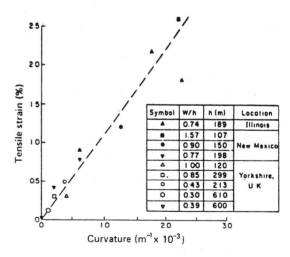

FIGURE 9.10 Horizontal tensile strain versus convex curvature for US and UK longwall panels. (*After O'Rourke and Turner, 1981.*)

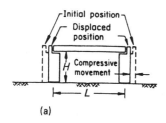

(a)

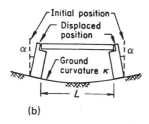

(b)

FIGURE 9.11 Structural response to horizontal strain and curvature at the ground surface. (*After Jones and O'Rourke, 1988.*)

It is significant that observations of US longwall subsidence show higher curvatures and strains than those typical of the British coal fields. Curvatures and horizontal strains can be approximately four times larger for US mining conditions.

Ewy and Hood (1984) have indicated that the relationship between curvature κ and horizontal strain ε_h appears to be in the form

$$\kappa = K\varepsilon_h{}^n \qquad (9.1)$$

in which K and n are empirical constants, with $n = 1$ being appropriate for the US data. Ewy and Hood suggest values of K between 0.049 and 0.082 m^{-1} for both convex and concave curvatures, where $n = 1$.

9.6.3 Structural deformation

Figure 9.11 presents a profile view of a beam simply supported on piers subjected to both horizontal compressive strain and concave curvature. If it is assumed that the curvature is more or less constant over the length L of the structure and that α is small, it can be shown that the rigid body rotation α of the piers is related to the curvature κ by

$$\alpha = \frac{\kappa L}{2} \qquad (9.2)$$

For the simplified model illustrated in Fig. 9.11, the horizontal displacement of the beam relative to the end piers is critically important with respect to stresses imposed in the beam and walls of the piers. Tolerable deformation is therefore related to a critical horizontal displacement δ_c above which structural damage will occur. If δ_c is taken as the horizontal displacement which accumulates across the entire structure, it can be related to the strain ε_h and rotation α by

$$\delta = \varepsilon_h L + 2H \tan \alpha \qquad (9.3)$$

in which H is the height of the structure. By combining equations (9.1)–(9.3) for small α, the horizontal strain ε_h is obtained as a function of δ_c, the structural geometry and the empirical constant K:

$$\varepsilon_h = \frac{\delta_c}{L(1 + KH)} \qquad (9.4)$$

In the case of masonry or brick arch structures behaviour is complex and three dimensional, and the model in Fig. 9.11 provides only a simplification. Nevertheless, some important aspects of structural response to mining subsidence are represented by equation (9.4). In particular, the permissible strain is related not only to horizontal displacement but additionally to changes in surface gradient.

Figure 9.12(a) shows the horizontal strain plotted relative to the length of the structure according to the relationship given by equation (9.4), with K taken as 0.075 m^{-1}. The plots were developed for various critical displacements and heights; relative rotation has been modelled as a rigid body phenomenon with structural tilting

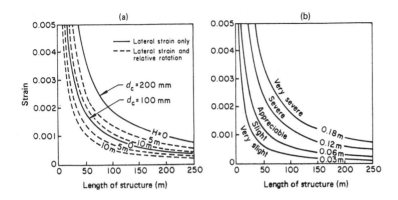

FIGURE 9.12 Relationship between strain and length of structure for various levels of deformation: (a) analytical trends; (b) empirical relationships. (*After Jones and O'Rourke, 1988; National Coal Board, 1975.*)

equal to the change in surface grade. If the relative rotation is expected to be minimal, H can be taken as zero when referring to Fig. 9.12(a). The empirical relationships for estimating building damage developed by the National Coal Board (1975) are shown in Fig. 9.12(b). The hyperbolic relationship between strain and length is evident in both the analytical and empirical approaches depicted in Fig. 9.12.

9.6.4 Cuttings and faults

The development of railways required the excavation of numerous cuttings, which are often steep sided and deep. It has been found from observation that these cuttings have an influence on potential mining subsidence strains in that the compressive strains are more acute than those predicted by conventional systems. A multiplying factor of 2.5 may be required to determine the true compressive strain which will develop and which may be imparted to any adjacent masonry or brick arch bridge (Institution of Civil Engineers, 1977). Similarly the presence of ground faults may cause serious additional distortion to occur (Lee, 1966).

9.6.5 Prediction of mining subsidence

A number of methods is available to predict the effects of mining subsidence on masonry structures. The method developed by the National

Coal Board is widely known and has been adopted for use outside the United Kingdom. The technique is simple and robust but requires an assumption of the position and layout of the mine workings (National Coal Board, 1975). A more general method which does not rely upon mine layout and which can be used to produce parametric predictions has been developed by Sims and Bridle (1966). This method has been used successfully in developing structural designs for the M1 and M62 motorways. If the mine layout is known, accurate computer predictions of subsidence and ground strains can be determined using the system reported by Jones and Bellamy (1973).

Burton (1977) devised three-dimensional systems to replace the static accounts given by subsidence and strain profiles in lines across panels. Such systems calculate the difference in

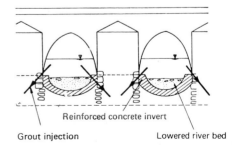

FIGURE 9.13 Propped piers and concrete invert. (*After Corte and Levillain, 1981.*)

horizontal/vertical movements of any surface points at any time and are used for large structures and installations. The calculations are complex, especially when taking into account the geology and positions of earlier mine workings or when analysing records of past events, and so a modern system, as used by British Rail for example, must be managed by computers. The introduction of computer programs to aid the prediction of mining subsidence was reported by Burton(1980).

9.7 Counter-measures

9.7.1 Ground strain

Ground strain resulting in forward movement of abutments (Fig. 9.2) may be particularly serious. Investigations are required to establish whether the movements are continuous; if they are, emergency procedures to stabilize the structure before collapse may be necessary. The cause of the movements should be investigated and established. The results may suggest the direction and mode of the counter-measures. Typically these might include the removal and replacement of saturated back-fill and the repair of fractured mains or services. Strengthening from underneath the arch using the methods described in Chapter 10, Section 10.4.5., is also a proven technique.

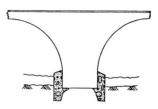

FIGURE 9.15 Use of concrete apron. (*After Ridings and Jones, 1981.*)

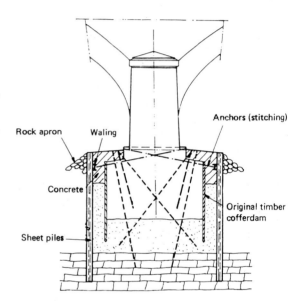

FIGURE 9.16 Repair of Mont Louis Viaduct, France, showing the position of the grout tubes. (*After Caumes, 1981.*)

Backward movement of abutments, if continuous, similarly demands urgent attention. The cause may be a reduction in the stiffness of the supporting soil or outward rotation of the toe. The latter can be checked by propping between abutment faces or between abutment and pier. Permanent remedial work may entail the construction of permanent props and include improvements to the water channel (Fig. 9.13). Alternative methods are possible using sheet piles positioned in front of the toe, although these may be difficult to install. Whatever counter-measure is selected, it must be based upon a sound site

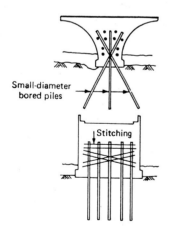

FIGURE 9.14 Small-diameter bored piling and stitching. (*After Ridings and Jones, 1981.*)

investigation and soil mechanics principles; remedial works undertaken without an understanding of the failure mechanism are likely to be uneconomic and unsuccessful.

9.7.2 Settlement and scour

Localized settlements and scour, resulting in defects such as those shown in Figs 9.1–9.5, have to be stabilized. Classical remedial measures involve underpinning, the use of needle piles, sheet piling, the construction of cofferdams around piers, the installation of concrete or rock aprons and grouting. Several methods can be combined and may include the stitching of cracked structures. Figures 9.14 and 9.15 show idealized remedial measures, whilst Fig. 9.16 illustrates the repair used on the Mont Louis Viaduct over the River Loire in France (Caumes, 1981).

In all cases the remedial works are to improve, protect or replace the existing foundation strata. An example of replacement is underpinning, which is used to transfer the existing (or additional) loads to a more competent stratum or to improve the existing one. The basic underpinning technique consists of excavating and replacing the material beneath the base of the structure, working in short bays, normally 1–2 m long, to a 'hit-and-miss' pattern. The bays are excavated to the required depth and filled with mass concrete to within 75–100 mm of the underside of the foundation; the gap is filled with a dry cement/sand mortar. A more sophisticated variation of the method involves the use of reinforced concrete pads connected to concrete beams. There are various proprietary systems, including the use of mini piles which can be categorized as having a nominal diameter of 100–250 mm.

Underpinning by grouting is a technique which seeks to improve the load-carrying capacity of foundations by treating the soil itself to enhance its shear strength. Soil grouting can be categorized as follows:

1. penetration grouting in which a cement or chemical grout is injected into the voids within the soil matrix
2. displacement grouting in which a grout is injected into the soil under pressure, thereby increasing compaction
3. replacement jet grouting in which grout is injected to replace soil eroded by high pressure jets.

Additional details of these techniques are provided in Chapters 15 and 17.

9.7.3 Mining subsidence

Counter-measures to ameliorate the effects of mining subsidence on masonry structures, particularly arch bridges, are limited. Support to the arch barrel in the form of timber centring or steel colliery arches can be used (Figs. 9.17 and 9.18). In the case of small structures carrying minor roads this may be an acceptable long-term solution. However, more radical measures are required for structures carrying important roads.

Accurate predictions of the effects of mining subsidence and the strains which will be imparted to the structure are possible. However, with modern mining the strains may be excessive and the counter-measures could involve demolition of the structure and the building of a replacement before mining occurs. Costs involved in precautionary works, including the demolition and reconstruction of affected structures, may be recoverable from British Coal in accordance with the

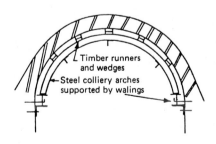

FIGURE 9.17 Use of colliery arches. (*After Ridings and Jones, 1981.*)

FIGURE 9.18 Effects of mining subsidence on a tunnel portal showing centring: (a) before mining commenced; (b) in 1934, after mining of the first seam; (c) in 1940, during mining of the second seam; (d) in 1941, after mining of the second seam. (*Courtesy of British Rail.*)

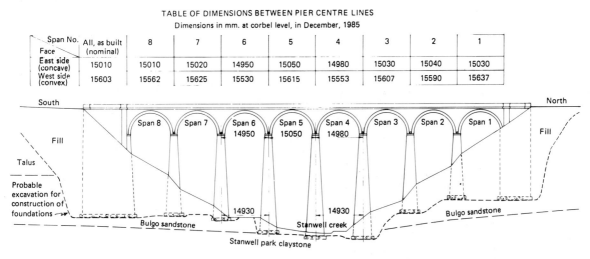

TABLE OF DIMENSIONS BETWEEN PIER CENTRE LINES

Dimensions in mm. at corbel level, in December, 1985

Span No. Face	All, as built (nominal)	8	7	6	5	4	3	2	1
East side (concave)	15010	15010	15020	14950	15050	14980	15030	15040	15030
West side (convex)	15603	15562	15625	15530	15615	15553	15607	15590	15637

FIGURE 9.19 Stanwell Park Viaduct: elevation of the east face before removal of span 6. (*Courtesy of State Rail Authority of New South Wales*).

provisions of the Coal Mining Subsidence Act 1957, the Coal Industry Act 1975 or specific Agreements (such as that of January 1982 between the Coal and Railways Boards).

9.8 Case History: viaduct affected by the release of rock stress

This account is based on information provided by the State Rail Authority of New South Wales.

9.8.1 The structure

Stanwell Park Viaduct was constructed in 1924 to carry a railway across a narrow valley near the east coast of New South Wales, Australia. Built of hard bricks in cement mortar to a radius of 240 m, it carries passenger and heavy freight traffic on two tracks at a height of 43 m above the Stanwell Creek. With an overall length of 148 m, it comprises eight semicircular arches, each of eight rings (Fig. 9.19). The massive box wing abutments and tapered piers are founded on, or just above, Stanwell Park claystone, which is generally sound and strong and which controls foundation behaviour.

Coal mining has long been undertaken in the area, but at a sufficient distance not to have affected the viaduct. Apart from minor cracking noted during regular bridge inspections, no defect of structural significance was observed prior to 1983. Mining in the immediate vicinity commenced at about that time.

9.8.2 The development of defects

In August 1985 a circumferential crack, up to 160 mm wide, occurred in arch 3. Transverse tie-rods were fitted, the track level was lowered to reduce the thrust on the spandrels and the crack was epoxy grouted. Small movements were also noticed throughout the viaduct and frequent measurement was commenced to identify any developing pattern. Enquiry of the Department of Mineral Resources revealed that a panel of coal 2 m thick had been worked, with 80% extraction, up to the 120 m horizontal reserve limit at a depth of 150 m below rail level during the period from May to September 1985.

By mid-November arch 6 showed local cracking and crushing, with several spandrels cracking along the line of the arch extrados and concrete

153

FIGURE 9.20 Stanwell Park Viaduct, span 6: east face, showing disruption of the spandrel wall at the level of the haunching. (*Courtesy of D. Walker.*)

haunching. Between December 7 and December 9 there was severe disruption of arch 6, with change of shape and partial loss of the lower three rings of brickwork. Cracking of spandrels and humping of parapets also increased between spans 4 and 8.

The line was then closed to traffic to permit remedial works to proceed without interference.

9.8.3 Investigation

The State Rail Authority's engineering geologist had previously investigated ongoing cracking and deformation of a tunnel some 2 km south of the viaduct which had also been affected by anomalous mining effects. It was his opinion that residual rock stress had been released by the effect of the mining on the regional stress field and had produced the movement and forces that were acting upon the viaduct.

To establish what deformations the viaduct and the valley had suffered and to monitor continuing movements, a range of activities was initiated. They included the following:

1. measurement of the offset of the piers from vertical and the distances between pier centre lines
2. a detailed survey of the entire viaduct, including arch profiles of spans 4 and 5 with monitoring of an array of pins on the down-side face of the structure
3. precise levelling of bars in the parapets and piers
4. monitoring of pier tilts by electronic tiltmeter and of changes in span lengths by tape extensometer
5. mechanical extensometer monitoring of crack dimensions in the superstructure

FIGURE 9.21 Stanwell Park Viaduct, span 6: east face, showing failure of the arch. (*Courtesy of D. Walker.*)

6. monitoring of tilts and shearing in the foundation bedrock and within pier 1 and the southern abutment by borehole inclinometer
7. monitoring of a mine subsidence traverse along the creek alignment

Horizontal compression of the foundation bedrock along the line of the viaduct had resulted in a shortening of the viaduct by up to 240 mm, with closure concentrated at spans 4 and 6 in the arch and the spandrels and the damage being most pronounced at span 6 (Figs. 9.19, 9.20 and 9.21). Failure of that span reduced the longitudinal compression on adjacent spans.

The compression of the valley profile was also accompanied by differential uplift at span 4, and monitoring revealed that deformation and cracking in the viaduct was occurring in concert with observed mining subsidence effects from the 1985 mining operations.

9.8.4 Remedial works

Because span 6 was considered to be beyond repair, it was decided to remove it and replace it with a steel span, with a precast concrete deck, that would permit continued movement between piers 5 and 6. A steel support tower on piled foundations was erected under span 6 to support 12 horizontal struts with a 90 tonne hydraulic jack at one end of each strut to give controlled release of strain. Steel arch ribs and timber lagging were supported on the struts to facilitate removal of the arch. Four 36 mm high tensile steel tie-rods were installed between piers at springing level, initially in each of spans 4, 5 and 7, to maintain as much of the function of the structure as possible; strain gauges were fitted to measure force changes in the rods. All these works were completed and the track was restored for reopening at the beginning of February 1986.

Later works included additional tie-rods installed in spans 1, 2, 3 and 8 and scour protection to the watercourse. On the removal of span 6 the dimension at springing level reduced by 20 mm, with some change in the inclination of piers 4, 5, 6 and 7. The force recorded per jack during removal was about 30 tonnes. Since February 1986 there has been a shortening of span 6 by about 60 mm, although the overall shortening between abutments has amounted to only some 20 mm, indicating internal relief in the super-structure, principally in spans 4 and 5. Support by a reinforced shotcrete skin has been provided under span 4.

9.9 References and further reading

Addy, N. (1985) Buried services associated with transport systems including mining subsidence. *Proc. ICE Symp. on Failures in Earthworks,* Pap. 2, Thomas Telford, London.

Anon, (1988) Using high PFA concrete. *Civ. Eng.,* June, 30–32.

Burton, D. (1977) A three-dimensional system for the prediction of surface movements due to mining. *Proc. 1st Int. Conf. on Large Ground Movements and Structures* (ed. J. D. Geddes), p. 209–28, Pentech Press.

Burton, D. (1980a) Advanced systems used in mining subsidence predictions. *Chart. Land Surv./Chart. Miner. Surv.,* **2** (1), 36–48.

Burton, D. (1980b) The introduction of mathematical models for the purpose of predicting surface movements due to mining. *Proc. 2nd Int. Conf. on Large Ground Movements and Structures* (ed. J. D. Geddes), p. 50–64, Pentech Press.

Caumes, A. (1981) The foundation repair of the Montlouis viaduct over the River Loire. *Colloque Inter. sur la Gestion des Ouvrages d'Art, Maintenance and Repair of Road and Railway Bridges,* Brussels/Paris. Presses de l'Ecole Nationale des Ponts et Chausées, Paris and Editions Ancient ENPC, Brussels, vol. I, pp. 279–84.

Cornet, D. and Corte, J. F. (1981) Classification of defects observed in masonry bridges. *Colloque Inter. sur la Gestion des Ouvrages d'Art, Maintenance and Repair of Road and Railway Bridges,* Brussels/Paris, Presses de l'Ecole Nationale des Ponts et Chausées, Paris and Editions Ancient ENPC, Brussels, vol. I.

Corte, J. F. and Levillain, J. P. (1981) Foundation repairs of small masonry bridges. *Colloque Inter. sur la Gestion des Ouvrages d'Art, Maintenance and Repair of Road and Railway Bridges,* Brussels/Paris. Presses de l'Ecole Nationale des Ponts et Chaussées, Paris and Editions Ancient

ENPC, Brussels, vol. I, pp. 311–16.

Department of Transport (1984) The assessment of highway bridges and structures. *Departmental Standard BD 21/84: Advice Note BA 16/84.*

Ewy, R. T. and Hood, M. (1984) Surface strain over longwall coal mines: its relation to the subsidence trough curvature and to surface topography. *Int. Rock Mech. Min. Sci.* **21** (3), 155–60.

Greenwood, D. (1987) Underpinning by grouting. *Ground Eng.,* **20** (3), 21–32.

Hunter, L. E. (1952) Underpinning and strengthening of structures. *Contract. Rec. Munic. Eng.*

Institution of Civil Engineers (1977) *Ground Subsidence,* Thomas Telford, London.

Jones, C. J. F. P. and Bellamy, J. B. (1973) Computer prediction of ground movement due to mining subsidence. *Geotechnique,* **23** (4), 515–530.

Jones, C. J. F. P. and O'Rourke, T. D. (1988) Mining subsidence effects on transportation facilities. *ASCE Symp. on Mine Induced Subsidence: Effects on Engineered Structures, Geotech. Spec. Publ.* **19.**

Jones, C. J. F. P. and Sims, F. A. (1981) The maintenance of bridges in areas of mining subsidence. *Colloque Inter. sur la Gestion des Ouvrages d'Art, Maintenance and Repair of Road and Railway Bridges,* Brussels/Paris. Presses de l'Ecole Nationale des Ponts et Chausses, Paris and Editions Ancient ENPC, Brussels, vol. II.

Leadbeater, A. D. (1986) *The Practical Use of Bridge Assessments,* Institution of Highways and Transportation National Workshop, Leamington Spa.

Lee, A. J. (1966) The effect of faulting on mining subsidence. *Min. Eng.* **125** (71), 735–45.

National Coal Board (1975) *Subsidence Engineers' Handbook,* 2nd edn.

O'Rourke, T. D. and Turner, S. M. (1981) Empirical methods for estimating subsidence in U.S. coalfields. *Proc. 22nd US Symp. on Rock Mechanics,* pp. 322–27, Amer. Soc. Civil Engrs, New York.

Piggott, R. J. and Eynon, P. (1977) Ground movements arising from the presence of shallow abandoned mine workings. *Proc. 1st Int. Conf. on Large Ground Movements and Structures* (ed. J. D. Geddes), pp.749–80, Pentech Press.

Ridings, I. D. and Jones, C. J.F. P. (1981) The load carrying capacity and maintenance of existing arch bridges. *Colloque Inter. sur la Gestion des Ouvrages d'Art, Maintenance and Repair of Road and Railway Bridges,* Brussels/Paris. Presses de l'Ecole Nationale des Ponts et Chaussées, Paris and Editions Ancient ENPC, Brussels, vol. I, pp. 101–7.

Sims, F. A. and Bridle, R. J. (1966) Bridge design in areas of mining subsidence. *J. Inst. Highway Engrs.* Vol. XIII November, 19–34.

Thorburn, S. and Hutchinson, J. F. (1985) *Underpinning,* Surrey University Press, Guildford.

Chapter ten

Defects due to wear and tear

A.M. Sowden and C.J.F.P. Jones

10.1 Sources of trouble

Wear and tear is taken here to cover the natural deterioration of a structure due to time-related effects of climatic and environmental conditions, as well as the adverse effects of normal use, particularly the loading to which it may be subjected.

Masonry structures comprise units connected by mortar joints; if either the units or the joints degrade, the structure tends to lose its integrity. Bricks and stones are often porous to a degree and may contain inherent fissures; the spaces between units are rarely completely filled, even in new construction. Whether falling as rain or flowing through the ground or derived from rivers and tides, water which has access to masonry will percolate into and through it, filling pores, fissures and voids, and so initiating the problems described in other chapters.

In particular, the loss of jointing material leaves units loose and able to drop out; it thus destroys the proper transmission of forces throughout the structure and may reduce the transfer of loads between units to a series of point loads, inducing high local stresses and possibly cracking of individual units.

If inadequate provision has been made for thermal expansion and contraction, forces are generated which will disrupt masonry. This effect is most marked where girderwork is built in or where bearings, intended to be moveable, have become seized or have worn into the supporting structure (Fig. 10.1). The corrosion of embedded metalwork will cause it to expand, with a similar outcome.

Old masonry structures were commonly founded on timber rafts and/or timber piles. If these are entirely and constantly submerged in water, they can enjoy very long lives, but any parts exposed to air will rot in circumstances where the first indication of such decay is likely to be settlement and fracturing of the masonry (Fig. 10.2).

Increases in the amount and density of traffic loading on structures is recognized as accelerating their deterioration. Under live loading, flexure occurs in masonry; cracking may ensue and fissures may open, allowing the ingress of water and debris from mechanical attrition, thus accelerating the weathering and fragmentation processes. If the flexure is not entirely elastic or the debris prevents full recovery from taking

FIGURE 10.1 Abutment fractured by girder movement and inability of the bearing to cope with it. (*Courtesy of British Rail.*)

place, a 'ratchet' form of progressive development of defects is likely. In the case of arched bridges, particularly under railways, differential deflection between the loaded and unloaded parts of the vault or contrary distortion as traffic passes on parallel tracks often gives rise to longitudinal cracking (normal to the axis of the vault). Similarly, the circumferential fractures which are so common a short distance in from the end face of a vault (Fig. 10.3) result, at least in part, from the differential deflection between the statically and dynamically loaded parts of the arch, and they may be made worse if the ends of a brick vault are stiffened by virtue of the provision of

stone voussoir facings or by the effect of the spandrel wall above.

10.2 Stability of spandrel walls

Specific mention should be made of the outward movement of spandrel walls, to which so many masonry bridges are susceptible; indeed, unserviceability arises more commonly from the collapse of such walls than from the condition of the arch barrel. Despite this, the Department of Transport *Departmental Standard BD 21/84* gives scant advice on the stability of spandrel walls, beyond stating that it is not amenable to calculation and 'must be assessed qualitatively by considering the condition of the structure and the significance of any defects'. In fact, this aspect of bridge assessment has been largely neglected and the mechanism of failure is not fully understood, and so the relative significance of defects is obscure.

The movement of spandrel walls may take the form of tilting or bulging or sliding over the extrados of the vault, possibly accompanied by horizontal shearing at different levels above it, or the above-mentioned fracture through the end of the arch barrel may open up (Fig. 10.4). There are various potential causes of this movement. Vertical live loading on the arch may initiate separation between its extrados and the stiffer spandrel wall and parapet acting together as a beam, or it may lead to a longitudinal fracture near the end of the vault as already described. Lateral forces tending to push the spandrel wall outwards may be exerted by the fill, especially if it becomes saturated and freezes or if road/railway track levels are raised appreciably. The forces may be due to traffic, either vertical live loads generating horizontal pressures from a 'Poisson ratio' effect on the fill or centrifugal forces on curves being transmitted through the fill. It is noticeable that the position of traffic relative to the spandrel wall is often significant; the presence of a footpath, restraining vehicles from passing close to the wall, leads to less damage or displacement. Direct vehicular impact with the parapet is, of course,

FIGURE 10.2 Effect of settlement of a pier supported on timber piles which have rotted. (*Courtesy of British Rail.*)

likely to have adverse effects, as may, in the case of railway underbridges, the operation of certain types of track maintenance machinery.

Instability may well result from any of these factors or some combination of them, but few studies have been undertaken to determine which are most likely to be critical. Accurate analysis of the lateral pressures generated within the fill is difficult, and may be impossible, owing to the wide range of infill materials encountered and the fact that adequate records are rarely available; investigations into the characteristics of the fill material are equally rare. It is noticeable, however, that, when spandrel walls collapse, the exposed faces often stand almost vertically (Figs. 10.5 and 10.6), suggesting that as the fill compacts and its voids become filled with fines it

behaves as a cohesive mass which tends to retain its shape despite a complete lack of side support.

It is apparent that, at present, any assessment of the stability of spandrel walls is subject to a wide range of reliability and the most effective approach for preventive maintenance is similarly indeterminate. In general, it seems that a significant horizontal movement of such a wall over the arch extrados may be tolerable, but any longitudinal fracture **through** the arch barrel, near its end and extending across a major part of the span, should be very closely monitored; if, as it develops towards the springing, it runs outwards towards the face, it should be regarded with concern and early counter-measures should be initiated.

FIGURE 10.3 Circumferential fracture in arch vault near its end face. (*Courtesy of British Rail.*)

10.3 Preventive maintenance

Since wear and tear are progressive with time and use, it follows that preventive maintenance will pay dividends. Thus early action should be taken over such matters as the regulation of live loading (e.g. by imposing weight restrictions or restricting the width of the carriageway), the reduction of impact effects (e.g. by repairing pot-holes or eliminating rail joints), the provision for expansion and contraction, the diversion or improvement of drainage, and the repair of leaking services or the prevention of water percolation.

This last, however, is not particularly easy to achieve. Conventional and effective methods of waterproofing are described in Chapters 11 and

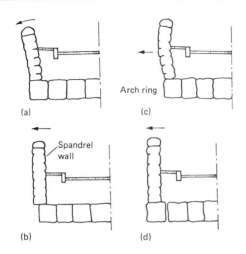

FIGURE 10.4 Movement of spandrel walls: (a) outward rotation; (b) sliding; (c) bulging; (d) outward movement and cracked arch. (*Courtesy of C.J.F.P. Jones.*)

20. However, most involve excavation and access to the non-exposed faces of the masonry; their implementation may therefore prove inconvenient, relatively expensive or disruptive of the normal use of the structure. Other measures leave the masonry in a saturated state. There is obvious scope for a technique which, by injection from the exposed face, will create an impermeable barrier preventing the entry of moisture into the rear face of the masonry. As yet, none has fully proved itself. Ground-stabilizing resin injected into voids in the back-fill should reduce its permeability and might produce such a membrane *in situ;* but most resin grouts are water-based gels which may irreversibly dry out and become ineffective. There appear to be possibilities in using water-reactive polyurethane foaming resins. The most intractable problems are likely to lie in controlling the

FIGURE 10.5 Collapsed spandrel wall of a road underbridge. (*Courtesy of C.J.F.P. Jones.*)

FIGURE 10.6 Collapsed spandrel wall of a railway underbridge. (*Courtesy of British Rail.*)

spread of the resin, especially in the light of the variable nature of most back-fill material.

Some attempts have been made to reduce the permeability of the masonry itself by the injection of epoxy resin, microsilica – cement or water-reactive polyurethane foaming resin, but the injection pressure considered safe to avoid damaging the brickwork was insufficient to achieve adequate penetration and the results were disappointing.

If a successful technique of waterproofing in this way can be developed, it will be essential to ensure relief of water pressure by providing effective and strategically located weep holes.

10.4 Remedial action

10.4.1 Effectiveness

Repair work must address the nature of the defect and its cause. There can be only cosmetic reasons for spending resources on dealing with a fault which has no adverse effect on the integrity of a structure. For instance, arches, particularly if relatively flat, may suffer from distortion under load as a result of the spreading of inadequately rigid abutments and this may lead to ring separation or transverse cracking (parallel to the axis of the arch); both these defects call for remedial action. However, observed distortion may have been due to premature striking of centring at the time of construction – a circumstance which can be inferred if there is no matching misalignment

FIGURE 10.7 The distortion in the haunch of the nearest opening probably occurred when the centring of this eight-ring arch was struck. (The concrete supporting structure was provided as a precaution, in place of brick diaphragm walls, as in the furthest span, when the way under the nearest span was opened up to provide an additional carriageway.) (*Courtesy of British Rail.*)

of the parapet or visible cracking. If monitoring verifies that the situation is stable, there is no reason to attempt rectification (Fig. 10.7). Similarly, the longitudinal cracking of the arch of a railway underbridge between adjacent tracks, as described above (p. 158), may in fact relieve the structure and allow it to act more efficiently and naturally; however, since it may facilitate water percolation and cause attrition to spread on each side of the defect, some countermeasure is clearly called for. Merely making good on a like-for-like basis or stitching across the fracture would be unlikely to be effective for very long. Rationally, it would be better to cut a wider chase along the crack and seal it with a resilient material, however difficult that might be in practice. An alternative course of action would be

to improve the load distribution on the arch, possibly by constructing a substantial saddle over it (see Section 10.4.6). Where masonry has been crushed under points of concentrated load (such as from a girder bearing), the initial action should likewise be to improve the load distribution. In the process it may be possible to transfer the loading from a superstructure to a position such that the stability of the substructure is improved – perhaps by moving a girder bearing back on an abutment or away from an acute angle in the masonry.

Generally, no sizeable repair should be carried out without an attempt to assess its effectiveness. Visual observation will suffice in many cases, but in others there is need for some quantitative confirmation or measure of the degree of improvement achieved. Traditionally, the approach has been to measure deflections under load, on the premise that repairs increase stiffness and reduce live-load distortion. Limited British Rail data on grouted repairs to masonry arches seem not to confirm this supposition. Reference has already been made (p. 41) to the fact that arch deflections are an insensitive measure of condition. That they are apparently equally unreliable indicators, on a comparative basis, of improvements achieved by repair emerges from such examples as measurements taken on 22 arches of a viaduct which showed negligible differences between deflections before and after repairs (by cement grouting with stitching). Also, a comparison of 73 pairs of 'before' and 'after' readings on a variety of other bridges showed an overall average improvement of about 25%, but that figure derives from very mixed results and conceals the fact that some arches deflected more after repair than they had before it! The measurement of dynamic deflections at a number of points using accelerometers would appear to hold promise; a detached area of brickwork should give a characteristic frequency of vibration which would change after repair. However, such readings on an arch before and after repair by resin injection showed a reduction in train-induced acceleration levels but no change in natural frequency values (in the range considered) as would be expected if the repair had effectively stiffened the structure.

It therefore seems that there are no proven techniques currently available for use on a routine basis to evaluate the efficacy of a repair or to compare the effectiveness of different treatments. There is considerable scope for development here.

10.4.2 Grouting

If all that is necessary is to restore the integrity of masonry by replacing lost jointing material or filling cracks and excluding water, the injection of a suitable grout into the body of the structure (internal grouting) should provide the solution (see also Chapter 15, Section 15.3.1). However,

this is an operation fraught with uncertainty; moreover, it is usually considered ineffective in masonry less than about 500 mm thick. Considerable expertise is certainly needed to achieve a satisfactory degree of penetration, and it is difficult to verify the extent of such achievement (although core sampling gives some indication). The subsequent demolition of structures which have been treated in this manner has sometimes revealed very variable dispersion of the grout, with poor adhesion to the brickwork where this was covered by a film of old lime mortar. (It may be, of course, that this variability is the cause, at least in part, of the lack of correlation between grouting repairs and deflection measurements described in Section 10.4.1). Problems will undoubtedly arise if the voids are small or clogged with debris, preventing adequate dispersion, and so a useful technique is to flush water through the injection holes prior to grouting; compressed air is sometimes introduced for the same purpose. Water flushing also aids the removal of drilling debris which may be smeared on the sides of the holes. Penetration depends, to an extent, upon injection pressures, but in order to avoid damage to old masonry these may have to be limited, with the holes located at sufficiently close centres and left open during grouting so that there is no undue build-up of force behind the face. With injection points spaced at 300–500 mm, grouting pressures are typically restricted to 0.3 N/mm^2 (3 bar or about 50 lbf/in^2) for the first stage and 0.5 or 0.6 N/mm^2 for the second.

For the impregnation of masonry it is usual to use a suspension grout based on hydraulic cement with the addition of 4% – 6% of bentonite and possibly pulverized fuel ash (PFA) to promote stability and reduce cost. Improved flow and quicker curing characteristics are claimed from the use of microsilica-modified cement grouts.

More expensive, but with the ability to penetrate fine fissures, are various resin grouts. Epoxy resins are strong and tough, are water tolerant, have good adhesion properties and will usually bond to wet masonry (but not to surfaces con-

FIGURE 10.8 Banding of a defective brick column. (*Courtesy of British Rail.*)

FIGURE 10.9 Strapping of a defective stone pier. (*Courtesy of British Rail.*)

taminated by oil or silt); they undergo low shrinkage during curing and have good resistance to most environments and most chemicals. Phenolic resins also hold promise, having similar qualities apart from higher shrinkage; their curing characteristics can more readily be varied to suit circumstances. Curing time and the rapid development of adequate strength may be critical in order to allow the early restoration of traffic. Formulations which impart thixotropic properties allow the injected material to flow readily into all voids and remain there until it sets.

Epoxy resin grouting of masonry has been practised for over 15 years on British Rail to arrest the development of cracking, to fill open joints and to stabilize loose masonry in appropriate situations. To all appearances, it has been consistently successful.

The cementitious grouting of poor quality masonry hearting to bridge piers, abutments and the like, where voids are often larger, is reliably effective. Grouting of the fill over arches has the potential to increase the capacity of the latter by improving load distribution and raising the fill factor (see Chapter 9, Table 9.2) in the modified MEXE method of assessment; it should also reduce water percolation through the fill and so improve the stability of the spandrel walls. However, the variable nature of fill materials detracts from the dependability of the results.

10.4.3 Stitching

Stitching (see Chapter 18) also serves to unify and strengthen deteriorated masonry. Once a fractured structure has been stabilized, stitching acts in conjunction with grouting to restore coherence. It is particularly effective in overcoming problems of ring separation in arches and the detachment of the face from a wall. It can also be used to improve the stability of detached spandrel walls by anchoring them to the arch backing (provided that the latter is of adequate quality). Building Research Establishment *Digest 329* (1988) describes some alternative systems using wall ties for stabilizing faces which are becoming detached from their backing.

10.4.4 Face patching

Face patching is more common than it deserves to be; it often serves only to obscure more deep-seated deficiencies, and that only temporarily. Refacing on a like-for-like basis may be desirable for cosmetic reasons; otherwise it should be confined to surfaces which have become severely eroded or spalled due to atmospheric or chemical action and it should have the objective of averting further deterioration which might hazard the integrity of the inner ring of an arch or the face wythe of a wall. In such circumstances, it must utilize materials of appropriate resistance, such as bricks complying with the frost resistant–low soluble salt content (FL) durability designation of BS 3921 or mortar made with sulphate-resisting

cement. However, there are problems of reconciling the adoption of strong materials (in the interests of optimum durability) with the normally advocated aim of matching existing strengths (in order to avoid hard spots).

The quality of this type of repair is dependent upon the efficiency of the bonding of the new work into the backing material. It is a labour-intensive operation and only economical for relatively small or scattered areas.

10.4.5 Reinforcement

The reinforcement of disintegrating structures can take a variety of forms. An additional 'skin' of masonry may be provided but, like face patching, it serves little purpose other than to arrest environmental action on the original structure and improve its appearance. Strapping of columns and chimneys using flat steel bands and of piers using steel beams and tie-bars is often effective but unsightly (Figs. 10.8 and 10.9). An encasing reinforced concrete shell, similarly used, will serve both to strengthen and protect but may change the appearance of the structure unacceptably.

Where clearances permit, an arch can be strengthened from underneath by the construction of a relieving vault using one of the methods shown in Fig. 10.10, or some variant of them, involving masonry (Fig. 10.11), steel or concrete. Reinforcement by steel liners may utilize either plain or corrugated steel sheets, pre-formed to the profile of the intrados, with the intervening annular space filled with grout or pumped concrete. With a minimal thickness of infilling, such underlining would normally be regarded as no more than a 'holding' operation to maintain the integrity of the structure pending its reconstruction within a reasonable time. However, that shown in Fig. 10.12 continues to perform its intended function satisfactorily after more than 25 years! A greater degree of support is achievable with a series of steel universal beam or column sections, bent about their x–x axis to the required radius and used in the manner of colliery arches; the old

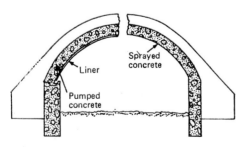

FIGURE 10.10 Arch strengthening from underneath. (*After Ridings and Jones, 1981.*)

FIGURE 10.11 Arch strengthening in brickwork. (NB The diaphragm wall under construction was for the purpose of stabilizing the outer arch of three while the centre arch was demolished and reconstructed.) (*Courtesy of British Rail.*)

masonry is lagged, on a temporary basis, with timber (Figs. 10.13 and 10.14) or, for a more permanent solution, with reinforced concrete, possibly acting compositely with the steel (see the case history in Section 10.5). The facility now exists commercially to bend, by cold rolling, a very wide range of structural sections to match existing profiles, however deformed, with considerable accuracy (see Chapter 21, Fig. 21.8). Alternatively, permanent repair from below may take the form of a relieving arch in conventional reinforced concrete or in sprayed concrete reinforced with steel mesh or steel or polypropylene fibres (see Chapter 16).

A feature of all forms of arch underlining concerns the transference of the thrust of the relieving element into the substructure. This may necessitate cutting a skewback into the face of the abutment, providing corbels or needles of adequate strength or carrying the new work down to the footings or even to a new foundation (see Chapter 12, Fig. 12.1). Although the construction of a relieving arch underneath does not involve obstructing the way over the bridge, it is preferable to close the bridge to traffic during the critical stages of the operation so as to relieve the new work of live load while 'green'. Except in the case of sprayed concrete treatment, the interface between the old and the new work must finally be grouted up to fill any voids left or created by

FIGURE 10.12 Arch strengthening using corrugated steel sheets. (*Courtesy of British Rail.*)

shrinkage or elastic compression and to ensure the intimate contact necessary for proper load transference. It is, of course, impossible to evaluate the degree of load sharing, and so it is customary to ignore it and to assume that the existing masonry will have been restored to at least its original capacity or, alternatively, to design the new relieving structure to carry the whole of the live load. In all cases of underlining, provision must be made to drain away any water percolating through the old masonry.

10.4.6 Saddle/relieving slab

A more effective, but often more expensive and disruptive, way of reducing the loading on a masonry arch or improving its distribution involves the construction of a 'saddle' or relieving slab over the extrados (Figs. 10.15–10.17). This technique has the advantages of neither reducing

the size of the opening under the bridge nor changing the external appearance. Here again, the choice lies between designing the saddle to carry the whole of the live load or regarding it as purely nominal strengthening of the original structure. In the former case, the existing arch is disregarded except as providing permanent shuttering; therefore it is common to introduce a compressible material over the extrados in the crown area to avoid live-load transference, or at least to interpose a debonding membrane between the old and the new. It is essential however, to ensure that the end reactions from the saddle/slab are properly transmitted to the abutments. The second approach is to dimension the saddle/slab to suit the construction depth available and to provide nominal/distribution reinforcement only. In this case, a saddle must be cast directly onto the existing extrados to ensure composite action with the original masonry, which may, additionally, be

FIGURE 10.13 Relieving arch using steel beam sections. (*Courtesy of The Angle Ring Co.*)

stitched to the new concrete. Any form of saddle reduces the lateral loading on the spandrel walls and provides anchorage for stitching them. Saddles and relieving slabs afford the opportunity and a stable foundation for conventional waterproofing (as described in Chapter 20); this opportunity should be siezed. Both techniques inevitably involve temporary interruption of traffic over the arch and removal of some or all of the fill, but they can be undertaken in sections transversely, allowing staged partial closure. In the interests of minimizing the duration of closure, the technique of vacuum dewatering the freshly laid concrete has been successfully employed to accelerate curing. It should be specially noted that all operations in connection with the removal of the fill and its replacement should be undertaken uniformly and symmetrically about the axis of the arch to avoid unbalanced loading on the vault.

In the case of viaducts with voided spandrels (as in the example shown in the Glossary, Fig. 5) the voids are often spanned by flat stone slabs or by somewhat insubstantial arches, both of which are a potential source of trouble (Fig. 10.18), and this may be compounded by their relative inaccessibility. Remedial measures have included strengthening by casting concrete relieving slabs over the vaults or by substituting precast concrete units (Fig. 10.19). Such measures do not overcome the difficulties of access for inspection, and a better solution could involve filling the voids with weak concrete. Other problems may then arise through the addition of weight which the voids were intended to obviate; weight added at a high level of a viaduct can critically affect the natural frequency of the structure, increasing its lateral oscillation under traffic. It may therefore be necessary to use lightweight concrete as infilling

FIGURE 10.14 Relieving arch using steel beam sections. (*Courtesy of The Angle Ring Co.*)

and/or to lower the formation level, i.e. to reduce the depth of filling.

10.4.7 Stabilization of retaining walls

Methods of stabilizing retaining walls against sliding or tilting under load include the following.

1. Ground anchor tie backs (see Chapter 19).

2. Lines of piling at the rear or in front of the toe to relieve or absorb some of the horizontal pressures.

3. For parallel walls retaining the two sides of a cutting, struts below the right of way and/or (if clearances permit) above it.

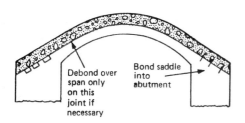

FIGURE 10.15 Concrete saddle. (*After Ridings and Jones, 1981.*)

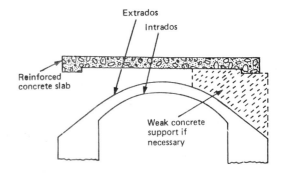

FIGURE 10.16 Concrete relieving slab. (*After Riding and Jones, 1981.*)

FIGURE 10.17 Saddle reinforced with curved steel beam sections. (*Courtesy of The Angle Ring Co.*)

4. For parallel walls retaining the two sides of an embankment (e.g. box wing walls to a bridge) ties below the right of way.

5. Buttressing. Buttresses need to be used with considerable care, however. To be effective they must be relatively massive, and so if they are founded on compressible ground they will be prone to settlement, whereas the ground under the existing wall will already have been consolidated. The result is all too often a buttress which parts company from the wall it is intended to support (Fig. 10.20) or hangs from the face of the wall, dragging it over rather than supporting it (Fig. 10.21). Therefore, if used, buttresses should be well founded, possibly on piles, and keyed, rather than bonded into the walls that they are intended to stabilize.

FIGURE 10.18 Distorted vault over spandrel void. (*Courtesy of British Rail.*)

6. Gabions stacked against the front face or, even more economically, material of any suitable nature, tipped and graded to an appropriate profile. Such a solution presupposes that available space exists in front of the wall.

7. Counterforts at the rear. Here, in contrast with buttressing, settlement of the new work, if the latter is properly bonded in, could improve the resistance to overturning. However, the practical problems of constructing counterforts are

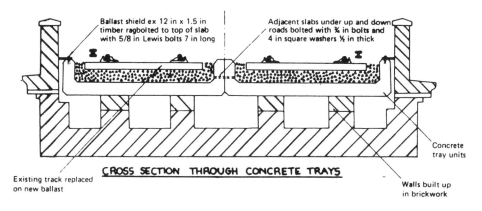

FIGURE 10.19 Precast units over spandrel voids. (*Courtesy of B. Towse.*)

FIGURE 10.20 Failed buttress. (*Courtesy of British Rail.*)

such as to make this a relatively costly stratagem in most cases.

10.4.8 Stabilization of spandrel walls

As discussed in Section 10.2, special mention needs to be made of the prevailing problem of distress in spandrel walls of bridges and viaducts. Diverse techniques are available to cope with this. The traditional method of increasing the resistance of such walls to outward movement involves the use of tie-bars with appropriate methods of distributing the stabilizing effect; the results are reasonably effective, if inelegant (Figs 10.22 and 10.23). The technique is still extensively used, and the appearance is improved

by taking advantage of the pattress plates now commercially available in either plain or self-aligning forms (Fig. 10.24). Owing to the difficulty of calculating the stabilizing forces required, it is usual to tighten tie-bars to nominal pre-loads only, but load-indicating end plates are obtainable if needed. It is normal to locate tie-bars in positions where they can be grouted into holes drilled from the face through the arch backing or into polyvinyl chloride (PVC) sleeves laid on the backing (involving excavation from above). If laid directly in the fill material, the problem of corrosion is critical and the life of the ties may be limited unless they are considerably over-dimensioned (e.g. by using old railway rails) or well protected (e.g. by wrapping in waterproof tape). Additional information on the durability of steel tension members buried in soil is given by Jones (1988). It is also essential to provide some ready means of identification of the ties as a precaution against damage by subsequent public utility excavations or highway/track maintenance operations. High degrees of skew present problems which are not always overcome by the use of self-aligning pattress plates. In such cases, an alternative approach could comprise a reinforced concrete beam on the outside face of the structure to extend the distribution of the effect of the tie-bars, but it would be necessary to give careful consideration to the aesthetic aspects of the beam in order to make it a visually acceptable feature.

An effective alternative way of strengthening spandrel walls is by thickening them on their inner faces – usually in reinforced concrete – so as to add to their mass and stability and increase their potential to act as horizontal beams. This work has to be undertaken with care and with the bridge at least partially closed to traffic. An excessive head of newly placed concrete has been known to cause the collapse of the spandrel wall it was being provided to support! It should therefore be limited to about a metre, with accelerating agents added to reduce the intervals between pours and the overall duration of the operation. The top level of the concrete thickening should not be

FIGURE 10.21 Failed buttress. (*Courtesy of British Rail.*)

FIGURE 10.22 Tie-bars and spreader beams used to stabilize spandrel walls. (*Courtesy of British Rail.*)

FIGURE 10.23 Tie-bars and spreader beams used to stabilize spandrel walls. (*Courtesy of British Rail.*)

brought up so high that it impedes the working of track maintenance machinery over railway under-bridges, while on road underbridges it may be limited by the necessity to accommodate public utility services. Some positive anchorage against horizontal displacement of such beams is desirable – not by keying into the extrados of the arch or the fill over the haunches, but possibly by ties between the ends of the beams or preferably, in the case of individual spans, by raising the wing walls to form buttresses against the outward reactions (Fig. 10.25).

Methods of reducing the effective lateral pressures acting on spandrel walls include the construction of a concrete saddle (as described above). The concrete can be brought up to provide a complete infill over the arch. This provides a permanent solution to the problem but may have a number of drawbacks, including increasing the loading on the foundations and/or temporarily hazarding the stability of the spandrel walls by an excessive head of newly placed concrete (see above); ducts may be needed to avoid the perma-nent encasement of services. Lightweight concrete can be used and the costs can be reduced by adopting high PFA concrete to the specifica-tion given in Table 10.1, laid in layers 100 mm thick.

Reinforced soil can be used as the infill between the spandrel walls (Fig. 10.26). The advantages of this technique are twofold: total relief is provided to the walls, and the method is economical as the primary construction material is soil. Com-prehensive information concerning the theory and

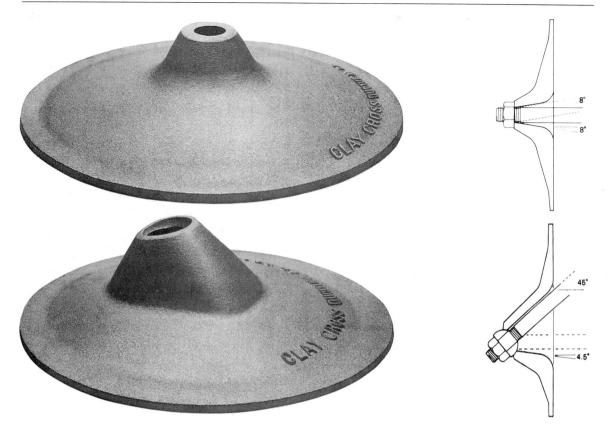

FIGURE 10.24 Modern pattress plates: (a) plain; (b) self-aligning. (*Courtesy of The Clay Cross Co.*)

construction of reinforced soil structures is given by Jones (1988).

The problems associated with vehicular traffic passing close to the spandrel walls can be relieved by changing the infill material. However, this does not obviate direct impact from road vehicles. The provision of a cantilevered slab deck has been used to resolve both exigencies (Fig. 10.27).

10.5 Case History: repair of bridge by relieving arches

Bridge 95 comprises three segmental arches, each of 7.5 m clear span, in brickwork with stone voussoirs and stone substructure; it carries two lines of railway over the River Whitting near Chesterfield. The river normally occupies only the central opening, but the side spans are necessary for flood relief.

Table 10.1 High PFA concrete infill between spandrel walls

	Quantity (kg/m^3)
Cement (OPC)	76
PFA (BS 3092: Part 2, type A)	212
Aggregate	
20 mm	723
10 mm	480
'Fine'	781

Average 28 day cube strength, 11.5 N/mm^2.

FIGURE 10.25 Wing wall raised to buttress spandrel wall. (*Courtesy of British Rail.*)

10.5.1 Defects and proposed remedial measures

As a result of weathering and water infiltration, the soft handmade bricks and lime mortar joints had suffered serious deterioration. The arches had cracked longitudinally and there was considerable

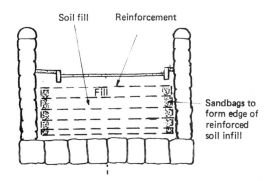

FIGURE 10.26 Reinforced soil infill used to relieve lateral pressures on spandrel walls. (*After North Yorks County Council.*)

permanent deflection; the degree of vertical movement under the heavily loaded freight trains using the line was such that the middle span had to be supported by timber centres and laggings as a precaution against possible collapse.

A study of the viability of reconstruction as compared with various forms of repair resulted in a decision to provide new relieving arches under all three spans, using reinforced concrete acting compositely with curved steel sections. The latter could be erected between the existing temporary centres in the middle span and be used to maintain

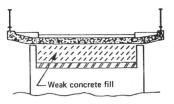

FIGURE 10.27 Cantilevered deck slab. (*After Ridings and Jones, 1981.*)

177

support to the arch until immediately before concreting; their strength would reduce the thickness of reinforced concrete required and so minimize the additional loading on the foundations. To ensure equal arch stiffness on completion and consistency of appearance, all three spans would be treated identically, with the outer spans being dealt with first.

10.5.2 Design

It was intended that the new relieving arches should be capable of carrying all the live loading on the basis of three-pin jointed structures. Various symmetric and asymmetric loading conditions were investigated to reflect all stages of construction and use. In service, the steel is already stressed by the weight of the concrete, but the concrete remains unstressed until the live load is applied. Under asymmetric live loading, part of the upper surface of the concrete is in tension, but it was considered that, in practice, cracking would be unlikely because of the restraint provided by the mass of the existing arch.

The design was based on the following method of construction in the middle span:

1. install needles (old rails and dowel bars) in piers
2. cast new reinforced concrete springer beams
3. erect new steel ribs, supported by the springer beams (point 2), between the existing timber centres
4. transfer support for the timber lagging from the timber centres to the new steel ribs
5. remove the timber centres
6. pressure grout the existing brick arch
7. fix reinforcement (threaded through holes pre-drilled in the steel ribs) and shuttering (supported by packing off the lower flanges of the steel ribs)
8. under possession of the tracks, remove the timber lagging and concrete the new arch
9. grout the interface between the new arch and existing brickwork
10. provide new weep holes

10.5.3 Construction

The work was undertaken largely as designed, but the following practical points are worthy of note.

1. The steel ribs were formed of 254 mm x 146 mm x 43 kg/m universal beam sections with shelf angles welded to their webs. Each was in two halves with a bolted connection to form a pin joint at the crown. They were curved to a nominal radius of 4.5 m and set about 150 mm clear of the existing intrados. They were lifted into place by block and tackle from adjacent timber centres. The close proximity of these centres hindered erection to line and level, and so, as the work progressed through the span, the timber centres were removed in turn and the support to the lagging was immediately transferred to the steel ribs using folding timber wedges.

2. To minimize the risk of further brick movement, particularly whilst transferring loads from the timber to the steel centres, the existing brickwork was grouted, using cementitious grout, before erection of the steel ribs. It proved difficult because the presence of the timber lagging and centres prevented effective pointing in advance. In the event, only a relatively small amount of grout was accepted by the brickwork.

3. Face shutters were erected flush with and bolted to the faces of the existing voussoirs. A better finish would have been obtained if these shutters had been set back slightly so as to avoid grout loss up the face of the arch.

4. Grade 40 concrete, with a nominal thickness of 265 mm, was specified for the new arches. It was placed through three holes in each span drilled down through the crown of the existing arch, one under each track and one under the space between them. The mix design was critical, and quality control was essential to achieve placing and compaction as effective vibration was impracticable. Proportions (by weight) were approximately 1 : 1¾ : 2⅞ OPC : sand : aggregate, with a water : cement ratio

of 0.49. A super-plasticizer (1% by weight of cement) was added to give a 'flowing' mix which could be placed by concrete pump (target workability, flow table reading of 550 – 620 mm). Trials showed that a concrete strength of 10 N/mm² could be achieved within the limited curing time available within the possession. With the shutters remaining in place, this was considered acceptable.

5. Final grouting utilized injection points at 1.5 m centres.
6. Four weep holes of 150 mm diameter were formed in each of the side spans through the reinforced concrete and the brickwork, with ductile iron spigots fixed in place using non-shrink cementitious grout.

The work was undertaken in 1986 by contract (A. Monk and Co.) at a total cost of some

FIGURE 10.28 Relieving arches to bridge over the River Whitting. (*Courtesy of British Rail.*)

£113 000 and appears to have been so successful in extending the life of the spans without significant disruption to traffic that other bridges are being similarly treated (Fig. 10.28).

10.6 Case History: repairs to Kincardine Viaduct by grouting, stitching and waterproofing

Kincardine Viaduct, near Auchterarder, Scotland, comprises six main arches each of 18.2 m span. The rail level is some 28.5 m above the valley bottom. It was built of sandstone masonry in lime mortar. In 1926 much of the spandrel fill material was replaced by a concrete slab 600 mm thick, but no waterproofing was provided.

Over the years, cracks in the piers had been repaired by iron cramps or straps, but there had been no sign of any foundation or structural failure. During the 1960s and 1970s, cracking and spalling of the lower portions of the piers spread progressively and fractures developed between the face voussoirs and the arch vaults. There was considerable water percolation through the arches and out of the piers at various places.

Coring and the cutting of windows in the piers revealed that they were each formed of a peripheral wall or casing, approximately 0.7 m thick, well built from large blocks of stone, but the hearting was composed of handplaced rubble of similar stone, varying considerably in size and set in poured lime mortar which was up to 150 mm thick in places. The lime mortar in the hearting had seriously deteriorated, virtually to sand. There was little evidence of an earlier attempt to grout the lower portion of one pier.

A consultant geologist reported that the rock was as capable of load bearing as when first quarried, but noted that most of the cracks in it were parallel to the likely direction of maximum load; he postulated that the outer stones could expand laterally under the compression loading and could therefore fail longitudinally along tensile fractures. Moreover, because of the irregularities in stone dressing, mortar composition and

thickness, that compression was probably being transmitted as a series of point loadings. The strength of sandstone is commonly reduced by about a third when the sandstone is water saturated and is further reduced when the loading rate is low. On the assumption that the condition of the pier cores meant that the peripheral walls were carrying the majority of the load, the stresses in the lower portion of the piers (of the order of 1.5 N/mm^2) were comparable with the failure loads derived from point-loading tests on samples after making adjustments for saturation and long-term loading.

Accordingly, it was deduced that the deterioration of the rubble core of the piers, as a result of long-standing water percolation, had led to overloading of the stones forming the pier casing. The main objective of the chosen repair scheme was therefore to re-create the ability of the hearting to carry its reasonable share of the load by grouting it over its full height, and to resist possible bursting forces by stitching the casing to the core.

On the basis of tests, it was found that a high degree of penetration could be achieved with a two-stage injection of a 1:3 OPC:PFA grout, with a water:cement ratio of about 0.5 and a grouting pressure not exceeding 0.4 N/mm^2. Because of the tightness of the joints in the peripheral walls, it was possible to inject the grout without preliminary pointing.

Grouting was undertaken from pier base to springing level in a succession of 0.5 m lifts. Primary holes 38 mm in diameter were drilled oblique to the pier faces and to the horizontal, followed by secondary holes at intermediate locations. When grout penetration had been adjudged acceptable by coring, the primary holes were re-drilled and 16 mm high tensile steel bars 2.5 m long were inserted and grouted in using the same mix.

Associated repairs to the remainder of the structure included the following:

1. grouting of the spandrel walls, in two stages,

FIGURE 10.29 Grouting repairs to Kincardine Viaduct. (*Courtesy of British Rail.*)

to depths of 2.5 and 5.0 m

2. grouting of the arch soffits, in two stages, to depths of 1.0 – 2.5 m (Fig. 10.29)

3. insertion of 20 steel anchor bolts, 25 mm in diameter and 2.5 m long, to dowel the voussoirs to the main arch vault in each span Anchorage was achieved with Selfix resin capsules and the remaining annulus was filled with neat cement grout. Pattress plates and nuts were given a protective bitumastic coating.

4. hand pointing of selected areas and some replacement of damaged masonry

5. waterproofing of the 1926 concrete deck using trowelled mastic asphalt in two layers, protected by Servipak

6. surface cleaning with high pressure water jets to remove grout stains and plant growth.

Grouting and stitching, by Messrs Colcrete, took ten months and used nearly 400 tonnes of grout (of which about 10% was for the second-stage grouting); it cost £143 000 (at 1979 prices), excluding access scaffolding. Waterproofing, undertaken in six track possessions, cost about £33 per square metre.

Nearly nine years after this work was completed, close inspection revealed no areas of deterioration and the repairs have clearly been effective.

10.7 References and further reading

Building Research Establishment (1988) *Digest 329, Installing Wall Ties in Existing Construction.*

Department of Transport (1984) *The Assessment of Highway Bridges and Structures. Departmental Standard BD 21/84; Advice Note BA 16/84.*

Jones, C. J. F. P. (1988) *Earth Reinforcement and Soil Structures,* Butterworths, Guildford.

Ridings, I. D. and Jones, C. J. F. P. (1981) The load carrying capacity and maintenance of existing arch bridges. *Colloque Inter. sur la Gestion des Ouvrages d'Art, Maintenance and Repair of Road and Railway Bridges,* Brussels/Paris. Presses de l'Ecole Nationale des Ponts et Chaussées, Paris and Editions Ancient ENPC, Brussels, vol. I, pp. 101–7.

Sawko, F. (1986) Rehabilitation of Masonry Arch Bridges – Procedure and Theory. *Proc. 1st Intern. Masonry Confer.* (ed. H. W. H. West). Pp. 73–5. British Masonry Society, London.

Turton, F. (1972) *Railway Bridge Maintenance,* Hutchinson Educational, London.

West, J. D. (1956) *Some Methods of Extending the Life of Bridges by Major Repairs or Strengthening.* Railway Paper 63, Institution of Civil Engineers, London.

Defects due to water

J. Powell

11.1 Introduction

In this chapter we identify problems commonly occurring in masonry which are created by the action of water. Preventive and corrective measures which can be taken are discussed, and these techniques are illustrated with practical examples. It should be noted from the practical examples that very often the defect has arisen owing to a combination of several unfavourable events and therefore caution should be exercised in diagnosing defects to a single cause.

11.2 Groundwater

It is essential in the construction of brick and masonry structures that a full site investigation is undertaken which determines whether groundwater is present. If adequate measures are incorporated within the structural solution to deal with the problems created by groundwater then the structure will give satisfactory service.

11.2.1 Problems created by groundwater

(a) Structural instability

Allowance should be made in the design of the structure to cater for the hydrostatic pressures generated by groundwater, particularly with regard to allowable bearing pressures, settlement and overall stability.

(b) Piping

Hydrostatic pressure on earth-retaining structures can cause water seepage under pressure from the increased head of water, resulting in undermining of the foundation and leading to possible collapse of the structure.

(c) Loss of jointing material

If insufficient consideration is given to water-proofing the structure then the ingress of groundwater can cause deterioration of the mortar and in severe cases the complete loss of the jointing material, particularly if it is lime based. As the mortar disintegrates, water movement from the rear of the wall pushes out the old crumbly mortar and deposits fines from the back-fill as a replacement. The wall has no tensile strength and needs to be reconstructed.

(d) Environmental effects

Groundwater penetration of the structure can lead to water dripping onto pedestrians and unsightly staining.

11.2.2 Preventive action

(a) Rear wall and formation drains

The groundwater-table can be effectively controlled by drains installed at formation level in conjunction with a rear wall drainage system which is positively connected to an outfall. To allow for blockages to the rear wall drainage system and to prevent a build-up of hydrostatic water pressure a series of weep pipes is installed approximately 0.5 m above ground level. The weep pipes are inclined to fall away from the front face of the structure to prevent the normal water movement within the rear wall drainage layer from entering the weep pipe and causing unsightly staining. As the weep pipes are a relief back-up to the primary drainage system, if water is seen to be issuing from them an investigation should be undertaken forthwith to determine the cause of the build-up of hydrostatic pressure.

(b) Well pointing and pumping

Groundwater can also be lowered by installing a system of well pointing and pumping the water to an outfall. This system is mainly used during the construction stage but can be adopted as a permanent solution where appropriate as it is very effective and not unduly expensive.

(c) Waterproofing

An alternative to groundwater lowering is to protect the structure against the ingress of water using mastic asphalt tanking or a proprietary waterproofing system (see Chapter 20). This method can also be used in conjunction with groundwater lowering.

11.2.3 Corrective action

(a) Pressure pointing and grouting

Pressure pointing and grouting is probably the most popular method of preventing the ingress of water into a brick or stone masonry structure. All large cracks over 5 mm wide, open joints and voids within the brick or stonework should be grouted with thixotropic cementitious grout injected at low pressure to avoid damage to the structure. Cracks up to 5 mm wide should be injected with epoxy resin, again at low pressure. All exposed surfaces should be cleaned by dry blast cleaning using a dry air–abrasive system and then washed down using a high pressure water lance in accordance with the recommendations of BS 6270: Part 1. All exposed joints which are to be repointed should be raked out to a minimum depth of 30 mm, removing all loose or soft mortar, before pointing with a cement mortar which should be struck flush as the work proceeds and given a bucket handle finish.

(b) Well pointing and pumping

Well pointing and pumping is an effective corrective solution inasmuch as the groundwater can be quickly lowered, but continual pumping of the water is required.

(c) Relieving drains

Installation of a system of relieving drains can effectively lower the groundwater but has a limited application for deep foundations.

(d) Waterproofing

Waterproofing is cost effective for the superstructure of masonry arch bridges but is impractical to install for existing substructures.

11.2.4 Case History: Farnhill Bridge

Farnhill Bridge is a stone masonry arch bridge 40 m long which carries a canal over an unclassified road in the Yorkshire Dales. The canal is contained within a reinforced concrete channel and crosses the bridge at a skew angle of 45°. The western approaches to the bridge entrance are flanked by tall free-standing masonry retaining walls.

There were many water leakages throughout the whole length of the arch barrel and in particular a steady flow of water issued at the

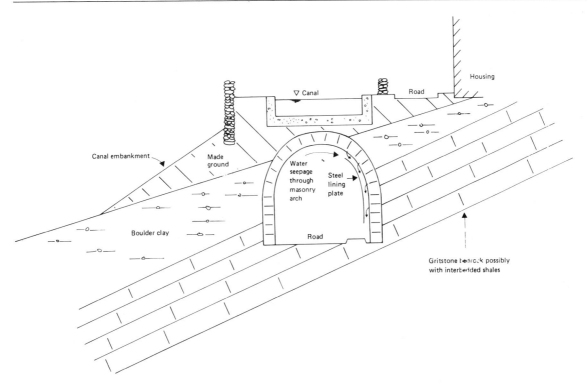

FIGURE 11.1 Farnhill Bridge. (*Courtesy of British Waterways Board.*)

junction of the south abutment and the footpath. A channel had been formed to collect the water issue which discharged into a road gully situated halfway along the bridge. A radial steel lining plate had been erected immediately above the footway to protect pedestrians from the many water leaks; water percolating through the arch soffit dripped onto the back of the steel plates and was discharged into the collecting channel. This situation had been in existence for many years. However, the steel lining plates were beginning to collapse because of severe corrosion, particularly at the connections, and the masonry was also showing signs of deterioration together with loss of mortar from the joints.

It was decided that a more effective repair should be undertaken and an investigation was undertaken to determine the source of the water. Initially the canal was dewatered and the concrete lining and joints were examined but no faults or obvious leakage paths were found. Penetrant dye testing was undertaken on the canal water but no observed leakage into the bridge or road drainage system was found. Having determined that the water was not leakage from the canal, an investigation into the adjoining ground conditions was undertaken, including test drillings to determine the bridge construction and the nature of the surrounding fill.

Figure 11.1 shows that the bridge is located in a sensitive area, with sidelong ground, high retaining walls, adjacent housing and a canal embankment formed on the hillside. It was determined that the arch bridge had been built hard up against the bedrock and that no provision had been made for rear wall drainage. The bedrock was a water-bearing gritstone with interbedded shales. Because of the long combined length of the bridge and retaining walls, the structure was forming an effective dam to the groundwater and this was the prime cause of the water leakage into the arch structure.

A number of alternative forms of remedial works were considered.

1. Gunite lining was discounted because of the risk of water pressures building up and blowing it off;
2. Grouting behind the arch would be the normally accepted solution but it was felt that an impermeable layer would be formed that would force the water to escape, possibly over the top of the arch crown, causing instability to the canal embankment or, in the event of blocking of the drainage path, causing damage to the adjacent housing and/or instability to the hillside above the canal;
3. Intercepting drains would provide a sound solution to overcome the problems of redirecting the water flow but unfortunately, owing to the close proximity of the housing and associated services, such drains were not a practical proposition;
4. The preferred solution was to accept the water flow through the structure but to control it to cause the least damage. The following action was undertaken:
(a) Twenty-two weep holes 55 mm in diameter were drilled at springing level and metal downpipes were connected into them such that the water discharged via the downpipes into the footpath drainage channel;
(b) Three or four percussive jack hammer holes 25 mm in diameter were drilled on a line between the canal and the wall on the west side and two were drilled in the canal towpath. Mild steel perforated tubes 25 mm in diameter with screw caps were inserted and water levels were monitored both before and after pointing;
(c) The masonry joints were raked out with an air disc cutter and repointed with waterproof mortar; all joints were struck flush. Flowing water was plugged with waterstop material. The pointing mortar (3:1 sand:OPC) included

a waterproofing agent (Setcrete). The repointing proved to be a difficult operation because of the narrow (1–2 mm) joint width, but would be more successful where the joints are, say, 10 mm wide.

The above work was completed in April 1986 at a cost of only £8 600, and although it was not completely successful in stopping all water leaks it was considered to be a satisfactory solution to the problems.

11.2.5 Case History: Bridge 198

A single-span bridge was to be reconstructed by providing a new superstructure but utilizing the existing brick abutments by backing them with mass concrete and linking them together with a new reinforced concrete invert slab. A full soil and structural investigation had been undertaken which indicated that the existing brick abutments were founded below the proposed invert slab formation level on a clay stratum which in turn overlays a water-bearing sand stratum.

Contract works commenced in January 1984. Flexible dams were erected between the canal banks and the canal water was pumped out. Excavation for the invert slab was undertaken which revealed that the wet abutment was founded at a higher level and also on a water-bearing sand layer instead of on the clay stratum predicted by the soils investigation.

The excavation within the canal and into the water-bearing sand resulted in an unbalanced water pressure and a flow of water from under the brick abutment at foundation level which washed out the fine sand deposits from under the foundation. The abutment immediately settled and the brickwork cracked along the bed joints because of the loss of ground support, as shown in Fig. 11.2. In the first instance the brick abutments were temporarily made safe by substantial timber props and two alternative methods of alleviating the water problem were considered.

FIGURE 11.2 Bridge 198: abutment settlement. (*Courtesy of British Waterways Board.*)

1. An interlocking steel sheet piled cofferdam was a technically feasible solution but access was a problem for installation and eventually the solution was discounted on cost grounds;
2. Well pointing was the preferred solution both from an economical viewpoint and because of speed of installation.

Vertical riser pipes were installed at about 1.5 m centres around the perimeter of the abutment. The water was drawn up the tubes by vacuum to a header pipe and discharged through the pump. The well pointing completely lowered the water-table, the sand stratum was stabilized as shown in Fig. 11.3 and no further settlement occurred. Consequently the bridge was reconstructed as originally planned without further mishap.

11.3 Scour/washout and loss of fines

Scour and washout are generally associated with rivers subject to flooding and/or tidal conditions. Substructures located in rivers can affect the bed regime, resulting in movement of deep-water channels and changes to silting deposits. In the case of major river crossings it is advisable to undertake hydrological studies. Loss of fines can occur as a result of changes in the river and groundwater regime and also through leaking drainage systems. Water leaking from a motorway road drainage system resulted in loss of fines from the PFA back-fill to the abutment of a motorway underbridge, causing a large cavity to be formed which eventually collapsed forming a large depression.

FIGURE 11.3 Bridge 198: stabilized sand stratum. (*Courtesy of British Waterways Board.*)

11.3.1 Problems created by scour/washout and loss of fines

(a) Structural instability

Scour and washout can undercut the substructure foundations, resulting in instability due to the formation of cavities and leading eventually to collapse of the structure.

(b) Impact damage

Under flood conditions large amounts of debris can be carried at speed along river channels causing structural damage to submerged piers and intermediate supports, particularly those formed of discrete columns. In severe flooding, water levels can rise to such levels that debris can collect under the superstructure, forming a dam.

11.3.2 Preventive action

(a) Inspections

Regular inspections and probing of the foundations of river-bank structures, particularly after flooding, can identify the early formation of cavities such that remedial measures can be undertaken before serious damage occurs. In the case of piers within the river channel the only effective inspection is by the use of underwater diving techniques.

(b) Fendering and cutwaters

Cutwaters and fendering structures can offer protection against impact damage during normal water flows, and floating fendering systems have been developed which can be of assistance under flood conditions.

11.3.3 Corrective measures

(a) Concreting

Concreting is an effective solution in filling voids and can be effectively placed under water, in still conditions, by the use of tremie tube. However, in fast-flowing water extensive temporary works are required to place the concrete in the dry.

(b) Sheet piling

Scour to river and canal banks can be effectively and economically formed by driving sheet piling which will generally require to be tied back to firm strata.

(c) Underpinning

Large voids under brick and masonry structures can be underpinned by the use of mini piling.

(d) Rip-rap

Slopes close to the water-line of rivers, lakes, or dams can be protected by rock rip-rap. The weight of the stone blocks prevents displacement by the action of water.

11.3.4 Case History: Chain Bridge

Chain Bridge is a single-span brick arch bridge with typical curved spandrel walls and brick parapet walls 1.2 m high. The bridge deck had been reinforced with a reinforced concrete saddle 250 mm thick and oversurfaced with tarmacadam.

A breach, situated a quarter of a mile to the north of the bridge, had occurred; to effect a repair stop planks were inserted in the specially formed grooves set into the masonry canal bank walls adjacent to the bridge and the canal was dewatered by running off through the next flight of locks to the north of the bridge.

The canal was successfully dewatered, but problems were experienced in getting the planks to make an adequate seal against the sill beam set in the canal bed. It was noted that the clay lining to the canal bed had been eroded away, possibly by overladen barges negotiating the sharp bend in the canal adjacent to the bridge, exposing a sandy gravel stratum.

Initially sandbags were used to effect a seal but it was noted that piping was occurring underneath the canal bed sill beam. It was therefore decided to drive a line of steel sheet piles across the canal on the upstream side close to the stop planks. The piles were driven to a sufficient depth to form an effective cut-off to prevent a recurrence of the piping. However, difficulty was then experienced in effecting a seal at the junction of the piles with the old masonry canal bank walls; the joint was sealed with puddle clay which was successful in stemming most of the leakage.

Having stabilized the situation, work was able to proceed on repairing the breach. The bridge was subjected to regular daily inspections during this period, and it was noticed that a crack had developed in the arch soffit. The road was closed to traffic as a precautionary measure when suddenly the bridge partially collapsed. The complete spandrel wall, part of the parapet wall and a small portion of the arch barrel collapsed together with the adjacent canal bank wall, as shown in Fig. 11.4.

The spandrel wall collapse exposed the backfill material behind the arch bridge, and it was found to be composed of a medium dense slightly orange brown pebbly sand. The fill had also partially collapsed, undercutting the road surfacing. Figure 11.5 shows the collapsed bridge.

An investigation into the collapse concluded that canal water was seeping through the unpointed masonry canal bank wall and underlying gravel bed and gradually washing away the fine material in the sandy gravel stratum under the spandrel wall foundation, resulting in a gradual softening and undercutting of the foundation to the bridge. After initially making safe, it was decided that the bridge was beyond repair and must be completely demolished. A new arch bridge was then constructed on piled foundations.

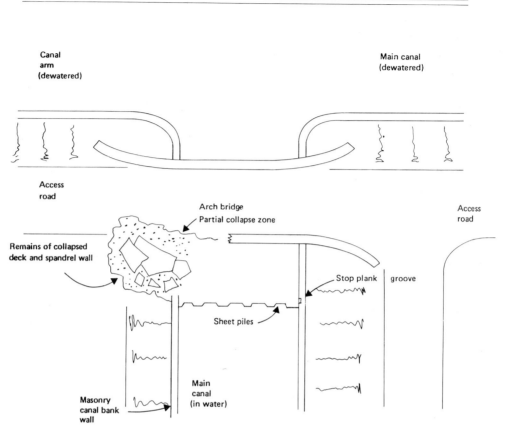

FIGURE 11.4 Chain Bridge. (*Courtesy of British Waterways Board.*)

11.4 Weathering, freeze-thaw and spalling

The durability of brick and stone masonry will be seriously affected when it remains saturated by water for long periods of time. Exposed masonry walls can become saturated by rainfall, by upward water movements from the foundations, laterally from retained back-fill in the case of undrained retaining walls, spandrels etc. and by downward water movements from the exposed tops of the wall.

11.4.1 Problems created by water penetration

(a) Frost action

Free-standing walls, parapets and retaining walls are susceptible to frost damage when the masonry is very wet. The characteristic effect of frost damage is spalling of the face and disintegration of the mortar. Frost can also accelerate sulphate attack damage.

(b) Atmospheric pollution

Sulphur compounds in the air or rainwater may react with the masonry or be deposited on its surface, leading to deterioration. Mortars may be attacked by sulphates derived from the bricks

FIGURE 11.5 Chain Bridge collapse. (*Courtesy of British Waterways Board.*)

themselves. Attack is gradual and occurs when the brickwork remains wet for long periods.

(c) Rainwater

Rainwater penetrates the masonry itself or the mortar carrying the salts when the salts crystallize within the pores of the masonry. Stresses then develop which can cause local fragmentation of the stone. This is the major cause of stone decay as shown in Fig. 11.6.

11.4.2 Preventive action

(a) Waterproofing

In the case of bridge decks a proprietary water-proofing system should be used which will prevent the ingress of water into the corpus of the structure.

(b) Copings

The use of copings to the top of the parapet and retaining walls will prevent the downward penetration of water and additionally will direct water clear of the wall below. Cappings such as brick on edge do not direct water clear of the wall as shown in Fig. 11.7, but are less liable to vandalism because of the lack of overhang.

(c) Mortar

The mortar should not be stronger than the stone and should contain the recommended amount of lime to give it some movement tolerance. The mortar should be struck flush or given a bucket handle finish. Recessed joints are not recommended as they form a ledge on which water can be retained.

11.4.3 Corrective action

(a) Stone preservative solutions

Water-repellent solutions based on silicones, silicurates or silicester solutions can be applied. On sandstone surfaces the water repellence of silicone treatment is relatively long lasting but it has no appreciable preservative effect. On limestones the effect is transient. Treatment can accelerate decay, particularly if the water is contaminated with aggressive soluble salts which pass through the treated layer (in the case of silicones this is only 2 – 3 mm thick). The water evaporates from behind the treated layer and the salts in solution crystallize, leading to spalling of the treated layer.

A more promising approach to preservation is in the deep impregnation of stonework (25–50 mm) with a silane resin treatment. This deeply penetrating treatment can secure the decaying surface to the underlying sound material, and the frequency of damaging crystallic cycles for soluble salts is considerably reduced or completely eliminated.

FIGURE 11.6 Stone decay. (*Courtesy of British Waterways Board.*)

FIGURE 11.7 Wall capping which has failed to direct water clear of the brickwork. (*Courtesy of British Waterways Board.*)

It is recommended that specialist advice is taken before any solutions are specified for water-repellent preservation.

(b) Masonry repairs

Decayed stonework can be repaired by descaling or limited dressing back of the decayed area and replaced with new stone indents or repaired with special mortars (Figs. 11.8 and 11.9).

(c) Painting

In the case of limestones the 'lime-wash' treatment is a simple and most effective paint treatment as it attempts to replace the calcium lost through decay.

11.4.4 Case History: Glen Bridge

Glen Bridge is a 14 m single-span parabolic arch

FIGURE 11.8 Spalled voussoirs and an earlier misguided attempt at repair. (*Courtesy of British Rail.*)

FIGURE 11.9 Voussoirs repaired by *in situ* concrete (with stitching bars to be installed). (*Courtesy of British Rail.*)

bridge with a height from crown level to water level of 6 m. The bridge masonry was in poor condition in several areas with extremely heavy spalling to the southwest voussoirs in particular. The spalling extended approximately from the crown to the south springing level. A large crack of variable width was present between the west elevation springing levels 600–900 mm from the arch face. In places the spalled masonry had broken away to the crack which resulted in loss of support to the parapet and the edge of the roadway.

The bridge carried a water main which had a history of leaking over a considerable period of time. The road surfacing drainage across the bridge was also inadequate, in that the road gullies were frequently blocked with silt resulting in water ponding for long periods of time and leading to the ingress of surface water into the corpus of the bridge. The bridge was not waterproofed and therefore the continuously saturated masonry was constantly damaged by the repetitive freeze–thaw cycle. (NB It was subsequently discovered, on undertaking the remedial works, that the road gullies had no pots so that the road surface water drained directly into the bridge fill).

The arch soffit above the towpath was very damp with water seepage, and spalling had taken

FIGURE 11.10 Glen Bridge: masonry repairs. (*Courtesy of British Waterways Board.*)

place on the surface of the masonry. The severe spalling of the masonry on the west arch face and the associated longitudinal crack was well advanced, and therefore in the interest of public safety temporary single-lane working was introduced to move the live loading away from

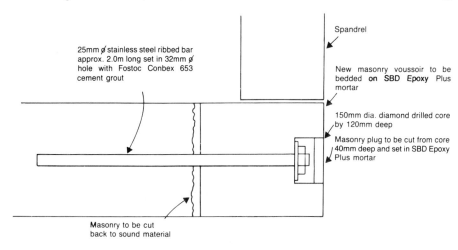

FIGURE 11.11 Glen Bridge: voussoir fixing detail. (*Courtesy of British Waterways Board.*)

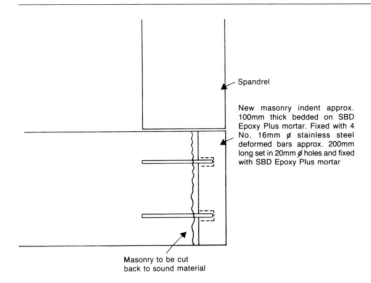

Spandrel

New masonry indent approx. 100mm thick bedded on SBD Epoxy Plus mortar. Fixed with 4 No. 16mm ⌀ stainless steel deformed bars approx. 200mm long set in 20mm ⌀ holes and fixed with SBD Epoxy Plus mortar

Masonry to be cut back to sound material

FIGURE 11.12 Glen Bridge: indent fixing detail. (*Courtesy of British Waterways Board.*)

the damaged area. Several voussoirs had fallen into the disused canal, and so the arch soffit was temporarily supported by a scaffold structure.

When the leaking water main had been repaired, it was considered that the remaining water problems stemmed from two sources: the surface water ponding on the bridge deck and thus percolating through the fill and masonry, and groundwater seepage from the hillside to the north and leakage from Glen Burn to the south. It was therefore decided to improve the surface water drainage by realigning the carriageway and incorporating a proprietary kerb–drainage system (Beany Block) together with new road gullies. In order to prevent the groundwater from reaching the bridge it was proposed to construct cut-off drains across the carriageway at both ends. In order to ensure that the structure was fully waterproofed a concrete slab was cast over the arch extrados and waterproofed with a proprietary membrane.

Masonry repairs consisted of completely renewing all the southwest voussoirs with new masonry, cutting back all weathered and perished masonry to sound material, and bonding new masonry indents into the arch barrel as shown in

Fig. 11.10. Because of the presence of the crack in the arch barrel it was necessary to stabilize the voussoirs using cement-grouted stainless steel anchors; the holes were drilled through the centre of every fifth or sixth voussoir into the main arch barrel. The ends of the bars were then blanked off with a masonry plug bedded with suitably coloured mortar as shown in Fig. 11.11. The longitudinal cracking in the arch soffit was sealed using a cement-based non-shrink grout, and the spalled masonry on the arch soffit was broken back and refaced with new masonry indents fixed as shown in Fig. 11.12.

The work was completed in August 1987 at a cost of £124 400.

11.5 References and further reading

British Standards Institution
BS 6270 *Code of Practice for Cleaning and Surface Repair of Buildings.* Part 1:1982 *Natural Stone, Cast Stone and Clay and Calcium Silicate Brick Masonry.*
Ministère des Transports: Direction Générale des Transports Intérieurs (1980) *Fondations de Ponts en Site Aquatique en Etat Précaire. Guide pour la Surveillance et le Confortement,* Laboratoire Central des Ponts et Chaussées, Paris.

Chapter twelve

Defects due to human actions and accidents

A.M. Sowden

In this chapter we deal largely with the results of acts of carelessness, misjudgement, ignorance and wilful damage. Many of the events are avoidable, but they continue to occur and the Maintenance Engineer is obliged to deal with them, often with a high degree of priority.

12.1 Overloading

Live loading in excess of that for which a structure was designed or which it can safely sustain may arise from the increased weight of modern vehicles (on both road and rail) or as a result of a disregard of weight restrictions. Retaining walls and bridge spandrel walls may be subject to additional loading when traffic is brought nearer to them by widening a carriageway or when greater speed of traffic on curves increases the centrifugal forces on them. Poor surfacing of a roadway, such as the presence of pot-holes or the inadequate reinstatement of excavations made in connection with installing or exposing public utility services, often causes undue impact effects, as can badly maintained rail joints. Inadequate cover can lead to unacceptably concentrated loads from traffic, especially if heavy individual wheel loads can be applied almost directly through large stones or pipe flanges.

The deposition of additional material, possibly in connection with raising road or track levels, may increase the surcharge effects of live loads on retaining and spandrel walls. It will also increase dead loading on bridge structures without a concommitant improvement in strength. Building works or tipping behind retaining walls may adversely affect their stability. The Maintenance Engineer must be constantly aware of all such possibilities when developments occur over or adjacent to structures for which he is responsible, and he must liaise with those concerned in the hope of inducing them to mitigate the adverse effects of their actions by varying their proposals or incorporating suitable precautions.

Many structures, particularly bridges, are regularly overloaded by the traffic using them, with the result that their maintenance costs are inflated and the margins of safety against their becoming unserviceable are reduced. In fact, the Bridge Census and Sample Survey undertaken by the Department of Transport (1987) indicated that some 10% of the masonry bridges carrying public roads in Great Britain fail to meet the standards of strength required for the vehicles having virtually unrestricted use of those roads, i.e. vehicles complying with the Construction and Use Regulations, including those of 38 tonnes laden weight. *Departmental Standard BD 21/84* (1984) is quite specific as to the action regarded as necessary in respect of any highway bridge assessed to be inadequate for the traffic using it:

1. Vehicle weight and/or lane restrictions should be applied;
2. The condition of the bridge should be monitored by special inspections at intervals not exceeding six months if it is considered that further deterioration of the structure may occur despite vehicle weight and/or lane restrictions;
3. If incapable of carrying even the lowest level of reduced loading defined in the Standard, the bridge will have to be closed to all traffic;
4. Replacement, or strengthening of the structure to carry the full loading, should be undertaken without undue delay.

Similar action should clearly be taken in respect of any structure which is inherently too weak to sustain the loads imposed upon it. It will generally fall to the Maintenance Engineer to initiate such action and ensure that it is pursued to a conclusion. Realistically, it must be accepted that replacement will take a considerable time to complete, bearing in mind the numbers involved and the financial resources likely to be available.

The effectiveness of the first of the courses of action mentioned above – traffic regulation – will depend upon its practical and commercial acceptability and the degree of discipline which can be imposed on the traffic in question. Reductions in the number of road lanes or railway tracks may incur intolerable operational delays. In the case of highways, they will not be reliably achieved by signs and markings on the road surface, but must be regulated by traffic lights or substantial physical barriers; on railways, signalling alterations will be needed.

The success of the application of weight restric-

tions is directly related to the effectiveness of their enforcement. In this respect there are particular difficulties with road traffic. At present, weight restrictions on roads are usually displayed by signs which indicate the maximum permissible vehicle weight or axle weight, with reference to the actual laden weights at the time of passage. They are legally enforceable, but in reality it is virtually impracticable to enforce them. This is because regulatory action depends upon proof that laden weights are in excess of those permitted; it is therefore necessary that a suitable weighbridge should be conveniently available, and this is rarely the case. As a result, the sanctions which exist are seldom invoked.

However, the proposals in the Department of Transport *Departmental Standard BD 21/84* (1984) define loading limits in terms of maximum gross vehicle weight, which is itself defined in Regulation 4(i) of Statutory Instrument 1981 No. 859 as the gross weight recorded on the Ministry plate attached to goods vehicles, irrespective of the load actually being carried. There are proposals to introduce a new restriction sign, applicable to all road traffic and referring to the maximum gross (plated) weight, with possible dispensation for empty vehicles. Infringement of a weight restriction would then be verifiable simply by comparing the plated weight of a laden vehicle with the limit shown on the sign. It would, of course, be essential that such a change should be accompanied by an effective publicity exercise, so that it is fully understood by drivers, traffic managers and enforcement authorities. The introduction of this or some similar workable system for enforcing weight restrictions is long overdue. Nevertheless, the imposition of restrictions must be regarded as a negative approach, with adverse effects which should be avoided if at all possible. Therefore, if early reconstruction is unlikely, attention must be concentrated on the feasibility of strengthening.

Most techniques of strengthening are akin to those of repair described elsewhere in this book: excess foundation pressures can be reduced by

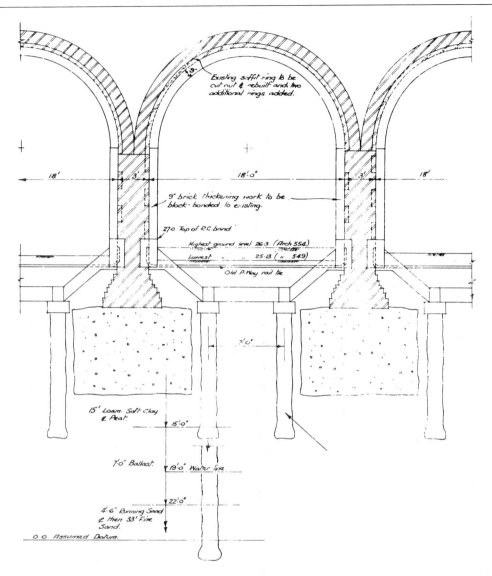

FIGURE 12.1 Scheme for countering overloaded foundations. (In the event, 9 in of reinforced concrete was substituted for the two additional rings of brickwork.) (*Courtesy of British Rail.*)

various methods of underpinning; abutments and retaining walls can be reinforced by being thickened or by the installation of a line of piling behind them; their stability can be improved by ground anchors, while that of spandrel walls can be increased by methods described in Chapter 10, Section 10.4.8; arches can be strengthened by grouting the fill over them, by thickening applied to either the intrados or the extrados or by the provision of a relieving slab; voids can be concrete filled; the live-load surcharge on abutments can be reduced by 'running-on' slabs; more rarely, unacceptable tensile stresses in masonry can be countered by modest prestressing using tensioned bars. Circumstances will influence the selection of what is practical and economic. Thus, thickening

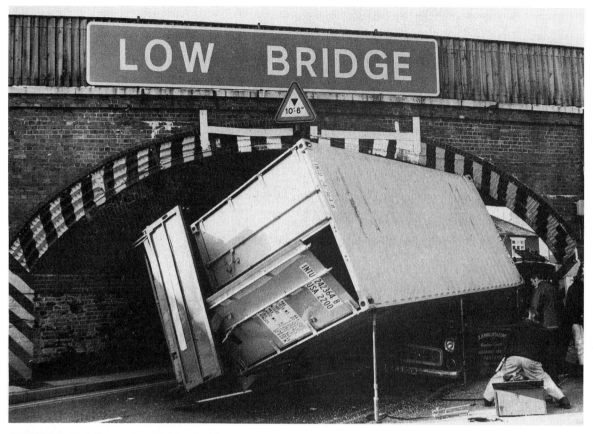

FIGURE 12.2 Example of bridge bashing. (*Courtesy of British Rail.*)

on exposed front faces may impinge on acceptable clearances, while the addition of material to unexposed rear faces is likely to involve costly excavation and possible interference with operational use. Furthermore, it will be virtually impossible for the new work to match the physical characteristics of the existing fabric of the structure, leading to design problems of assessing the degree of load sharing. It should also be borne in mind that it is rare for such new work to be pre-loaded, so that it will only assist in carrying the live load. Unless the new work is properly bonded into or anchored to the existing structure, a contraction or elastic deformation crack may develop at the junction; if it is so connected, settlement or long-term creep may cause the more recent work to drag on the old, thus increasing

dead-load stresses in the latter. Also, with any form of strengthening involving the addition of dead loads, some form of reinforcement of the foundation may be required in order to sustain them.

An interesting example of dealing with overloaded foundations is illustrated in Fig. 12.1. Here, one of the arches of a railway viaduct had been used for the storage of a considerable weight of metal, much of it on racking adjacent to the piers; this was more than the existing foundations could withstand and it resulted in settlement. The method shown for strengthening the brickwork and relieving the load on the foundations proved entirely effective.

Conversely, carrying capacity and/or stability can be improved by the reduction of dead loading

FIGURE 12.3 Example of bridge bashing. (*Courtesy of Bob Lord.*)

which makes no positive contribution. Thus, fill material can be removed by regrading to a lower profile or lightweight concrete can be substituted.

12.2 Bridge bashing

Bridge bashing is the common term for an all too common cause of trouble, i.e. structures being struck and damaged by traffic, usually as a result of lack of care or discipline. Occasionally river, canal or railway traffic is responsible, but the overwhelming majority of the damage is caused by road vehicles which are too large or unmanoeuvrable to negotiate the available openings. As is apparent from Fig. 12.2, fatalities have resulted and many incidents are potentially disastrous

(Fig. 12.3) in that they are such as to be capable of derailing a train or breaching an aqueduct. Repeated warnings by the Institution of Civil Engineers/Institution of Structural Engineers Standing Committee on Structural Safety and by the Chief Inspecting Officer of Railways do not seem to have led to any significant reduction in the incidence or severity of such accidents. This is not for want of effort on the part of some authorities, but there remains a degree of indifference elsewhere. Furthermore, every increase in permitted vehicle weight effectively increases the potential severity of impact, and every increase in length which adversely affects manoeuvrability adds to the risk of impact.

In the case of bridges over roads, it has long

FIGURE 12.4 Restricted headroom sign with chord marking (sign 532.1) and road markings. (*Courtesy of British Rail.*)

been the practice to erect warning signs to indicate substandard headroom (less than 16ft 6in). However, a survey by Galer (1980) showed that one driver in five did not have even a reasonably accurate knowledge of the height of his vehicle or its load, while many did not properly understand the relevant road signs. Other studies established that engineering plant (especially when being transported on low loaders), container lorries and skips were the 'arch enemies' in that they were involved in most of the serious incidents. Therefore in 1979 regulations were made requiring the heights of such vehicles (including their loads) to be marked in their cabs. There is no statutory height limit for road traffic in the United Kingdom, and at various times recommendations have been made that there should be one. Most European countries have a 4 m limit and the Armitage Report of 1980 suggested 4.2 m for this country. However, this would have only limited effect, as

there are so many existing bridges with headrooms less than any tolerable limit of vehicle height. Much effort has been put into devising ways of reducing the risk of occurrence of such impacts: trials have been made of various types of height gauges to warn drivers, physical height gauges to act as barriers have been considered and experimental traffic control systems have been tried. It may be that significant improvements will be achieved not so much by these means as by motivating drivers to exercise greater care and responsibility and to appreciate the dangers to themselves and others if they take risks in this respect. The publicity initiatives taken to date have not achieved this.

These are not matters for the Maintenance Engineer. His responsibility is to consider what preventive action he can take at each vulnerable location. The measures suggested by the Department of Transport *Circular Roads 5/87* include the following.

FIGURE 12.5 Spandrel wall of railway underbridge 'hooked off' by a road vehicle. (*Courtesy of British Rail.*)

1. Ensure that any road overbridge with a substandard headroom is marked with the appropriate warning signs indicating the available clearance (signs 530 and 532.1 of the Traffic Sign Regulations). Where a traffic regulation order prohibits vehicles over the specified height from passing under the bridge, the regulatory sign (sign 629.2) is substituted and action should be taken in respect of infringements;

2. Check the visibility and effectiveness of the relevant signs to ensure that they are of adequate size and condition, properly sited, not obscured (e.g. by foliage or street furniture), clear of distracting advertisements and conspicuous at all times (clean and illuminated or reflectorized);

3. Verify periodically the accuracy of the headroom indicated, since it may have been reduced by road resurfacing;

4. Consider the possibility of a road diversion avoiding the bridge;

5. At those structures most at risk, paint the more vulnerable parts (e.g. the arch face of a bridge over a road or the parapets of a bridge under a road) so as to make them more conspicuous. The normal form of such painting comprises alternate black and yellow diagonal stripes, each 150–250 mm wide (see Fig. 12.2);

6. At arch bridges, the signed clearance may be available over only the centre part of the width of the carriageway and in such circumstances 'chord marking' on the bridge face (sign 532.1) is appropriate. Drivers of high vehicles find that it is helpful for these signs to be supplemented by road markings, similar to those shown in Fig. 12.4, to guide them through the highest part of the bridge. If, in these circumstances, there is a risk that vehicles attempting to use the centre of the road could be forced to

FIGURE 12.6 Protective reinforcement caught by emerging traffic. (*Courtesy of British Rail.*)

one side to avoid oncoming traffic, the only effective course is to enforce single-lane operation by the installation of traffic lights.

Most of the above measures are the responsibility of the Highway Authorities and can be implemented only by them, but the engineer responsible for the maintenance of vulnerable bridges should be aware of the possibilities. He may have to provide the stimulus for their implementation and initiate action to ensure their continued effectiveness.

The damage sustained by arched masonry overbridges, despite such precautions, is invariably concentrated in the haunches and is cumulative rather than of immediately serious effect. Damage to the

parts facing oncoming traffic is usually superficial. However, the type of abrasive contact which progressively carves a chase along the intrados until the integrity of the inner ring is hazarded is more serious. Most at risk is the face on the exit side of the bridge, which can be 'hooked off' in a single incident, especially if there is already a fracture in the arch running parallel with that face (see Chapter 10, Fig. 10.3). In these circumstances, individual voussoirs may be dislodged or a major part of the end of the arch may be brought down, taking with it the spandrel wall above; the road or railway carried by the bridge is thus left without lateral support (Fig. 12.5).

Repairs which merely replace the damaged or demolished parts on a like-for-like basis do little

FIGURE 12.7 Protective plating to the haunch of an arch. (*Courtesy of British Rail.*)

to obviate a recurrence. Substantial protective reinforcement of the vulnerable leading edge of the soffit (i.e. at the haunch on the entry face) is often effective, but in the case of a narrow arch this is at risk of being caught by emerging traffic (Fig. 12.6). A generally successful countermeasure involves lining the arch intrados, at the critical level, with steel plates curved to the arch profile and anchored to it with countersunk bolts (Fig. 12.7). It is essential to leave no protrusions which could foul passing traffic and, to prevent the plates being 'sprung', their abutting edges should be welded together and any annular spaces between the plates and the masonry must be grouted.

In the case of vulnerable abutments, piers and similar roadside structures, consideration should be given to the provision of raised kerbs or barriers which will preclude vehicles from being driven too close to them. Reflective studs can be used to supplement the distinctive black and yellow painted stripes advocated above. Where parapets to retaining walls and underbridges are demolished by vehicular contact, there is a prima facie case for their redesign to 'high containment' standards in accordance with the recommendations of Department of Transport *Technical Memorandum (Bridges) BE 5*. This can be achieved without unduly drastic alterations to the appearance of the structure by the use of masonry-faced reinforced concrete, although it may require some ingenuity to ensure adequate provision to react the moments and shear forces at the base of such parapets.

12.3 Vandalism

Vandalism is an affliction of society; therefore, as in the case of various other aspects of this chapter, the Maintenance Engineer can be absolved from the obligation of curing the root cause rather than the symptoms. There was a time, now long gone, when he could have effective recourse to the deterrent of a sign reading:

Any person wilfully INJURING any part of this COUNTY BRIDGE will be guilty of FELONY and upon conviction liable to be TRANSPORTED FOR LIFE

However, he still has to cope with the results of vandalism, which, in respect of masonry structures, generally manifest themselves as graffiti and the removal of individual units, leading to progressive disintegration.

It is sometimes possible to deter the more offensive and mindless graffiti by accepting a form of more disciplined decoration: local artists and even schoolchildren under proper control can be allowed to paint murals in suitable locations. However, there are potential problems in establishing responsibility for appropriate action in the event of defacement, peeling off or other degradation.

The rough surfaces of most brickwork and stonework make them less prone to disfigurement by pens, but they are particularly vulnerable to the ubiquitous aerosol paint sprays. Various solvent-based water-rinsable cleaners to BS 3761 are effective in removing paint from glazed bricks and tiles and from polished stone surfaces, but on other masonry they may leave only a smeary mess or, at best, ghosting. Clifton and Godette (1986), describing performance testing of graffiti removers, reported that, of 99 potential removal treatments identified, 24 were considered as being worthy of investigation. Standard graffiti were produced by applying aerosol paints, felt-tip pens, crayons and lipstick to brick, sandstone, limestone and aluminium surfaces. In the case of the brick substrates, for example, the effectiveness of each remover was tested on a total of 98 specimens. Each of the 24 treatments gave some individual results which were adjudged acceptable for general purposes (effectively 90% removal), but in every case the average results for each remover were

below that standard, ranging from 46% to 88%. Many of the treatments proved to be graffiti specific, i.e. more effective in the removal of certain types of mark than others, indicating that the effectiveness of any proposed treatment should be tested on the particular graffiti before use.

Wire brushing, grit-blasting, hydroblasting and similar methods of cleaning can cause damage by the removal of the masonry surface, exposing a softer substrate to the elements. Abrasive cleaning should therefore be avoided.

Clear moisture-cured polyurethane coatings for application to clean surfaces are being developed. Graffiti applied subsequent to coating can then be more readily removed using solvents. However effective these may be on painted surfaces and the like, their impermeability makes them unsuitable for masonry since any build-up of hydrostatic pressures in the substrate or freeze–thaw action is likely to cause them to peel or flake off. There remains a need for genuine microporous coatings of this type. The alternative of obliterating graffiti by overpainting is subject to a similar criticism as well as to aesthetic objections.

The ability of vandals to dislodge or remove masonry units largely depends upon the condition of the jointing material. If the mortar is allowed to deteriorate to the extent that units become loose, there is a real risk of their removal and possible use as missiles, even in apparently well-ordered communities. Such considerations emphasize the need for prompt identification of conditions conducive to this kind of trouble and equally prompt preventive maintenance in the form of repointing and the replacement of any missing units. The forcible breaching of boundary walls by irresponsible acts of wanton destruction are not uncommon; these are primarily matters for police action.

12.4 Exposure to fire - contributed by G. J. Edgell

Fires involving materials stored in arches under viaducts are common and in some circumstances, such as those described in the case history in Section 12.5, structures can be exposed to very high temperatures for relatively long periods. The Maintenance Engineer needs information as to the resultant effects on structures for which he is responsible. The information available regarding the performance of masonry in fires is relatively limited compared with that for other structural materials, e.g. steel or timber. This is not really surprising since, in the case of clay bricks, the production process involves firing to high temperatures and consequently their performance is good. It is understandable that, in the situation where walls 215 mm thick made from almost any type of clay brick or sand-lime brick will withstand a 6h fire test and suffer no damage, there is little pressure to carry out research on these materials.

The work which has been carried out divides broadly into research aimed at providing some information on how properties of units or mortars vary with temperature and testing to provide basic data which could be used to support provisions in various National Building Regulations. Much of the work was carried out a considerable time ago.

The research has been largely uncoordinated and too limited to provide anything more than a sketchy understanding of these fundamental properties of masonry. Nevertheless it is required if a fire engineering approach to the design of masonry structures is to be possible. The approach in fire engineering is that the probable severity of a fire is assessed and the structural response is computed taking into account loading conditions and changes in material properties with temperature. Clearly, with the lack of available data on masonry and the low level of current research in this area, it will be many years before a fire engineering approach can supersede the use of tabulated data based on fire tests.

Of greater interest to the Maintenance Engineer than the behaviour of materials at elevated temperature, i.e. during a fire, is the appraisal of a structure following a fire. Essentially the problem is that the engineer is faced with a structure with indeterminate structural properties. Fortunately some useful conclusions can be drawn from the results of American tests where bricks were retrieved from walls following either a fire test alone or a fire test followed by a hose stream test (Ingberg, 1954). The results available are from tests on walls that were nominally 300 mm thick and consequently are relevant to massive rather than slender brickwork. The tests involved heating the masonry from one side to a temperature of, in some cases, 1370°C and subsequently cooling, the whole cycle lasting as long as 10h. The hose stream test was applied for up to 5 min immediately following a period of exposure to the furnace of about 1h, during which time the furnace temperature reached approximately 1000°C.

The compressive strength results for half-bricks loaded on end provide the most relevant comparison. Bricks loaded on end usually have a different strength to that when they are loaded normal to the bed face, especially when they are perforated. However, in this case the bricks were all solid or had a small frog in them, and the conclusions drawn from the results should be reasonably reliable. The compressive strength of clay bricks in the exposed face of the wall fell by less than 20% and that of concrete or calcium silicate bricks fell by about 50%. The bricks that had been subjected to both fire and hose stream tests indicated resultant losses of strength of a similar order.

A number of comparisons of the strength of prisms cut from walls with that of some built at the same time as the test walls could be made. The mean loss of strength due to the fire test alone or to the fire and hose stream tests was 40%. As the results for individual bricks could well underestimate the true strength losses because any bricks failed by the test were unavailable, it seems that a 40% loss in strength might not be an unreasonable estimate for clay brickwork, and some higher figure of 50% or 60% might be appropriate for calcium silicate or concrete brickwork.

Substantial reductions in the modulus of rupture for bricks exposed to the fire, whether as headers or stretchers, were noted, on average 80% for stretchers and 60% for headers. For other unexposed bricks the reductions were lower, on average 23%. This sort of information can only be used with caution as it applied to bricks not damaged by fire; consequently, it would be necessary to examine the bricks in any work carefully to see whether such values were reasonable. However, the modulus of rupture is not generally used in masonry design, and these data may be relevant in special circumstances only.

Laboratory furnace tests on individual sand-lime bricks (Ernest 1910) have indicated a reduction in the residual compressive strength after the temperature had been raised to between about 800 and 1200 °C which was similar to that referred to above. At temperatures above 1200 °C the strength increases, probably because of the production of a new type of bond between lime and silica. The residual modulus of rupture showed a progressive and large decrease until the temperature to which the bricks had been exposed exceeded 1200 °C. These tests all indicate that the conclusions above are reasonable.

Bessey (1950) carried out an investigation of the visible changes that occur in naturally occurring stones, aggregates and mortars as a way of determining the temperature to which they had been heated. Some of the observations may be relevant here; for example, if the temperature reached is known from the melting of other materials, some conclusion can be drawn about the condition of the material in question. Sand, sandstone, flint and limestone which contain any hydrated iron oxide tend to go pink or reddish brown at 250 – 300 °C, and this seems to be due to the dehydration of the hydrated iron oxides at these temperatures. Sand and sandstones tend to weaken and fail at temperatures beyond 573 °C as there is a considerable expansion of quartz grains at this temperature. At temperatures above 700 °C limestone begins to calcine and this reaction becomes rapid at between 850 and 900 °C. When this happens the marked pink or red coloration developed at 250–300 °C dis-

appears and the specimen disintegrates slowly on exposure to moist air. Igneous rocks tend not to change colour on heating, but acid types, e.g. granite, may shatter above 573 °C as a result of quartz expansion and basic types, e.g. basalt, may expand above 900 °C.

Tests on mortar showed the development of a pink colour at 300 °C as might be expected, and there was no other effect until 500–600 °C when the mortar clearly became weak and friable. A similar strength loss at this temperature has also been demonstrated in concrete.

The detailed information given above is, admittedly, limited and based on fairly old data but may be helpful in some circumstances.

Guidance has been given elsewhere (Malhotra 1982) that, if brickwork walls are otherwise satisfactory following a fire, their load-bearing capacity can be assumed to be satisfactory. Clearly cracked and deformed structures are unlikely to be satisfactory, but the data from the results of bricks taken from the fire tests does indicate the order of strength loss which may occur, and may enable the Maintenance Engineer to make a more informed judgement as to whether the damaged structure retains a sufficient margin of safety for its intended use.

12.5 Case History: fire in a railway tunnel

It is unusual to be able to study brickwork in detail after a sustained high temperature fire. However, such an opportunity arose in the Summit Tunnel, which takes the Leeds – Manchester railway under the Pennines near Todmorden.

Constructed by George Stephenson in 1839–41, Summit Tunnel is some 2600 m long and brick lined, generally to a thickness of six rings (685 mm). The initial construction used locally made bricks, but considerable areas had subsequently been repaired with Accrington bricks, while some minor patching had involved undressed sandstone masonry, probably of local origin. Recent investigations had revealed that the brickwork was mostly tight, but with some ring

FIGURE 12.8 Summit Tunnel: effect of fire on locally made bricks. (*Courtesy of West Yorkshire Fire Service.*)

separation and local loose patches up to 300 mm deep. A void 200–500 mm wide between the lining and the surrounding rock was loosely filled with rubble.

In December 1984 a goods train transporting 1300 tonnes of four-star petrol became derailed; leaking petrol ignited and the resulting fire burned for three days. Although it is not known how much petrol was consumed in the fire and how much drained away into the ballast, it is possible to say with some confidence what sort of temperatures were experienced by the masonry. There is evidence from the melting of the rail, the buckling of the tank wagons and the degradation of the ballast, coupled with a knowledge of the burning behaviour of the fuel, as well as evidence from the bricks themselves to indicate that the temperature in the worst-affected areas at the crown of the

tunnel would have been above 1540 °C, and probably as high as 1700 °C, for several hours.

12.5.1 Effects of the fire on the masonry

Over the 400 m length of tunnel affected by the fire, some 20% of the inner ring of the brickwork lining had spalled, while locally, at the centre of the fire, up to three courses had been shed because of heat-induced ring separation.

A detailed survey revealed that the two types of brick had behaved very differently in the fire. The original locally made bricks had sweated, and part of the matrix had exuded as a dark glassy residue which was sufficiently fluid to run down the walls and solidify on cooling in the form shown in Fig. 12.8. It was determined by experiment that this 'melting' occurred at about 1250 °C. Analysis indicated that the major metallic components in the melt were aluminium and silicon, with little carbon. The Accrington bricks, which were of much more uniform composition had merely spalled; at some points, up to 75 mm had been lost from the face, probably in a series of successive layers. The sandstone masonry had similarly spalled in thin layers along weak micaceous bedding planes. The masonry behind the spalling, both brick and stone, appeared sound, while the effect on the surrounding rock (mostly shale and sandstone) was negligible.

The mortar was lime-based in the original construction but had been repointed in some places with materials based on ordinary Portland Cement (OPC) and in the areas where newer bricks had been used it was based on either OPC or high alumina cement. It generally reacted in sympathy with the surrounding bricks, but its function as a jointing material did not seem to have been significantly affected by the fire. The composition of the cementitious component had radically altered; over the years, the mortar had become sulphated due to the presence of smoke, but on heating this calcium sulphate dihydrate had converted into the drier bassanite or anhydrite form.

The experience showed that brickwork will fail progressively when exposed to high temperatures, but the failure will be limited to the immediately exposed brick. Hence, once a course has fallen away, the newly exposed course will then react as if it had previously been uninvolved in the fire. This demonstrates the insulating properties of brickwork and the advantage of using small stable building components bonded by a relatively weak elastic material.

By selecting controls from undamaged parts of the tunnel, it was possible to compare the compressive strengths and water absorption values of heated and unheated examples. The main findings were that, although the original compressive strengths of the two types of brick were very different, in neither case did the fire significantly alter them, despite the loss of material from the locally made bricks. The Accringtons maintained a strength of about 60 N/mm², while the strengths of the older bricks all fell within the range 7 – 29 N/mm², with no statistical shift as a result of the heating.

Water absorption tests gave more disturbing, but not unexpected, results. The Accringtons, with their better composition, maintained a water absorption of typically around 4%, with a dry density of about 2.4. However, the older bricks, with a much higher initial absorption value of 13% and a dry density of 1.9, showed an increase in the former to 24% and a decrease in the latter to 1.5. This was clearly because the loss of the exuded material resulted in a more porous structure, and it could cause durability problems in a wet tunnel.

12.5.2 Repairs to the tunnel

Construction shafts near each end of the fire had acted as flues through which flames had raged for many hours. Their post-fire condition was such as to cast doubts as to their safety and stability, to the extent that considerable remedial measures were undertaken (Duncan and Wilson, 1986). Away from the shafts, the only work considered necessary involved the application of sprayed concrete in layers to achieve a total lining thickness at least equivalent to a five-ring arch (about 570 mm) and to act as a safeguard against future durability problems. The sprayed coating, with a minimum thickness of 100 mm, was reinforced with steel mesh pinned to the brickwork and incorporated polyvinyl chloride (PVC) drains to relieve hydraulic pressure. The repairs were undertaken in 1985 by contract (Whitley Moran and Co. and Colcrete Ltd) at a cost of about £1 million.

12.6 Adjacent works

Works are often undertaken in the vicinity of a structure with insufficient regard for their effect on its stability. Examples include the following:

1. works which lead to overloading (see Section 12.1)
2. the removal of passive resistance by the lowering of ground levels in front of retaining walls, such as by action to increase headroom under overbridges, by railway track blanketing or by deepening watercourses
3. the removal of subsoil support by over-enthusiastic excavation, often in ignorance of just how shallow are the foundations of many older structures
4. the lowering of the water-table and resultant soil shrinkage
5. work which may alter the course or regime of a river, causing scour at abutment and pier foundations, such as dredging or reclamation work or any construction which increases the speed or line of flow of a watercourse past a structure
6. vibration from pile driving or similar activities
7. excavations by service authorities, such as
   (a) deep continuous trenches having the same effect as described in point 2
   (b) careless excavation over bridges, which has been known to sever tie-bars, puncture water-proofing and even cut away parts of the structure itself
   (c) inadequate reinstatement over bridges, which may reduce load distribution, increase water percolation and impact effects, or leave hard spots
8. mining activities (see Chapter 9, Section 9.6).

Apart from fostering an increased awareness of the potentially adverse effects of such activities, the actions required of the Maintenance Engineer involve the following:

1. vigilance for their incidence, since he cannot rely upon receiving prior notice
2. ensuring that they are undertaken in such a way and under such control as to minimize their adverse effects, e.g. excavations taken out in short lengths, possibly by hand rather than with a machine, support maintained by strut-

ting and reinstatement properly completed
3. temporarily supporting or strengthening the affected structure
4. monitoring the condition of the structure by surveying existing defects and observing their development or the appearance of new ones, by check measurements of levels, verticality, convergence, profile distortion etc., and by crack surveys, tell-tales, photographs and more frequent detailed examinations
5. continued monitoring of the situation for some time after completion

In connection with the activities by or on behalf of public utility undertakings there should be strong support for many of the recommendations of the Horne Committee on behalf of the Department of Transport (1985), particularly those to the effect that 'there should be a national training scheme in excavation and reinstatement work, which should apply to all organisations which have general powers to excavate in the highway, should cover both direct labour organisations and contractors, should embrace different levels of staff from manual workers to senior supervisors and engineers, . . ., with uniform standards of assessment . . . complemented by a system of certification . . . Every job should be supervised by someone with a certificate of appropriate training at supervisory level.'

12.7 Case History: effects of river improvement works on a railway viaduct

This section is based mainly on information provided by the consulting engineers (C. H. Dobbie and Partners) with the permission of Thames Water and British Rail.

Following a period of prolonged rainfall in 1968, severe flooding occurred in areas adjacent to the River Mole in Surrey. The flood alleviation works undertaken in consequence included improvements to the channel where it passes under four lines of railway at Royal Mills Viaduct, Esher. This viaduct comprises four brick

FIGURE 12.9 Royal Mills Viaduct: upstream (south) face showing new piled foundations, cracks which developed during piling and temporary shoring in one arch. (*Courtesy of British Rail.*)

arches; originally built (in 1838) to carry two tracks, it was subsequently widened on each side on separate foundations.

The improvement scheme required the lowering of the invert under the most westerly arch, which hitherto had accommodated the river, and the lowering of ground levels under the other arches to enable the river to utilize all four spans. Trial holes revealed that the actual foundation levels were not as deep as indicated on the drawings and were separated from the London Clay below by a layer of loose gravel, fine sand and silt. The particle size of the finer soils was considered too small for successful treatment by geotechnical methods.

It was apparent that underpinning was necessary. The preferred system was one proposed by Cementation Projects Ltd which involved surrounding the piers and abutments with a curtain wall of contiguous bored piles carried down into the London Clay. Heavy concrete capping beams were cast around the tops of the piles and the weight of the viaduct was transferred to the piling using Macalloy high tensile bars to clamp the bridge foundations securely between capping beams. It had been intended to drill holes for the Macalloy bars into the concrete foundations under the brick footings, but it was found that the concrete was of a loose and friable nature as it had been made using Blue Lias lime, a naturally occurring material used prior to the manufacture of Portland cement. In view of the condition of the foundations and the poor strength expected of them, the positions of the

FIGURE 12.10 Royal Mills Viaduct: preparations for chemical grouting repairs to brickwork. (*Courtesy of British Rail.*)

bars were raised to pass through the brickwork above the level of the concrete. New reinforced concrete inverts were constructed in each span. To maintain the river flow, fluming was first undertaken through the two most easterly arches and the river was then diverted to them while work proceeded on the two western arches.

Settlement occurred during the first-stage piling operation, apparently as a result of reorientation of the granular materials in the gravel, and therefore compaction, caused by vibration during boring. Cracks, varying from fine hair lines to a width of 6 mm, appeared in the brickwork of the arches and spandrel walls (Fig. 12.9). Temporary shoring was erected as a precaution to safeguard

the continued operation of the train services. It was not considered practicable to repair the brickwork with cement grout, and so a chemical grout was used. It consisted of a resin and a hardening agent pumped from separate tanks through special metering pumps to a mixing head, from which it was injected into the cracked structure (Fig. 12.10). Core samples later indicated a penetration of grout to a depth of 312 mm.

Although the earlier permeability tests had indicated that uniform penetration of grout under the foundations would be difficult to achieve, it was decided, in the light of the settlement in the eastern half, to seek some improvement in the soil strength under the western pier and abutment by injecting a silicate bicarbonate grout (with a viscosity only slightly above that of water) before undertaking the second-stage underpinning operation. After setting, the grout would provide a small strength increase, of the order of 0.14 N/mm², and although its life expectancy would be only three to six months, such an increase would help to neutralize any groundwater effects during piling. A cement–bentonite mix was also injected to seal any unrevealed pockets or gravel layers.

Because of fears of possible frost damage if the cracking had penetrated through to the arch extrados, it was stipulated that a concrete under-lining be added to each arch opening. A comparison of methods and costs for placing this underlining favoured the use of sprayed concrete. The linings were designed to carry the total live loading and were 250 mm thick, with two layers of reinforcement. Initial problems of lack of bond and laminations of sand were identified with considerable rebound of the larger (10 mm) aggregate in the mix and the difficulty of spraying through two layers of reinforcement. The substitution of a gunite mix consisting of 1 part of cement to 3 parts of sand (graded 5 mm down) with a water:cement ratio of 0.45 gave very satisfactory results. Core samples gave 28 day cube strengths averaging about 60 N/mm², double that required by the design. To avoid the occurrence of hidden pockets ('shadowing') during spraying, this was undertaken in two stages with the inner and outer reinforcements being separately fixed. All existing drainage holes were maintained and extended through the sprayed concrete using plastic tubes.

On the downstream side of the viaduct, steel sheet piling with concrete capping beams and ground anchors was extended alongside the main channel of the river with a gabion bed and concrete block revetment. Permanent ground anchors consisted of 13 mm diameter strand, covered with polypropylene over the free length and encapsulated in resin over the fixed length, with standard VSL anchor heads. The anchorages

were formed in the London Clay. Temporary ground anchors were used to tie back the eastern abutment whilst placing the concrete invert slab; they were similar to the permanent anchors but without the encapsulated end anchorages.

In 1987, ten years after completion of the above work, the viaduct was noted to be in good condition, albeit with slight water percolation through the arch soffits.

12.8 References and further reading

12.8.1 Overloading

Department of Transport (1984) *The Assessment of Highway Bridges and Structures, Departmental Standard BD 21/84.*
Department of Transport (1987) *The Assessment of Highway Bridges and Structures: Bridge Census and Sample Survey.*
Motor Vehicles (Construction and Use) Regulations 1978 (S.I. 1978 No. 1017), as amended by S.I. 1982 No. 1576.
Traffic Signs Regulations and General Directions 1981 (S.I. 1981 No. 859).

12.8.2 Bridge bashing

Armitage, Sir A. (1980) *Report of the Inquiry into Lorries, People and the Environment,* HMSO, London.
British Standards Institution BS 6579: 1989 *Safety fences and barriers for highways.*
Department of Transport (1982) *The Design of Highway Bridge Parapets, Technical Memorandum (Bridges) BE5.*
Department of Transport (1987) *Circular Roads 5/87.*
Department of Transport (1988) *Report of Working Party on a Strategy for the Reduction of Bridge Bashing,* HMSO, London.
Galer, M. (1980) An ergonomics approach to the problem of high vehicles striking low bridges. *Appl. Ergon.* March.
Mallet, G. P. (1981) Protection of low bridges. *Colloque Inter. sur la Gestion des Ouvrages d'Art, Maintenance and Repair of Road and Railway Bridges,* Brussels/Paris. Presses de l'Ecole Nationale des Ponts et Chaussees, Paris and Editions Ancient ENPC, Brussels, vol. II, pp.603–7.
Rose, C. F. and Mallett, G. P. (1982) 'Bridge bashing' – damage to bridges by road vehicles. *Proc. Inst. Civ. Eng. Part 1,* 72 495–7.
Traffic Signs Regulations and General Directions 1981 (S.I. 1981 No. 859).

12.8.3 Vandalism

Brick Development Association (1976) *Building Note No. 2, Cleaning of Brickwork.*

British Standards Institution
BS 3761: 1986 *Specification for Solvent-based Paint Removers.*
BS 6270: *Code of Practice for Cleaning and Surface Repair of Buildings.*
Part 1: 1982. *Natural Stone, Cast Stone and Clay and Calcium Silicate Brick Masonry.*
Clifton, J. R. and Godette, M. (1986) Performance tests for graffiti removers. *Proc. Symp. Cleaning Stone and Masonry* (ed. J. R. Clifton), Special Technical Publication 935, American Society for Testing and Materials, Philadelphia, PA, pp. 14–24.

12.8.4 Exposure to fire

American Society for Testing Materials (1969) ASTM Standard E119–69: *Standard Method of Fire Tests of Building Construction and Materials.*
Anon (1926) Common brick withstands fire tests. *Brick Clay Rec.,* **68** (12), 939.
Bessey, G. E. (1934) *Sand Lime Bricks,* DSIR Build. Res. Spec. Rep. 21, HMSO, London.
Bessey, G. E. (1950) *Investigation of Building Fires.* Part 2: *The Visible Changes in Concrete or Mortar Exposed to High Temperature.* Nat. Build. Stud. Tech. Pap. N4, Part 2.
Brick Institute of America (1974) *Fire Resistance. Tech. Notes Brick Construc.* 10.
British Standards Institution BS 476: 1972. *Fire Tests on Building Materials and Structures.* Part 8: *Test Methods and Criteria for the Fire Resistance of Elements of Building Construction.*
Danish Standards Institution Danish Standard 1051.1: 1979. *Fire Resistance Tests. Elements of Building Construction.*
Davey, N. and Ashton, L. A. (1953) *Fire Tests on Structural Elements.* Nat. Build. Studies Res. Pap. 12, HMSO, London.
Duncan, S. D. and Wilson, W. (1988) Summit fire - Post fire remedial works. *Proc. 5th Inter. Symp. Tunelling '88,* organized by the Institute of Mining and Metallurgy, London, April, pp.87–95.

Edgell, G. J. (1982) *The Effect of Fire on Masonry and Masonry Structures: A Review.* BCRA Tech. Note 333.
Ernest, T. R. (1910) Fire tests on sand lime bricks. *Trans. Am. Ceram. Soc.,* **12** 83–9.
Fisher, K. (1982) Fire resistance of brickwork: regulatory requirements and test performance. *Proc. Brit. Ceram. Soc.,* **30** 65–80.
Foster, H. D. (1939) Fire resistance of ceramic building materials. *Ohio State Univ. Eng. Exp. Stn. News,* **11**(5), 11.
German Standards Committee DIN 4102: 1965. *Behaviour of Building Materials and Structures in Fire – Classification with Respect to Definitions.*
Ingberg, S. H. (1954) *Fire Tests of Brick Walls,* Nat. Bur. Stand. Build. Mater. Struct. Rep. 143.
Malhotra, H. L. (1962) Fire resistance of perforated brick walls. *Builder,* March, **202**, 513–15.
Malhotra, H. L. (1966) *Fire Resistance of Brick and Block Walls,* F. R. S. Fire Note 6, HMSO, London.
Malhotra, H. L. (1982) *Design of Fire Resisting Structures.* Surrey University Press.
Monk, C. B., Jr. Goldenberg, J. E. and Jearkjirm, V. (1971) Structural action of brick bearing walls exposed to fire temperatures. *Proc. 2nd Int. Brick Masonry Conf. Stoke-on-Trent,* British Ceramic Society.
Saemann, J. C. and Washa, G. W. (1975) Variation of mortar and concrete properties with temperature. *J. Am. Concrete Inst.,* **29** (5), 385.
Sullivan, P. J. and Poucher, M. P. (1971) *The Influence of Temperature on the Physical Properties of Concrete and Mortar in the Range 20 °C to 400 °C,* ACI Spec. Publ. 25.
Underwriters' Laboratories (1970) *Fire Tests of Buildings, Construction and Materials,* Underwriters' Labs (US) Stand. UL 263.

12.8.5 Adjacent works

Department of Transport (1985) *Roads and the Utilities: Review of the Public Utilities Street Works Act, 1950,* p.6, HMSO, London.

Chapter thirteen

Defects due to vegetation

A.M. Sowden

The deterioration of individual stones due to bacterial attack and the biological effects of algae, lichens, fungi and mosses were described in Chapter 8, Section 8.2. Larger plants – creepers, shrubs and trees – can have a more deep-seated disruptive effect on masonry.

Vegetation growth on and around structures is often considered to enhance their appearance, softening their lines and setting them in an attractive environment of ever-changing colour. However, the physical effects of the growths of roots and suckers, as they penetrate joints and cracks or withdraw moisture from the soil under foundations, may be major contributors to the disintegration of those structures.

Grasses, weeds, wall flowers and the like growing on ledges do relatively little harm; plants begin to cause concern when they develop significant root systems which initiate mechanical damage. Roots have extraordinary penetrative powers (Fig. 13.1) and destructive capabilities (Fig. 13.2). They infiltrate any available crevice and, as they grow, force it further open and allow the access of water and air and hence promote weathering (see also Chapter 8, Section 8.3.4). Creepers and ivy may have the advantage of insulating masonry surfaces from extremes of weather, but equally they inhibit evaporation and tend to maintain those surfaces in a permanently moist condition; possibly even more serious is the extent to which they prevent proper examination and may conceal incipient defects.

Certain species of trees, particularly elms, poplars and willows, make high moisture demands on the subsoil and thereby can affect foundations. Clay soils undergo considerable changes in volume with varying moisture content, shrinking as they dry out and expanding as water is absorbed. Therefore root systems which extend into such soils and remove water from the ground under a structure can cause settlement damage, which will be exacerbated by seasonal variations.

FIGURE 13.1 Tree roots found within a structure during its demolition. (*Courtesy of British Rail.*)

FIGURE 13.2 The effect of root growth on a wing wall. (*Courtesy of British Rail.*)

The effect is most marked when foundations are shallow, as they frequently are for older structures. However, deeper foundations are not necessarily immune, as root penetration to considerable depths is possible and for mature trees drying influences can extend to a depth of 5 m or more. Conversely, the removal of established trees may reverse the process, allowing such soils to re-absorb moisture, swell and cause heave, which may continue for a decade or more with insidiously adverse effects on foundations. Autumn leaf fall may block surface drains at a critical time; root growth penetrating a buried drain at a joint or crack can readily develop to the stage where it forms an effective plug. Over-mature, diseased or unstable trees may fall and damage nearby structures.

It follows from the above that the proper control of vegetation on, over and adjacent to structures is an important aspect of their preventive maintenance.

It may be possible to dispose of small shrubs and creepers by the application of a herbicidal spray to the foliage during dry weather in the summer growing season (June – August). Alternatively, or as a follow-up treatment during winter to ensure that the roots are killed off, the growth should be cut back to just above ground level and a herbicide applied immediately to the stump. Scrub and young trees should similarly be cut down and the stump treated with herbicide. Mature trees should be felled if they are a potential source of trouble, but it would be unwise to drag out stumps which are close to structures. A preferred alternative could well be coppicing, whereby the tree is cut down to about 5–10 cm above ground level, but the stump is not treated and so regrowth produces a low, sturdy, multistemmed plant.

Suitable herbicides currently include glyphosate (available commercially as 'Roundup'), which will kill or damage all plants it touches, and triclopyr (available commercially as 'Garlon'), which will kill or damage all plants except grasses. For spraying, the herbicide should be diluted (1 part herbicide to about 5 parts water) and applied using a low pressure nozzle. It can also be applied with a sponge held in a rubber glove or using a brush. For stump treatment the herbicide solution should be less dilute or even undiluted and it can be mixed with wallpaper paste or oil. It is painted thickly onto the cut face and sides of the stump. Drilling holes in the stump will increase the contact area. A few hours of dry weather is needed after treatment for herbicides to penetrate.

Appropriate precautions should, of course, be taken and the proper protective clothing worn when using herbicides, bill hooks and chain saws. Specialist advisers or experienced tree surgeons should be retained in the event of suitably skilled staff not being available in-house to ensure that vegetation growth is not unnecessarily damaged. Equally, precautions against injury to adjacent trees and shrubs should be taken during maintenance operations on structures; care is necessary over the location of fires and every effort should be made to avoid the spillage of, for example, fuel oil or cement-based materials within the probable range of root spread. The possibility that trees are the subject of Preservation Orders or are located within conservation areas should not be overlooked.

From a more positive point of view, there need be no objection to planting for landscaping pur-poses provided that care is exercised over the choice of trees and shrubs in relation to the type of soil and the proximity of structures. Shrubs which are relatively slow growing and easy to control include bush ivy* (*Hedera helix congesta*), dogwood (*Cornus sanguinea*), guelder rose (*Viburnum opulus*), hazel (*Corylus avellana*), holly* (*Ilex aquifolium*), juniper* (*Juniperus communis*), privet* (*Ligustrum vulgare*) and yew* (*Taxus baccata*) (the asterisks indicate evergreens). These form good natural ground cover and so inhibit less acceptable varieties from colonizing. Suitable attractive small stable trees include the cherry (*Prunus padus*), rowan (*Sorbus aucuparia*) and whitebeam (*Sorbus aria*). As a rough guide, roots extend horizontally for a distance equal to the height of a tree. Single trees should therefore be located no nearer to an existing structure than their mature height. For groups or rows of trees competing for water over a limited area, the distance should be increased to some 1.5 times the mature height. If leaf fall is a potential problem, evergreens should be chosen.

Appropriate vegetation can also be used constructively to stabilize slopes, to control water erosion and reduce scour, and to provide shelter (Coppin and Richards, 1990).

References and further reading

Ashurst, J. and Ashurst, N. (1988) Control of organic growth, Chapter 2 of Vol. 1. *Practical Building Conservation*, English Heritage Technical Handbook, Gower Technical Press.

British Standards Institution. BS 5837:1980. *Code of Practice for Trees in Relation to Construction.*

Coppin, N. and Richards, I. (1990) *The use of vegetation in Civil Engineering*, Butterworths, Guildford.

Cutler, D. F. and Richardson, I. B. K. (1981) *Tree Roots and Buildings*, Construction Press, London.

Ward, W. H. (1949) The effect of vegetation on the settlement of structures. *Proc. Conf. on Biology and Civil Engineering*, Institution of Civil Engineering, London.

PART FOUR

Remedial techniques

Chapter fourteen

Mechanical (or pressure) pointing

D.J. Ayres

14.1 Development

At the end of the Second World War the condition of railway tunnels in the United Kingdom was poor after six years of reduced maintenance; in particular, the mortar in the brickwork joints had seriously deteriorated owing to attack by the sulphur dioxide from steam locomotives. A method of mechanizing the cleaning and repointing operations was developed, starting in 1951 with an aerated mortar which had good flow properties because of its air content of 20%. This was fairly successful except at locations where the locomotive exhaust was sufficiently severe to attack the new mortar before it had cured properly. A better mortar was then developed, using pulverized fuel ash (PFA) and a non-ionic wetting agent, with minimum air entrainment to give good flow and minimal reaction between the ordinary Portland cement (OPC) and the wetting agent. With the mix now used (see Section 14.3) the likelihood of alkali–aggregate reaction is avoided, and tests on the mortar after immersion for five years in extremely high magnesium sulphate concentrations have shown it to be unaffected owing to its pozzolanic nature.

From the outset, pneumatic pressure pots with no moving parts were used to place the mortar. Each pot supplied mortar along one hose to one pointing gun, and just one prime mover, an air compressor, supplied six pots arranged on a staging on a rail wagon. A conventional reciprocating piston pump could not have dealt with the aerated mortars used initially because of their compressibility in the cylinder and abrasion problems with the stiff 2:1 sand: cement mix used. Elsewhere, British Rail has been using helical continuous flow pumps for over 30 years to place aerated 3:1 sand:cement grouts of lower viscosity. However, using a continuous flow pump to place a pointing mortar with an aggregate: cement ratio of 2:1 (or higher) would be very difficult, as it would arch in the hose in the event of any overlong stoppage. Improved flow could be achieved by reducing sand content, by adding clay or by extra aeration, but this would conflict with the objective of optimizing the qualities of the mortar in respect of durability, shrinkage etc.

14.2 Site preparation

Some investigation of the masonry is necessary before specifying the preparation, since the quality of the existing mortar can be very variable and it must not be removed to the extent that units become loose or dislodged. Sand–lime mixes, possibly with a little hydraulic cement, are likely to have deteriorated under conditions of water flow. Black lime mortars included ground ash which could produce strong material because of the pozzolanic reaction with the fine particles, but their properties varied according to the degree of ash grinding before use. White lime mortars were weaker with less pozzolanic action and greater dependence on the carbonate produced; they are more susceptible than the black mortars to breakdown by aggressive water. Portland cement mortars have been used in construction for over a century.

The inner joints of brickwork were often not completely filled with mortar and may, in fact, be channels for water flow. Previous repointing (by hand) will often be found to comprise strong mortar only 5 or 10 mm deep backed by empty joints up to 20 mm deep. This mortar should be removed by hand raking or mechanical chasing tools, as the use of high pressure water or air jets will remove the superficial mortar slowly but at the same time may remove soft mortar behind it, up to a depth of 100 or 200 mm, so loosening the facing course.

Nevertheless, the most effective and economic approach is removal by jetting. With a standard pump offering up to 0.7 N/mm² pressure and a jet of diameter 1 or 2 mm, all soft mortar is washed away to the required and observed depth. There is such a degree of splashing, however, that the operator, who should be applying the jet to within 150 mm of the joint, tends to stand away from the work, dissipating the jet energy. High pressure jetting at 14 N/mm² (2000 lbf/in²) or more is possible with modern equipment, and this requires less water for a given area to prepare the face quite adequately. A dual jet of compressed air with water at low pressure is similarly effective. Compressed air alone for jetting produces much dust, necessitating safety masks. In any case, track ballast, drainage systems and installed equipment must be protected against fouling by the debris which results from the preparatory operation.

If there are very wet areas of brickwork, holes should be drilled through it and pipes mortared in to conduct the water away before pointing. This should reduce the area requiring wet pointing but will also provide injection points for sequence grouting if general waterproofing is sought.

Wetting agents can be added to the jetting water to prepare masonry and to loosen sooty oily deposits. These should be of low foaming power, for otherwise the foam can carry the dirt for long distances and into the drainage system where it may cause blockages. If the outfall is into a natural watercourse, the additive must be biodegradable.

14.3 Normal mechanical pointing

The mortar mix is prepared in a drum mixer with blades to give a positive mixing action. OPC, sand and PFA (in proportions of about 1:2:0.5 by weight) are mixed with water and a non-ionic wetting agent (e.g. Nonidet LE); the water content is such as to produce a mortar of lower consistency than for trowel work but still exhibiting dilatancy. The sand should be of a sharp rendering quality, medium to fine (100% passing a 600 micron sieve). The mortar is transferred via a 2 mm sieve to a pressure pot and a pressure of 0.1 – 0.3 N/mm² is applied, depending on the viscosity of the mix, the length of mortar hose and the height of the pointing gun above the pot.

An auxiliary air line goes to the gun. Air enters the nozzle via four 1 mm tubes set at a critical shallow angle to the central mortar flow. An operator at the pot controls the rate of mortar flow on the instructions of the nozzle man; one pot man can serve up to six pots. As the mortar passes through the gun, the compressed air breaks it into a series of pellets and projects them into the joint. The back of the joint is thus filled first and the mortar builds up to the surface of the brickwork to give a better result than is possible by hand. An average of 8 m²/h is possible, and overhead work is as simple as vertical work (Fig. 14.2).

Joints up to 100 mm deep can be filled in

FIGURE 14.1 Normal mechanical pointing of a wall. (*Courtesy of British Rail.*)

normal brickwork, while in stonework the technique will deal with joints up to 300 mm deep, especially if they are wide. Obviously a higher speed is achieved with a shallower joint to be filled, but approximately 30 l of mortar is placed by one gun in 10 mins. The mortar surface which results is rough and generally proud of the brick surface. In the case of stone masonry the effect can be quite decorative, but it is less visually acceptable for brickwork (Fig. 14.1). However, a weathered face can be struck or a jointing tool applied if the mortar is first left for an hour to start its initial set. In the case of tunnels and structures not exposed to public view, the surface is generally left untreated and the extra mortar provides additional protection (Fig. 14.2).

14.4 Quick-set mechanical pointing

In quick-set mechanical pointing the pointing gun has an extra attachment by which a chemical line is introduced into the middle of the mortar flow before the mortar passes an air inlet. The quick-set chemical used is a sodium silicate with an optimum silica content in relation to its viscosity (84 grade); only 1%–2% by weight of mortar is needed. Mixing of the chemical and the mortar takes place when the air agitates the material, which is then projected into the joint as with normal mechanical pointing. The setting time is between 0.05 and 0.02 sec. The nozzle must be held very close to the joint, as otherwise the set mortar will bounce off. In any case, because of the

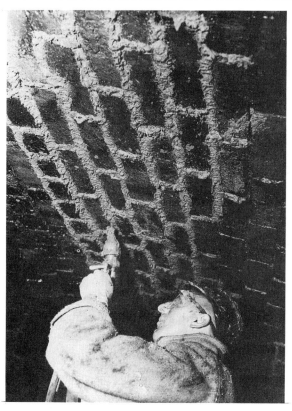

FIGURE 14.2 Normal mechanical pointing overhead. (*Courtesy of British Rail.*)

FIGURE 14.4 Quick-set pointing: normal mortar slumped on the left; adding quick-setting chemical produces the column in the centre (12 in rule on the right). (*Courtesy of British Rail.*)

FIGURE 14.3 Quick-set pointing of a wall with water running over its surface. (*Courtesy of British Rail.*)

caustic nature of the chemical, goggles should be worn.

As the mortar strikes the joint and sets, the shrinkage strain takes place before the next amount of mortar lands upon it, so that the shrinkage stresses associated with quick-setting systems are largely obviated. Brickwork with water running over its surface can be pointed easily using the quick-set system (Fig. 14.3), and water emanating at low pressure from behind the surface can be held back if required. The appli-cation of a trowel to give a weathered finish is more difficult, but the mortar can be trimmed flush with the surface with a sharp metal edge provided that this is done within 5 mins of pointing.

At the turn of a tap, the operator can revert to normal mechanical pointing; the chemical resistance is equally high for the mortars used in both systems. Operators can learn the techniques for both in a short time.

Grouting

W.B. Long

15.1 Introduction

Grouting is the term used throughout English-speaking countries for the introduction into a structure or into the ground of a material in liquid form which subsequently cures or sets into a durable solid or gel form. The technique has been used for the repair of masonry structures for over a century but remains a largely empirical art despite a much better understanding of the theoretical principles involved. By the skilled and experienced interpretation of geotechnical information and of observed performance, satisfactory results can be achieved from carefully devised and executed injection procedures.

Early grouting repair works were carried out using hydraulic limes and pozzolans, but today most work is done with cementitious materials in aqueous suspension, although resin and chemical solutions are often used in special applications.

15.2 Injection methods

15.2.1 Pressure injection

The most common and universally suitable method of injecting grout is to supply it under pressure to the point of injection from where it can penetrate and permeate voids and interstices by displacing the gases and liquids (normally air and water) therein or by compressing the gases or, in special cases, by disturbing and displacing the soil structure. If gases within the pores are compressed rather than displaced, some grout may be expelled when the injection pressure is released. By choosing grout of suitable viscosity and set or gelled characteristics and by skilful injection procedures, highly successful results can be obtained with this technique.

Injection pipes are inserted and sealed into holes pre-drilled in the positions and to the depths required for satisfactory spread of the grout; in the case of natural soil or loose fill, it may be possible to drive in the injection pipes without pre-drilling, using suitable points. The pipes can be withdrawn and resealed in stages, enabling a series of injections to be made. Tube-à-manchette or similar sleeve equipment allows injection at predetermined positions within the hole.

Since permeation of the grout can only take place as air and water are displaced, provision for their escape must be considered.

The rate and amount of injection at each injection point will depend on the nature and viscosity of the grout, the fissure size and the permeability of the structure or ground, and the pressure employed. Considerable experience is required in choosing, in the light of available information as to the nature of the structure or ground, the most suitable combination of hole spacing, grout material and pressure employed in a particular situation. Great care must always be taken to avoid excess pressure build-up, which could cause disruption of the structure, ground heave or even overturning.

Injection normally commences at the lowest level and progresses upwards and sideways in a systematic manner. It continues at each injection point until refusal is reached at a predetermined limiting pressure, until a predetermined maximum quantity has been injected at that point or until grout emerges freely at adjacent injection points which will function as tell-tales. In this last case, injection at the original point can be stopped and the hole plugged, and operations can be continued at the adjacent holes from which grout has flowed.

When grout is injected to refusal at any point, without evidence of spread to the adjacent holes, consideration should be given to the provision of an intermediate pattern of secondary injection holes to ensure satisfactory and complete grouting. A tertiary set of injection holes may also be required in some circumstances.

When an injection point accepts grout without refusal or build-up of back pressure, grout must be leaking away and steps may be required to deal with this situation. These may include the use of quick-setters (with great care, to avoid seizing up the plant and equipment), flocculating agents and fillers, or very thick grout in successive small injections to set in and block the leakage paths progressively.

Leakages of grout at adjacent injection points and through the structure, as well as expelled air bubbles and water, can give a useful indication as to how the grout is spreading. Leakages should be stemmed with quick-setting mortars or by caulking, if necessary. Repointing the surface joints of a masonry structure prior to grouting will be helpful in retaining the grout, but such repointing is seldom grout tight. In order to avoid unsightly staining of the surface masonry, all grout leakages should be cleaned off promptly before the grout can set.

Water flow through a structure being grouted should be controlled to avoid the washing away of unset grout. It should initially be channelled into a few defined points in which pipes have been set and water emerging from elsewhere should then be controlled by quick-setting mortars. Grout injection should commence away from the piped leakage points and work towards them, progressively reducing the flow and sealing the other parts of the structure until the piped leakage points themselves can finally be injected.

Water takes the line of least resistance, and it is common to find that, after apparently successful grouting to control percolation, leaks appear elsewhere after a period of days or weeks owing to pressure build-up. Further treatment may then be required in the new leakage areas. Percolation can often be reduced to acceptably insignificant amounts by grouting, or at least controlled, but can seldom be stopped completely, as grouting is not waterproofing. Some residual dampness is common.

On completion of grouting, all injection pipes should be removed and the points made good in an appropriate manner; the final cleaning up of the surfaces will include the removal of temporary caulking and spillages.

15.2.2 Vacuum injection

Pressure injection techniques have a number of limitations. The high pressures sometimes required to force grout into fine voids and fissures by expelling the gases and fluids within can be dangerously disruptive. Filling of the voids may generally be less than satisfactory, especially where confining the grout is difficult. Spread of grout beyond the area being treated may be a nuisance and incur additional costs.

The development from the 1970s of the patented vacuum injection technique (Balvac Whitley Moran Ltd) has allowed injection to be carried out where the use of pressure might be disruptive, ineffective or wasteful. A vacuum is first established within a portion of the structure to be treated, removing gases and fluids from within the pores and effectively holding the portion together for as long as it is applied. Injection grout, usually resin or chemical, is then introduced and drawn into the pores by the pressure differential between the atmosphere and the vacuum. On completion of one portion, work progresses to the next. No leakage or unwanted spread of grout occurs, as injection is confined to that portion where the vacuum has been induced. Any leakage paths away from this portion will have been sealed first so that the vacuum can be achieved.

Under suitable circumstances, a very high standard of filling of fine voids and fissures and of delaminations has been achieved with the vacuum injection technique.

15.3 Uses and applications of grout injection

All types of masonry structure covered in this book, whether surface structures or those underground or those subject to marine conditions have been successfully repaired by grout injection techniques. Some of the particular applications are described below.

15.3.1 Internal consolidation and strengthening

Great improvement in the strength and durability of a structure can be achieved by the systematic injection of grout to fill cracks, joints, voids, fissures and interstices. These may exist because the original construction, particularly the core, is of random or rubble masonry of inferior quality, or they may be due to poor quality mortar, or lack of it, in the joints, to deterioration from weathering, to stress or to movement.

Cementitious grouts would normally be used for such work on the grounds of costs and compatability whilst providing adequate strength and durability. They are usually injected by the pressure technique. However, chemical and resin grouts have been successfully used in special situations for the injection of fine voids and interstices by both the pressure and vacuum injection techniques. The vacuum technique has pro-

ved successful for filling and rebonding delaminations and filling internal voids in brickwork.

15.3.2 Control or reduction of water leakage

Control of water leakage is often one of the most successful uses of injection grouting. Although complete water tightness is not normally attempted or possible, control or reduction of leakage to an acceptable or negligible level is commonly achieved.

The grouting can be undertaken within the structure to fill the internal voids and interstices which form the leakage paths and/or into the ground or fill around the structure to reduce access of water to the back-face. The latter is often particularly effective.

Experience shows that the success of this treatment is determined by the efficiency and thoroughness of the injection, rather than by the materials used. In particular, skill is required in dealing with initial heavy percolation and groundwater flows. Also, persistence is needed in applying secondary or even tertiary injections, because after the initial leakage through lines of least resistance has been stemmed, back pressure may increase and other leakage paths develop. Positive methods of dealing with groundwater under pressure, such as relief drains or pumping, may be required in association with the grouting.

Cement grouts have been almost universally used until recent years for water-leakage control, with considerable success as they are compatible with wet conditions, economical and convenient to use; they will still be the choice in many situations where adequate penetration can be achieved. However, water-tolerant and water-reactive resins, which have a lower viscosity and better penetration into fine voids and fissures than cement grouts, are now available; although more expensive, they may be more effective in reducing or stopping water leakage in difficult situations.

15.3.3 Grout treatment of foundations and fills

The grout treatment of foundations and fills, which is highly specialized work that should only be undertaken by skilled and experienced specialists and on the basis of the conclusions of a thorough geotechnical survey, is well described in the British Standard *Code of Practice for Foundations* (BS 8004:1986, Section 6.7); limitations of space allow only the principles to be outlined here.

Permeation grouting is carried out to strengthen the soil and reduce its permeability, and therefore to reduce the flow of groundwater. It involves the injection of grout into the natural pore spaces of the ground or fill without disturbing the particle structure. Chemical grouts are usually employed; resins or cementitious grouts are used less commonly as they will only penetrate large interstices.

Displacement grouting involves the injection of thick cementitious grouts which disturb the ground structure rather than permeate the natural pores. By varying the injection technique and grout mix and pressure, extensive sheets of grout, 1–5 mm thick, can be formed along lines of principal stress (hydrofracture), thus reducing the flow of groundwater. Less extensive lenses of grout, 100–150 mm thick, compress the adjacent ground (squeeze grouting), or spherical bulbs of grout around the ends of the injection pipes compact high porosity soils (compaction grouting). The squeeze and compaction techniques consolidate the adjoining ground and improve its strength; they may be able to restore settlement.

Replacement grouting, commonly known as jet grouting, involves the insertion of a cementitious grout into spaces eroded in the ground by a high pressure water and air jet to form vertical columns which may interconnect, vertical panels or horizontal wedge-shaped wings, depending on the technique employed. These can provide ground support, reduce the risk of erosion or scour, or minimize groundwater movement. Specialists such as GKN Colcrete carry out well-proven jet grouting techniques.

15.4 Grout materials and mixes

15.4.1 Cementitious grout materials (aqueous suspensions)

Cementitious grout materials consist of cements, usually blended with other fine powders as fillers and with admixtures to improve or modify the liquid and hardened properties, which are thoroughly mixed with water to form a pumpable aqueous suspension. They can be batched and mixed on site from the individual components when there is sufficient expertise and quality control, or alternatively they are available from a number of manufacturers (e.g. Pozament) as factory batched, mixed and prepacked standard mixes, or as special mixes formulated to meet particular requirements. Use of prepacked mixes may eliminate many of the uncertainties associated with site storage, handling and batching.

All materials should be manufactured, supplied, stored, handled and mixed in accordance with the relevant British Standard or other appropriate specification.

Cements used in grouting can be of almost any type, but ordinary Portland cement (OPC) is the most usual, with rapid-hardening and sulphate-resisting cements often specified for appropriate conditions. A more finely ground cement will permeate fine fissures better and set more quickly.

Although neat cement grouts are sometimes used where high strength is required or the grout quantities are too small to justify the extra site complication of storing and handling fillers, it is more common to reduce the cost and modify the properties of the grout by adding an inert filler, such as fine sand, or a pozzolanic filler, such as pulverized fuel ash (PFA), whilst still achieving adequate hardened grout properties.

Sand is often cheap and readily available. If used as a filler in cement grout, it should be fine, with a maximum particle size not exceeding 0.5 mm, and mixed in proportions of up to 3:1 sand:cement. A greater proportion of sand is difficult to pump. Sand–cement grouts are of high strength, but will not penetrate fissures less than about three times the maximum particle size in width because of 'bridging' across the gap.

Pozzolans react with free lime in the presence of water to form a cementitious compound, and so when they are used in a cement grout they increase the strength of the mix as well as acting as fillers, improving the pumpability and flow characteristics and reducing bleeding. **PFA** is an excellent cheap artificial pozzolan for use with cement grout when available at the appropriate quality and fineness. **Ground granulated blast-furnace slag** is another cheap artificial pozzolan, and both it and PFA are much used as fillers for low strength large-volume cavity-filling cement-based grouts. **Natural pozzolans,** such as finely ground shale, pumicite or diatomite, are more expensive and are used for improving the cement grout properties in special situations.

Clays, especially **bentonite,** absorb water and form gel structures, thus stabilizing a cement grout and preventing 'bleeding' (the formation of a layer of water on the upper surface of a grout, which can cause cavities when the latter sets). Where strength is required in the set grout, bentonite up to 5% by weight of water may be used. At higher proportions, there will be little strength; the bentonite is used as a filler for reducing the cost.

A very wide variety of **admixtures** is commercially available, including accelerators/retarders to regulate the rate of set and hardening, fluidifiers and air entrainers to improve the handling properties, expanders and anti-bleed or anti-dispersant agents. Admixtures should be regarded as aids to good grouting practice but not as a substitute for it. Their suitability should be verified by trial mixes when necessary and the manufacturers' guidance should be obtained, especially where the use of two or more admixtures is proposed for a mix. Admixtures are usually used in relatively small quantities, requiring care in batching and mixing; the effect of overdosage and underdosage should be ascertained.

Water used in cement grouts should preferably be from a public water supply and of a quality fit for drinking. It should not contain any detrimental substances, but seawater has been successfully used for cement grouts when these were not being placed in contact with steel, which may suffer accelerated corrosion from the chlorides present.

15.4.2 Chemical grouts

Chemical grouts are solutions of chemically active liquids that are specially formulated for grout injection. They are available in a wide range of strength, viscosity, setting or gel time and other properties. They are generally more expensive than cementitious grouts, often very considerably so. They are therefore chosen for the repair of masonry structures when the grout characteristics required are not available in a cement mix, e.g. very low viscosity, particular chemical resistance, rapid set and hardening, and formation of a flexible gel rather than a rigid mortar.

Chemical grouts are available with a viscosity little greater than that of water; these therefore possess the ability to penetrate exceptionally fine pores. Such grouts may form a gel rather than harden, but they are well able to resist considerable hydraulic gradients when gelled and so to be effective in controlling water penetration. Their long-term durability, particularly if alternate wet and dry conditions occur, should be investigated.

Polyester or acrylic resin grouts are also available, with very low viscosity as well as excellent strength and durability; they are particularly useful for strengthening work. Water-tolerant resins are available for use in wet conditions.

All chemical grouts are supplied as two or more reactive components in solution. The earliest practical system, known as the Joosten process, used two chemicals such as sodium silicate and calcium chloride which reacted together immediately. It was therefore necessary to inject one first and then to follow up with the second, making a relatively expensive site operation with a lot of injection holes and two separate injections. Currently available commercial chemical grouts require only a single injection; the components have a variable range of setting or gelling times after mixing, allowing adequate time for site injection before the set or gel commences. Times of 10 – 90 min are usually chosen in practice; these are very sensitive to temperature, however.

Chemical grouts should be obtained only from experienced and reputable manufacturers, able to offer appropriate quality assurance and full information on the properties, handling characteristics and long-term durability of their products. Their advice and instructions on suitability, handling, mixing and injection techniques should be carefully followed.

15.5 Grout plant and equipment

Grout plant and equipment should be purpose designed or adapted and should form a balanced and integrated system capable of efficiently batching and mixing the materials to the required standard and of delivering them to the injection point at the required volume and pressure without segregation.

(a) Batchers

Unless pre-batched materials are used, all site batching should be by weight, except for added water or other fluids which can be measured by volume.

(b) Mixers and agitators

Since efficient mixing is essential, all mixers should be purpose designed to suit the materials. The use of conventional concrete and mortar mixers is not recommended.

For cementitious grouts, high shear colloidal-type mixers should be regarded as the norm, as they wet the particles more thoroughly and break down agglomerations, producing grout mixes with lower bleed, sedimentation and filtration characteristics. However, care may be necessary with grouts containing accelerators and some admixtures, as flash setting may be induced. Rotating-paddle mixers of lower efficiency and slower speed and other non-colloidal mixers should only be used after ascertaining that they will produce a grout meeting the requirements of the particular job, or where the materials and admixtures to be used are formulated so that they mix readily.

Mixers for chemical and resin grouts should comply with the recommendations of the manufacturer. Chemicals and resins generally mix readily in suitable plant, as they are supplied in the form of two or more liquids, with fillers in powder form being only occasionally used. Thorough mixing must be achieved, but too high a mixing speed may be harmful to some materials, causing a rise in temperature and premature setting.

All grout mixers should be designed such that the materials do not build up or clog in the machine. They should also be capable of easy cleaning and be readily dismantled for this purpose when necessary. Facilities for washing out with water or solvents will be needed, together with means of disposal of the waste arising.

Cementitious grouts which are not used immediately after mixing are normally transferred to agitating tanks, which are equipped with slow speed paddles or the like to provide sufficient movement to keep the mix in proper suspension until used. Mixed grout must not be kept in such tanks beyond the time of the initial set.

Mixers and agitators are available with petrol, diesel, electric and air power units. The choice of type and power used can be left to the specialist contractor, subject to his meeting the requirements of suitability for the particular grout mix. Air power is often preferred, as it is commonly available on construction sites, and compressed air itself may be useful for blowing out grout lines and needed for drilling injection holes.

(c) Pumps

Grout materials may be abrasive and viscous. Grout pumps and associated valves and delivery pipes should be designed to be suitable for handling the materials at the required volumes and pressures.

Specialist manufacturers supply various types of reciprocating and continuous pumps. Reciprocating, or positive displacement, pumps are of the piston or diaphragm type and give a pulsating pressure. This is thought by some to assist grout penetration by the ram effect. Piston-type reciprocating pumps are generally used for high pressure applications and diaphragm-type pumps are used for higher volumes at lower pressures. Diaphragm pumps may be simpler to maintain and service. Continuous pumps of the screw or rotary type are also used for high volumes at lower pressures; they can be particularly light and compact.

The choice of pump type and power unit can generally be left to the specialist contractor. Speed control for the power unit will enable a range of injection outputs to be obtained, and in all cases rapid control of the pump is required with the facility to stop the pump or release pressure quickly in the case of build-up of high back pressure.

Manually powered piston or diaphragm pumps may be appropriate for low volume applications where close control of injection pressure and volume is required.

(d) Water supply

For high volume cementitious grouting, a large amount of mixing water will be needed and an overhead water reservoir, with provision for rapid discharge of a measured quantity into the grout mixer, will be advantageous.

(e) Grout hoses, injection pipes and cocks

Flexible rubber or plastic hoses, of suitable diameter to transport the mixed grout over the distance required and able to withstand the proposed pressures, are normally used. Quick-action grout-tight couplings should be used, enabling the hoses to be readily disconnected for cleaning out.

Injection pipes, which are normally steel, are inserted into each injection hole and made grout tight by sealing with quick-setting mortar or caulking with suitable packing. Hose connections to the injection pipes should also be quick action and grout tight and provided with an on–off cock or other means of quickly stopping the injection and retaining the injected grout, once the pipe is filled to refusal or the required amount has been introduced. Taper plugs in the steel pipe or clamps on a short length of flexible hose can be used for this purpose.

It may be convenient to provide a return grout line to the agitating tank, with a two-way cock at the delivery end of the pipe leading to it, so that, on completion of injection at a point, grout can be recirculated whilst connection is made to the next injection point, thus minimizing the risk of grout blockages in the delivery hoses owing to its settling out and avoiding the need to stop and restart the pump each time.

(f) Vacuum injection plant and equipment

Vacuum injection is wholly carried out by licensed experienced specialists, who can be relied upon to supply appropriate plant and equipment. In general, this will be similar to that used in pressure injection as far as mixing, delivery etc. are concerned, with the vacuum being provided by portable pumps of suitable capacity.

15.6 Contract documentation and supervision

15.6.1 Contract documentation

There are really no standard specifications or codes of practice for remedial grout injection work, apart from those relating to the materials used. Advice should be sought when necessary from appropriate independent experts or reputable specialist contractors, as considerable acquired knowledge and experience is needed to carry out successful remedial grouting.

Documentation may then take the form of a specification for the materials, requiring compliance with relevant published standards, together with a performance specification for the injection grouting. The latter should stipulate, as appropriate, the characteristics required of the grout in both the fluid and hardened phases (i.e. viscosity, setting time, strength, limits of allowable mix proportions, water: cement ratio etc.) and the requirements of the injection technique (i.e. number, size and depth of injection holes, injection method and limiting pressures and quantities), as well as the result to be achieved (such as filling of voids, stemming water flow or leakage, strength of injected areas etc.).

15.6.2 Workmanship

(a) Materials

Cements, chemical grouts and prepacked materials should be identified, stored in dry and frost-free conditions and in accordance with the manufacturers' instructions, and used in the order of manufacture or delivery. Aggregates, pozzolans and other materials used as fillers should be obtained from an approved source and stored in dry and frost-free conditions. Water should be from an approved source and free from harmful matter. Admixtures should only be used with the approval of the engineer, subject to full details being supplied as to the reason for and method of use, the amount to be used, the chemical names and proportions of the main active ingredients, the means of batching and adding to the grout mix and the detrimental effect of overdosage and underdosage.

(b) Batching and mixing

All site-batched materials should be weigh batched to an accuracy of ±3%. Thorough mixing in an approved machine is necessary.

(c) Mix proportions and trial mixes

Unless the proportions are stipulated in the specification, the contractor should design the mix to the approval of the engineer so that the penetration and hardened properties of the grout are such as to achieve the objective of the grouting. Trial mixes and injections may be required, and suitably tested, to verify this before the work is allowed to proceed.

(d) Injection technique

The injection technique to be employed should be stipulated by the contractor prior to commencement of work, with sufficient detail as to number, type and depth of injection points, pressure(s) to be used, sequence of work etc. for the approval of the engineer. No variation from this approved technique, which may be considered necessary in the light of the results being achieved, should be made without the engineer's prior agreement.

(e) Weather

Extremes of weather may require appropriate precautions. Cement grouts will be damaged by low temperatures and should not normally be placed when the air temperature is under 5 °C or the temperature of the structure is under 2 °C. Chemical grouts may also be seriously affected by low temperatures and the setting or gelling may be delayed or stopped, and so the manufacturer's recommendations should be observed. High temperatures may cause extra-rapid setting or gelling of many materials, and even flash setting, and so care and precautions are again needed, including modifying the grout mix if necessary.

15.6.3 Quality control

Variations in grout properties and in injection success are most likely to arise from the quality of the materials, the accuracy of batching and adequacy of mixing, the injection technique and the effectiveness of testing. Therefore, these are the aspects which should receive close attention by those supervising grouting operations. Monitoring should ensure adherence to the requirements described in the preceding sections of this chapter. In particular, the grout injection procedures must accord with those approved by the engineer in respect of the location and depth of injection points, the order of injecting into them, the amount of grout take at each and the observance of pressure limits; accurate records of these must be kept. Observation should be maintained on the effectiveness of the operation, especially the spread achieved (as evidenced by undiluted grout appearing at adjacent holes). The supervising engineer must be alert to the possibility of distortion of the structure or of the ground as a result of pressure build-up, as well as to the possibility of uncontrolled leakage or wastage. If necessary, in the light of observed performance, the mix and/or the injection technique should be modified.

Testing of a cement grout mix for flow characteristics can be carried out on site using the Portland Cement Association Cone or Colcrete Flowmeter; chemical grouts can be tested using a viscometer. Set characteristics can be determined from 75 mm, 100 mm or 150 mm cubes. The degree of penetration and the properties of the grout *in situ* can be determined only from tests on samples removed by coring.

15.7 Safety

Some of the materials used in grouting are toxic and/or corrosive. Care should therefore be taken accordingly, especially when working in enclosed spaces or tunnels. Appropriate protective clothing, including eye protection, must be worn; good lighting and proper ventilation must be provided and precautions taken to control run-off or accidental spillage which may seep into drains and watercourses.

15.8 References and further reading

Anon (1969) Grouting design and practice. *Consult. Eng. Tech. Suppl.,* **33** (10), 1–39.

Bowen, R. (1975) *Grouting in Engineering Practice,* Applied Science Publishers.

Bradbury, H. W. (1978) Design of grouts. *Int. Semin. on Drilling and Grouting,* Cement and Concrete Association.

British Standards Institution.

BS 3892 *Pulverised Fuel Ash.* Part 2: 1984. *Specification for Pulverised Fuel Ash for Use in Grouts and for Miscellaneous Uses in Concrete.*

BS 6699: 1986. *Specification for Ground Granulated Blastfurnace Slag for Use with Portland Cement.*

BS 8004: 1986, *Code of Practice for Foundations.*

Bruce, D. A. (1987) Structural repair by grouting. *Proc. Int. Conf. on Structural Faults and Repair, London, July 1987,* Engineering Technics Press.

Burns, D. A. (1983) Resins in the repair of brickwork. *Conf. on Refurbishing and Renovation of Buildings – Materials and Techniques, September 1983,* Plastics and Rubber Institute, London.

Coomber, D. B. (1985) Tunnelling and soil stabilisation by jet grouting. *Proc. 4th Int. Symp. on Tunnelling,* Institute of Mining and Metallurgy.

Glossop, R. (1961) The invention and development of injection processes. *Geotechnique,* **10** (3); **11** (4), 255.

Gourlay, A. W. and Carson, C. S. (1982) Grouting plant and equipment. *Proc. Conf. on Grouting in Geotechnical Engineering,* American Society of Chemical Engineers, New Orleans.

Littlejohn, G. S. (1982) Design of cement based grouts. *Proc. Conf. on Grouting in Geotechnical Engineering,* American Society of Chemical Engineers, New Orleans.

Littlejohn, G. S. (1983) *Chemical Grouting,* South African Institution of Engineers (Geotechnical Division) (available from GKN Colcrete).

Littlejohn, G. S. (1985) Underpinning by chemical grouting. In *Underpinning* (eds S. Thorburn and J. F. Hutchison), Chapter 8, Surrey University Press.

McLeish, W. P. Apted, R. W. Chipps, P. N. and Bell A. L. (1987) Geotechnical remedial works for dams: some U.K. experiences. *Proc. Int. Conf. on Structural Faults and Repair, London, July 1987,* Engineering Technics Press.

Perelli Cippo, A. and Tornaghi, R. (1985) Soil improvement by jet grouting. In *Underpinning* (eds S. Thorburn and J. F. Hutchison), Chapter 9, Surrey University Press.

Warner, J. (1982) Compaction grouting – the first 30 years. *Proc. Conf. on Grouting in Geotechnical Engineering,* American Society of Chemical Engineers, New Orleans.

Sprayed concrete

W.B. Long

16.1 Introduction

Equipment for the spraying of cement mortar was first developed in the early 1900s, since when sprayed concrete has been widely used for many new construction and repair applications, including the repair and strengthening of masonry structures. The process has been given many different names in English-speaking countries, such as gunite, shotcrete, pneumatic concrete etc., but the term sprayed concrete has now been adopted in the United Kingdom.

A particular advantage of the technique is that only the flexible delivery hoses and the discharge nozzle, together with a nozzle operator, are required at the point of application. The nozzle operator may be able to work away from the point of application by using a remote-controlled nozzle in special situations, for safety reasons and the like. Spraying can be carried out from temporary access facilities which need only to carry the weight of the operatives and hoses. As the delivery equipment itself can be relatively light and transportable, sprayed concrete work can be carried out in many locations of difficult access,

such as inside mines, shafts, tunnels and culverts and on high structures.

16.2 Definition of terms

The following are the accepted United Kingdom definitions of some of the terms used specifically in connection with sprayed concrete.

Sprayed concrete is a mixture of cement, aggregate, water and admixtures, projected at high velocity from a nozzle into place against an existing structure, or formwork, where it is compacted by its own velocity to form a dense homogeneous mass.

Dry process is that technique whereby the mixed cement and aggregate are transported by a high velocity air stream to the discharge nozzle, in a dry state, and the water required for hydration and workability is added at the discharge nozzle.

Wet process is that technique whereby the cement, aggregate and water are mixed to a suitably workable consistency and pumped to the discharge nozzle, where a high velocity air stream is introduced to project the material into place.

There is also a **composite process**, whereby the cement, aggregate and water are mixed to a suitable consistency, and introduced into a high velocity air stream which transports them to the discharge nozzle and projects them into place.

Gunite is a term used in the United Kingdom and elsewhere for sprayed concrete with aggregate of size less than 10 mm.

Shotcrete is the complementary term for sprayed concrete using aggregate of size 10 mm or greater.

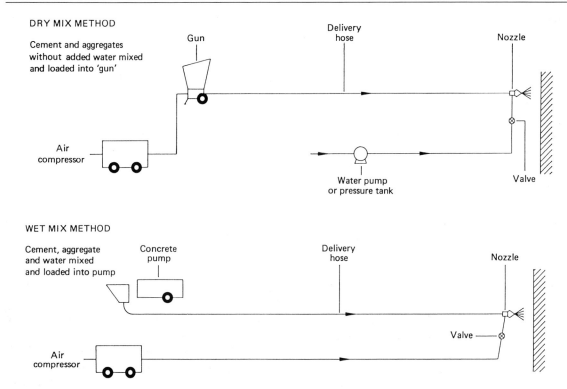

DRY MIX METHOD

Cement and aggregates without added water mixed and loaded into 'gun'

Gun

Delivery hose

Nozzle

Air compressor

Water pump or pressure tank

Valve

WET MIX METHOD

Cement, aggregate and water mixed and loaded into pump

Concrete pump

Delivery hose

Nozzle

Valve

Air compressor

FIGURE 16.1 Diagrammatic comparison of the dry and wet processes. (*Courtesy of Balvac Whitley Moran.*)

Layer is a discrete thickness of sprayed concrete built up at one time by a number of passes of the discharge nozzle and allowed to set.

Flash coat is a thin layer of sprayed concrete applied to protect, prime or finish the surface.

Rebound is all material which, having passed through the discharge nozzle, does not conform with the definition of sprayed concrete, i.e. has not been compacted in place into a dense homogeneous mass but has rebounded or dropped away from the surface against which it is being placed and fallen onto the ground or scaffolding.

Substrate or **interface** is the surface onto which sprayed concrete is being applied, i.e. the prepared surface of a structure or of a previously applied layer of sprayed concrete.

Profile guides are rigidly and securely fixed timbers, tensioned wires, pins or spacers etc. positioned and used as guides to the line and thickness of the sprayed concrete.

16.3 Sprayed concrete processes

The principal characteristics of the three application processes (dry, wet and composite) are as follows (Fig. 16.1).

16.3.1 The dry process

In the dry process cement and surface-dry aggregates are batched, mixed and loaded into a purpose-made machine. Here, the dry mix is pressurized and introduced evenly and without segregation into a high pressure, high velocity air stream which conveys it to a discharge nozzle through flexible hoses. At the discharge nozzle a finely atomized spray of water is introduced to hydrate the cement and provide the right consistency for placing and compaction.

As no excess water is required to provide workability during transporting and placing, dry process sprayed concrete can be placed at low water:cement ratios and has no-slump characteristics. It can be readily placed on vertical and overhead surfaces, subject to appropriate limitations on layer thickness. Admixtures for the purpose of modifying the hardened characteristics of the sprayed concrete or accelerating the setting time etc. can be introduced as dry powders to the pre-mix or as liquids added to the water introduced at the nozzle. Fibres of steel, polypropylene or other materials can be added to the dry pre-mix.

Equipment with a wide range of material throughput is available, enabling small quantities to be placed under close control in thin layers or on difficult or intricate shapes, and large quantities to be placed in thicker layers and over large areas with uninterrupted access. Delivery hose lengths in excess of 100 m can be used. The plant and equipment are light and readily transportable, quickly set up and simple for competent operatives to learn to use.

Although aggregates up to 20 mm size can be used with dry process plant, there may be little advantage in their use and aggregates of 10 mm maximum size are typically employed. Aggregate:cement ratios are typically in the range 3.5–4.0:1, although leaner mixes are possible. The proportion of material rebounding, which is mainly the coarser aggregate, can be high and the placed mix may have a significantly lower aggregate:cement ratio than the pre-mix.

Hardened characteristics of the dry process sprayed concrete include relatively high early strength, good 28 day strength (typically in the range 30–50 N/mm²), good density and good bond strength to substrate. Very low permeabilities can be achieved by careful selection of aggregates (the grading is particularly important), or by the addition of appropriate admixtures.

The properties of dry process sprayed concrete are more variable than in the case of other processes or of conventional concrete, but excellent results are achievable by skilled and experienced operatives, subject to good quality control; it is the process most commonly used in the

United Kingdom for repair and strengthening work.

16.3.2 The wet process

In the wet process the concrete is pre-mixed and pumped along flexible hoses to a discharge nozzle, in the same manner as in concrete pumping. The mixed materials are loaded into a hopper over a purpose-made or adapted concrete pump, which must be capable of delivering an even flow of concrete, without pulsation, to the discharge nozzle where high pressure air is introduced so as to provide enough velocity to project the concrete and compact it into place.

Because of the workability required for pumping through delivery hoses of relatively small diameter, admixtures to improve this are commonly incorporated in the pre-mix. The sprayed concrete can be satisfactorily placed on vertical and overhead surfaces subject to appropriate limitations on layer thickness. This build-up may be assisted by the incorporation of quick-setting admixtures, which can be added at the discharge nozzle if this is suitably adapted. Rebound may be significantly lower than with the dry process.

Any pumpable concrete mix can be used; the maximum aggregate size is normally 20 mm. Cement quantities are typically 350–450 kg/m³. Hardened characteristics will be similar to those of conventional concrete, with good 28 day strength (typically 30–50 N/mm²), good density and satisfactory bond to a prepared substrate.

Equipment available, purpose made or adapted for concrete spraying, has medium to high throughput with delivery hose lengths similar to those for concrete pumping. Output may be limited by the ability of the nozzle operative to control the placing effectively, rather than by plant capacity.

The process is most frequently employed in the United Kingdom for placing large volumes and large thicknesses of sprayed concrete and in medium to larger works, since the plant and equipment may be generally larger and heavier

FIGURE 16.2 Sprayed concrete strengthening of a masonry arch: reinforcement anchored to substrate. (*Courtesy of British Rail.*)

than dry process plant and less rapidly transportable. Because of the higher output achievable and the lower rebound, the costs of larger amounts of work may be lower than with the dry process. Better quality control of the concrete is achievable and excellent placed sprayed concrete work is practicable with skilled and experienced operatives.

16.3.3 The composite process

The composite process has been developed in Hungary, and the plant and expertise are now being made available elsewhere under licence. The process involves a purpose-designed train of plant, incorporating a batcher, mixer, elevator and placing machine, which can be rail mounted or set up statically. Concrete is pre-mixed by the plant, including all cement, aggregates and added water, and loaded into the placing machine as a wet mix. Here the mix is introduced into a high pressure, high velocity air stream and transported by this to the discharge nozzle, from which it is projected into place with very high compaction.

FIGURE 16.3 Sprayed concrete strengthening of a masonry arch: finished work. (*Courtesy of British Rail.*)

Advantages claimed for the process are the better control of concrete quality and water:cement ratio and the lower rebound associated with the wet process, together with the lower water:cement ratio, higher placing velocity and longer delivery hoses associated with the dry process. Furthermore, no admixtures are needed. Subject to further user experience and the selection of suitable available United Kingdom aggregates, the process should be particularly applicable to larger-scale projects such as tunnel linings, where the potential of the plant could be utilized economically.

16.4 Uses and applications

Whatever the purpose of the sprayed concrete, satisfactory results can probably be achieved with any of the three spraying processes. The choice in a particular situation should be based on considerations of previous satisfactory experience, plant and expertise available, the extent of the work, the thickness and quality of material to be applied, access and working space, the availability of suitable materials and the cost.

16.4.1 Surface protection and repair

The sprayed concrete technique can be used to apply a relatively thin layer of concrete onto the surface of a structure which has deteriorated due to weathering, erosion, mechanical damage, fire or chemical attack. Repairs can be carried out with a high quality concrete which will bond well, has high strength and durability and is speedy and economical to apply. Thicknesses of up to 100 mm are commonly used for this work, and the sprayed concrete can either follow the profile of the existing surface or be brought out to an improved line.

Experience has shown that sprayed concrete may develop cracks if unrestrained, and good practice requires either the incorporation of light steel fabric reinforcement or the inclusion of steel, polypropylene or other suitable fibres in the mix. Alternatively, a sufficient quantity of styrene butadiene rubber latex (SBR) or other suitable polymer admixture can be incorporated to modify the properties of the concrete and eliminate shrinkage cracking.

All types of masonry structure covered in this book have been repaired in the above manner. Whilst a tolerably good line and texture, acceptable on engineering structures, can be achieved

FIGURE 16.4 Sprayed concrete applied to brick culvert: work in progress. (*Courtesy of Balvac Whitley Moran.*)

FIGURE 16.5 Sprayed concrete strengthening: finished appearance. (*Courtesy of Balvac Whitley Moran.*)

with skill and care, and some degree of colour matching is possible, the technique is unlikely to be suitable for architecturally sensitive applications without further surface treatment.

16.4.2 Strengthening

Structural strengthening of tunnels, culverts, arches and the like is frequently undertaken by adding an appropriate thickness of sprayed concrete, incorporating steel reinforcement as indicated by normal design practice (Figs. 16.2–16.5). Special consideration needs to be given to the size and positioning of the steel reinforcement, in order that it can be properly encapsulated by the concrete. Surface finish and tolerance may be less critical in this type of work, but both will be below the standard of concrete cast in shuttering.

16.5 Contract documentation

Detailed guidance on specification, practice and recommended method of measurement in the United Kingdom is given by the Concrete Society (1979, 1980a, 1981) and the Sprayed Concrete Association (1987a). This can be incorporated in the contract documentation or referred to, as appropriate, for the particular job. The documents should clearly define the work to be carried out and the properties required of the placed sprayed concrete, so that compliance can be checked by the supervising engineer whilst a skilled and experienced contractor is allowed the flexibility to meet these requirements with the available plant and materials. In view of the limited general experience of such work, it may be helpful, before inviting tenders, to seek the advice of the Sprayed Concrete Association or some other specialist source so that all aspects of the documentation are clear and practicable.

The following may need particular consideration when drawing up the specification.

16.5.1 Materials

(a) Cement

Ordinary Portland Cement (OPC) is normally

245

used, but most types of cement are suitable and may be specified for particular conditions.

(b) Reinforcement

When reinforced sprayed concrete is to be used for structural strengthening, the reinforcement can be designed in accordance with normal practice. However, the detailing must take account of the particular requirements of the technique, especially the maximum bar size and the positioning, spacing and back-cover to the bars so that these can be properly encapsulated. Guidance is given by the Concrete Society (1978, 1980a) and the Sprayed Concrete Association (1987a).

For non-structural sprayed concrete, a light reinforcement to restrain cracking is usually incorporated. A welded steel fabric of 100 mm square mesh, weighing about 2 kg/m², or similar has been found to be satisfactory and is commonly specified. The incorporation of fibres in the mix (steel or polypropylene) in suitable proportions will also restrain cracking, as well as improving the flexural strength and impact and abrasion resistance by varying amounts depending on the type and quantity of fibre. Specialist guidance should be sought.

The choice of non-structural reinforcement type can be made on the basis of technical, cost and convenience considerations. The incorporation of fibres, which are relatively expensive, in the mix eliminates the separate operation of fixing steel reinforcement and may therefore speed up the whole repair work; also, it may allow a reduction in the sprayed concrete thickness, where this is otherwise determined by considerations of cover to steel reinforcement.

Flexural strength is very susceptible to variations in the fibre content of the concrete, and so it is important to note that the proportion of fibres which rebound during spraying may greatly exceed that of other ingredients. Allowance should be made in the design of the mix to ensure that the fibre content is adequate. The optimum fibre length for low rebound appears to be about 25 mm, while dispersion in the mix is improved by the use of non-magnetic fibres which do not 'ball up'. Tests and/or specialist guidance are advisable.

(c) Admixtures

Sprayed concrete of good strength and durability can be placed without admixtures in normal situations. Admixtures (SBR or acrylic polymers, silica fume etc.) may be specified, or proposed by the contractor, to achieve some specific improvement in the properties of the sprayed concrete, such as more rapid set, better flow characteristics or reduced water content for the wet process, or greater impermeability (see Section 16.4.1).

16.5.2 Requirements for hardened sprayed concrete

Sprayed concrete is usually specified in terms of the minimum 28 day crushing strength. 30 N/mm² will be adequate for normal purposes and is readily achievable. Higher strengths of 40 or 50 N/mm² may be required for particular applications and can be achieved by careful attention to the selection of aggregates and of mix design, together with good application technique. High early strengths may be desirable in certain situations, such as bridge strengthening, tidal work or structures subject to live loading, and may require the use of admixtures. Strengths of 4 N/mm² at 8 h and/or 10 N/mm² are typical requirements. Dry process sprayed concrete generally has a high early strength as a proportion of its 28 day strength.

Other properties which may be specified include flexural strength, permeability, bond strength and impact and abrasion resistance, although standard test methods have not yet been agreed for all of these.

16.5.3 Thickness of the sprayed concrete

The minimum thickness that can be applied in practice is about 15 mm, except in flash coats which are 5 – 10 mm thick. Other than in very uniform operating conditions, a thickness of 15 mm will tend to crack unless SBR or polymer admixtures are used. With fibre reinforcement, specialist advice should be sought as to the minimum thickness needed to incorporate the fibres. When welded fabric reinforcement is adopted, the minimum thickness is that needed to provide cover, which is 50 mm in the case of plain steel and less when galvanized or stainless steel is used, depending on the durability required and on operating conditions, in accordance with normal reinforced concrete practice.

Whilst theoretically any total thickness of sprayed concrete can be applied, this must be undertaken in layers of thickness limited to that which is possible in each application without slumping, sagging or debonding. The maximum layer thickness depends on the application technique, the concrete mix and admixture and whether the work is in a vertical or overhead configuration; it rarely exceeds 150 mm. In most cases, layer thickness should be specified on a 'performance' basis.

16.5.4 Guides to line and thickness

The importance of accurately fixed light timber profiles, tensioned 'piano' wires or other means of defining the line and thickness of the finished work cannot be overemphasized. Without these, unacceptable variations may occur. The thickness of the placed concrete can only be checked by drilling or coring.

16.6 Workmanship and quality control

Guidance on good working practice is given by the Concrete Society (1980a, 1981). The requirements of good reinforced concrete practice also apply. The following points require particular consideration.

16.6.1 Materials

Aggregates stockpiled for site batching and mixing for dry process sprayed concrete should be free to drain off excess moisture and protected from rain, so that they are at a suitable moisture content (normally less than 5%), as well as being afforded all other normal weather protection.

16.6.2 Plant and equipment

The type and quantity of plant and equipment to be used should be agreed by the engineer. It should all be purpose designed or adapted for spraying the proposed mix. It should be well maintained, working efficiently as designed, producing an even supply of well-mixed concrete material and transporting and placing this material evenly, without segregation, at the correct velocity and consistency.

The air supply to dry process plant and to the nozzle of wet process plant, as well as the water supply to the nozzle of the former, should be of sufficient volume and pressure and free from pressure fluctuations. The air supply should be dried to remove excess moisture.

All plant and equipment should be thoroughly cleaned out at the end of each working shift and when work is interrupted.

16.6.3 Batching and mixing

When ready-mixed concrete or bagged pre-mixed materials are used, the usual quality controls should be applied. For site batching and mixing, weigh batching with an accuracy to ±3% should normally be used. For small jobs using the dry process, volume batching can be used subject to the approval of the engineer. No material other than oven-dried bagged pre-mix should be used more than an hour after cement is added to the mix. Mix proportions should be designed by the contractor so that the specified properties are achieved.

16.6.4 Qualifications of operatives

All operatives employed in controlling the plant and equipment and placing the sprayed concrete should be approved by the engineer, who may require evidence of previous satisfactory experience and quality of work and/or the production of test panels to assess competence. In particular, a nozzle operator should be able to place uniform well-compacted concrete, without voids or trapped rebound, around reinforcement if appropriate and to the required tolerance and surface finish. This can be assessed by the examination of freshly placed panels and/or coring of cured panels.

16.6.5 Test panels

Test panels required to check compliance with the specified properties or to prove the competence of operatives should be sprayed using the same materials and plant as will be used in the contract, and in the same attitude (overhead, vertical etc.). Typically, test boxes 750 mm square by 100 mm thick are used. Panels should be made as close to the point of application as reasonably possible. After careful curing and protection from climatic extremes, they can be tested by taking cores 25 mm or 100 mm in diameter or sawn cubes of side 100 mm.

16.6.6 Interface preparation

The interface or substrate should be thoroughly prepared by the removal of all loose and unsound material and surface deposits, so as to provide a clean rough sound surface. This preparation may be more effectively undertaken by mechanical means such as abrasive blasting or high pressure water jetting after cutting away unsound material with light hand-held power tools.

Running water should be stemmed or diverted, so that cement is not leached from the freshly placed concrete; alternatively, a flash-setting admixture can be used in the first layer of sprayed concrete, which should then be ignored in con-

sidering the total thickness required for durability or structural strength. Weep pipes should be provided and/or permanent channels incorporated where necessary at the interface between the existing masonry and the new work in order to provide long-term drainage and prevent the build-up of unacceptable hydraulic pressures behind the placed and hardened sprayed concrete.

16.6.7 Reinforcement

Bar or fabric reinforcement should be securely fixed in the specified positions so that it cannot be dislodged by the force of the concrete spray, which may build up behind and force it outwards as well as inwards. Neither should the reinforcement be able to vibrate or spring, as this could result in cavities behind it or loss of bond.

16.6.8 Profile guides

Profile guides are the only means of determining that the correct thickness of sprayed concrete is applied and of achieving acceptable lines and tolerances. Light timbers or tensioned 'piano' wires should be securely fixed along main arises, and these or suitable pins or spacers should be located at appropriate intervals on faces in such a manner as not to interfere with the placing of the concrete.

16.6.9 Spraying procedure

If dry, the interface to be sprayed should be pre-wetted to reduce absorption of water from the sprayed concrete. Spraying should commence at the bottom, or to one side, with the nozzle being held at the optimum distance from the interface for the particular equipment, usually between 0.6 and 1.5 m, and at approximately right angles to the interface, so that the concrete impacts with enough velocity to compact itself but not so hard that previously sprayed concrete is displaced. 'Fanning' the nozzle will assist in the even build-up of placed concrete, and varying the angle of application will assist in compacting the concrete around and behind reinforcing bars. Care should be taken to avoid trapping rebound in corners and returns and behind reinforcement; the use of an air line to blow away accumulating rebound may be helpful.

When successive layers are to be applied, each layer should be lightly trimmed before it sets, and the next layer applied whilst the previous one is still 'green' and after pre-wetting the surface. Construction and other joints should be formed in accordance with guidance given by the Concrete Society (1980a). If the flow of concrete from the nozzle becomes uneven or intermittent, the operator should direct the material away from the placing surface until an even flow is restored.

16.6.10 Sprayed concrete finishing

Sprayed concrete is strongest and most durable when left as-sprayed, without surface treatment. The as-sprayed finish will be uneven in line, texture and colour, and some surface treatment may be necessary to improve these but should always be as little as possible. Weakening of the concrete itself and disturbance of the bond may result from trimming, floating or trowelling.

Light trimming to improve the line and profile can be done with the edge of a steel float or with purpose-made tools, while the texture can be improved by the subsequent application of a flash coat, 5–10 mm thick, placed while the original sprayed concrete is still 'green'. This flash coat may incorporate finer aggregate to improve the uniformity of texture, and if applied to substantial areas at a time may result in a more uniform colour. If a better standard of appearance and colour uniformity is required, this can be achieved by applying a render coat by hand, or a specially formulated cementitious protective coating or concrete paint, to the trimmed sprayed surface.

16.6.11 Curing and weather protection

Sprayed concrete is usually applied in relatively thin layers and has no protection from formwork, and so it is more susceptible than conventional concrete to drying out, freezing, rain damage and other weather conditions.

Curing and protection to a high standard are always most important. Curing should be by wetting by continuous or intermittent spraying (the latter at sufficiently close intervals), by covering with wetted hessian, impermeable polyethylene or other membrane, or by applying proprietary curing membranes (but not to intermediate layers where the subsequent bond may be affected). Protection against very high temperatures or direct hot sunlight, drying winds, and heavy rain should also be provided until the concrete is set.

Sprayed concrete should not be placed on frozen surfaces, nor when the air temperature is less than 3 °C unless special precautions are taken to the approval of the engineer. Care should also be taken when spraying onto very hot surfaces.

16.7 Practical considerations

16.7.1 Access

Access to the work, whether by scaffolding, cradles, hydraulic platforms or other means, must take account of the special requirements of concrete spraying, including safety and legal obligations.

For interface preparation, fixing of reinforcement and profiles and surface finishing, operatives should be able to reach the work face comfortably.

Whilst actually spraying, the nozzle operator should be able to stand back so that the discharge nozzle is at the optimum distance from the interface (up to 1.5 m or more), requiring a wider scaffold platform than normal. Also, to handle the nozzle with its weight and force of reaction and to move about as required to place the concrete (which is continuously discharging from the nozzle) to the required sequence, thickness and quality, the operator needs an unobstructed platform to serve adequate areas of the work or a platform capable of being readily moved about.

16.7.2 Artificial lighting

Lighting must be provided, especially for the nozzle operator, whenever the natural light, which may be reduced by the dust and cement fog arising from the work, is inadequate.

16.7.3 Artificial ventilation

Ventilation will certainly be needed in confined spaces which lack natural through ventilation to improve both visibility and the working conditions of the operatives.

16.7.4 Protection

Other parts of the structure, operatives, adjacent plant and equipment and the public may require protection against the concrete spray and rebound.

16.7.5 Sequence of work

Other operations need to be kept clear of the actual spraying because of the rebound, dust, overshoot and noise. Careful planning will ensure least disruption.

16.7.6 Debris and rebound

Debris will arise from the preparation of the interface, from material cut away, from spent abrasive from blast cleaning and from slurry from water jetting. Rebound and waste material will be produced by the concrete spraying and finishing. Appropriate measures may be required to confine, collect and dispose of all of this, and especially to exclude it from the drainage facilities.

16.7.7 Safety

Safety must be a prime consideration. In particular, means of access, lighting, ventilation and protective clothing, including eye protection, should be such as to allow safe and effective working.

16.8 Editor's Note on recent trials of the composite process by British Rail

Recent trials carried out by British Rail have confirmed the claims made by the originators of the composite process (Dorogcrete) to the extent that it produces a highly compacted semi-dry concrete, with good strength and durability potential. Strengths show high early gain and little variability.

With a 14 mm to dust crushed limestone aggregate and a water:cement ratio of 0.38, strengths averaged 65.6 N/mm² at 28 days (standard deviation of 2.8). Sprayed 200 mm thick, in a single pass onto a vertical face, the concrete showed excellent contact with the substrate and complete encasement of the reinforcement without shadowing. The process does not appear to be particularly susceptible to the grading of the aggregate, but elongated particles could pass the rejection screen and clog the screw feed to the placing machine. Costs appear competitive with other processes (particularly for larger jobs), but the throughput of 2 m³/h is lower and the air demand (up to 1000 ft³/min) requires considerable compressor capacity.

As yet there has been no opportunity to validate the claim that the process, as developed and carried out in Hungary, is not just a method of applying a better quality of sprayed concrete but is a total concept in which an analysis is undertaken of a structure's existing strength and a concrete mix is then designed and applied which will reliably meet a specified strength requirement.

16.9 References and further reading

American Concrete Institute (1977) *Specification for Materials, Proportioning and Application of Shotcrete.*

Concrete Society (1978) *Assessment of Fire-damaged Concrete Structures and Repair by Gunite.* Tech. Rep. 15, Concrete Society, London.

Concrete Society, (1979) *Specification for Sprayed Concrete.*

Concrete Society (1980a) *Code of Practice for Sprayed Concrete.*

Concrete Society (1980b) *Proc. Symp. on Sprayed Concrete; Construct. Int.* Pub. Construction Press, Lancaster.

Concrete Society (1981) *Guidance Notes on the Measurement of Sprayed Concrete.*

Cooke, T. H. (1989) *Concrete Pumping and Spraying. A practical guide.* Thomas Telford, London.

Ryan, T. F. (1973) *Gunite – A Handbook for Engineers,* Viewpoint Publications (Spon), London.

Sprayed Concrete Association (1987a) *Information, Design and Specification: Sprayed Concrete.*

Sprayed Concrete Association (1987b) *Fibre Reinforcement.* Tech. Data Sheet 1, Sprayed Concrete Association, London.

Chapter seventeen

Mini pile underpinning

I.W. Ellis

17.1 Introduction

Underpinning by the process of installing piles through existing structures was evolved by Dr F. Lizzi, of Fondedile SpA, Italy, in about 1952, for strengthening foundations subject to settlement or required to support additional load. Because the piles had to be installed through the existing masonry or concrete foundations, and occasionally old timber piles, with minimal disturbance, special piles were developed. These were of the bored injection grouted type and were originally called *pali radice* (root piles) by Fondedile because of their resemblance to the roots of trees. Today they are usually known as mini piles or micro piles to differentiate them from those in traditional use for structural support.

Typical examples of arrangements for piled underpinning are shown in Fig. 17.1 The piles achieve a direct connection between the existing foundation and competent ground immediately they are formed, without additional works such as needling or pile caps.

17.2 Methods of construction

The methods of installing mini piles vary from site to site and contractor to contractor, depending upon the condition of the structure, the subsoil and the plant available.

The method illustrated in Figs. 17.2 and 17.3 is typical for mini piles installed in non-cohesive soils. Boring is carried out with rotary drilling rigs by rotating and feeding down a temporary drill casing in short lengths. The leading length has a cutting bit at its lower end. Drilling fluid, usually water, is circulated through the casing to cool the cutting bit and to remove the drilling spoil from the bit to the surface (Fig. 17.2).

When the required depth has been drilled, all remaining spoil is removed by flushing with clean water and the borehole is filled with sand-cement grout via a tremie pipe extending to the bottom of the pile. Reinforcement, which may be a cage, a single bar or a tube, complete with spacers, is placed immediately the tremie pipe is removed. The temporary casing is then removed length by length, with grout being replenished to ground level after the removal of each length (Fig. 17.3).

In the method described above the drill casing provides temporary support to the soil until completion of the grouting operation. In cohesive soils and rock the method is similar except that the drill casing and cutting bit are removed before the grout is placed.

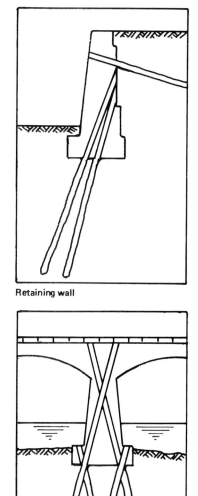

Retaining wall

Bridge

FIGURE 17.1 Typical pile arrangements for underpinning a retaining wall and a bridge. (*Courtesy of Fondedile Foundations Ltd.*)

17.3 Plant

The plant employed to construct mini piles can be divided into drilling equipment and grouting equipment.

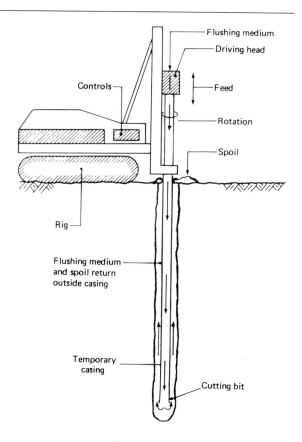

FIGURE 17.2 Drilling method for mini piles in non-cohesive soils. (*Courtesy of Fondedile Foundations Ltd.*)

The drilling equipment comprises a drilling rig, a high pressure water or mud pump and a drill string. The drilling rigs may be electrically, hydraulically or pneumatically operated and may be rotary or rotary–percussive. Most contractors use commercially manufactured machines made by companies including Wirth, Hands-England, Craelius and Klemm, but some larger contractors manufacture their own specially designed rigs. Drilling rigs vary considerably in shape and size to meet the wide range of situations encountered when underpinning. They may be skid mounted, track mounted (Fig. 17.4), lorry mounted or even mounted remotely from the power pack to facilitate easier handling in very restricted working areas.

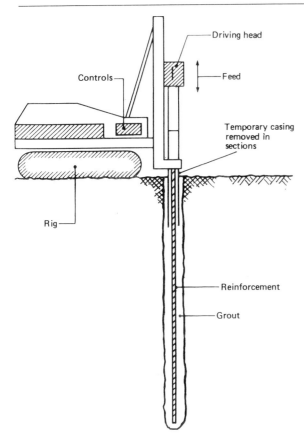

FIGURE 17.3 Mini pile grouting method. (*Courtesy of Fondedile Foundations Ltd.*)

tungsten carbide mounted on a drill casing of similar diameter. This allows the core to be retained in the drill casing for periodic extraction, usually every 1 or 2 m. In non-cohesive soils the drilling can be continued below the structure using the same equipment.

In cohesive soils the drilling method must be changed once the existing foundation has been core drilled – usually to a continuous flight auger or a drag bit with water flush. Rock encountered at a lower level may require yet another system, such as 'down-the-hole' pneumatic hammers or rock roller bits mounted on weighted drill rods. In the case of massive masonry retaining walls or bridge foundations down-the-hole and rock roller bit techniques can be used economically if large heavy drilling rigs can be accommodated. Where small or sensitive structures are to be underpinned it is advisable to use standard rotary coring techniques to keep vibration to a minimum. Coring techniques should always be used if steel or timber is anticipated.

The grouting equipment comprises a mixer, a storage tank and a pump. The mixer, usually of the collodial type, should have a high speed shearing action capable of mixing the sand and cement quickly and allowing it to be pumped without segregating. The storage tank, if used, should be fitted with an agitator. Grout pumps should be capable of pumping the grout several hundred metres in order that the equipment can be sited to facilitate easy unloading and storage of materials.

17.4 Pile sizes

Mini piles are normally 75–225 mm in diameter and 10–20 m long. Occasionally piles of larger diameter are used in underpinning and lengths of up to 40 m are possible when soil conditions warrant it.

The safe working loads of mini piles can vary considerably. Drilling fluids used in rotary drilling techniques produce irregular rough-surfaced bores which, coupled with the penetration of grout into

High pressure pumps are required to provide an efficient flushing system. If the drill casing is fed too quickly the cutting bit may become blocked, and considerable pressure will be required to clear the build-up of spoil. The velocity of the flushing fluid returning to the surface must be sufficient to carry the particles of soil in suspension without balling but not high enough to cause unnecessary erosion of the borehole and cavitation under the structure being underpinned.

The drill string includes the drill casing, (Fig. 17.5) drill rods and the cutting bit. Drilling methods vary considerably depending on the structure and soil to be drilled. The structure itself is normally core drilled using a crown tipped with

FIGURE 17.4 A track-mounted drilling rig. (*Courtesy of Fondedile Foundations Ltd.*)

the surrounding soil, result in piles of high frictional resistance with low settlement characteristics. The normal safe working load for typical mini pile diameters is 100–300 kN with loads of 500 kN possible when the piles are founded in rock or very dense gravels.

17.5 Design considerations

The piles must be designed to carry the specified loads, taking into account the required minimum factor of safety, settlement, pile spacing, downdrag, the overall bearing capacity of the ground beneath the piles, the bond between the piles the existing structure, and any other relevant factors.

When selecting piles for underpinning special consideration must be given to the following:

1. the possible continuing support that will be given by the original foundation
2. the transfer of load from the structure to the piles

FIGURE 17.5 Handling a length of 170mm diameter casing. (*Courtesy of Fondedile Foundations Ltd.*)

In piled underpinning the structure continues to rest on its original foundation and will only call on the piles to assist if further settlement occurs, and then only to the extent induced by that settlement. For this reason the original foundation is likely to continue to support a proportion of the total load and this fact should be borne in mind when determining the factor of safety.

When calculating the necessary bond length required to transfer the load from the structure to the piles, consideration must also be given to the effect of the stress concentration on the structure itself. In some instances the pile capacities may be limited by the stresses and settlement that can be accepted by the structure rather than by the ultimate bearing capacity of the piles themselves.

17.6 Pile testing

It is not normally practicable to proof-test piles constructed through existing foundations. For this reason, particularly where soil conditions are unknown, a preliminary test pile, or piles, should be installed to check the pile design.

Pile integrity testing, for the purpose of checking the structural soundness of piles, is unlikely to produce satisfactory results when used to test mini piles constructed using rotary drilling techniques.

This is because of the disproportionate variation in pile diameter along the pile length in different soil densities.

17.7 Specification

The *Specification for Piling* (Institution of Civil Engineers, 1988), although covering only traditional types of piling, is a very comprehensive document, which will assist Maintenance Engineers and others responsible for preparing piling specifications for major underpinning contracts.

The Federation of Piling Specialists has also issued advice on piling and pile testing, including *Specification for the Construction of Mini Piles.* These latter are not as comprehensive as the I.C.E. publication, but nevertheless they will be of invaluable assistance to all persons preparing contract documents, particularly for small to medium sized contracts.

17.8 Rate of working

The speed at which piled underpinning can be carried out depends on a number of factors, including the condition of the structure, the function of the structure and whether or not the structure is to be operational at the time that the work is to be undertaken.

For example, a large underbridge carrying a railway over a small stream or minor road can normally be underpinned using several drilling rigs working below bridge deck level. Provided that the structure is in fairly good condition, large rigs installing high capacity mini piles can carry out the work quickly and economically, with an average contract taking only a few weeks to complete. In contrast, a smaller bridge in a poor structural condition may take many months to complete if only one rig can be accommodated, and that only for a few hours each night when line or road closure is possible.

17.9 Advantages

In most situations, particularly when dealing with medium to large structures, piled underpinning is a fast and economical solution. Because the pile is effectively able to work as soon as the grout has set, strengthening of existing foundations is progressively obtained, thereby enhancing the already existing support of the ground itself.

Extremely sensitive structures, even during active settlement of several millimetres per day, can be successfully underpinned with mini piles provided that the work is undertaken by experienced contractors employing properly trained operatives, the correct plant and professional supervision.

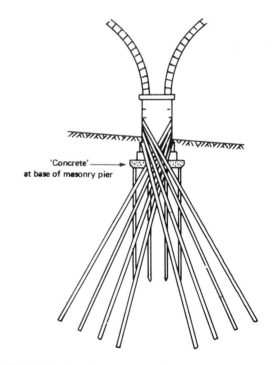

'Concrete' at base of masonry pier

FIGURE 17.6 Arrangement of mini piles at Northwich Viaduct (note the original timber piles). (*Courtesy of Fondedile Foundations Ltd.*)

17.10 Case Histories

17.10.1 Underpinning a railway viaduct

Northwich Viaduct is a multispan masonry underbridge carrying the Altrincham to Chester railway line. One of the piers, supporting a load of 1730 tonnes, settled approximately 50 mm over a number of years. Settlement was caused by decay at the top of existing timber piles supporting the pier, possibly because of a change in the water-table in the vicinity of the pier. The soil is predominantly glacial till.

Settlement was arrested by installing 70 mini piles 170 mm in diameter equally spaced in four rows along each face of the pier as illustrated in Fig. 17.6. The piles were approximately 15 m long and designed to support a safe working load of 250 kN each, with a factor of safety of 2.5. The contract was completed in six weeks in 1979 by Fondedile Foundations Ltd.

17.10.2 Stabilization of Delamere Dock wall, Weston Point Dock, Runcorn, Cheshire

(a) Introduction

Mini piles were used to stabilize a section of masonry wall at the north end of Delamere Dock which was sliding forward, probably because of excessive surcharge from materials stacked behind the wall at some time in the past. The unstable section was approximately 110 m long, 9.8 m high and 3.4 m wide at the base. Movement at either end had been minimal but the middle of the wall had moved forward about 700 mm and was unstable.

A study of old drawings and a site investigation revealed that the wall was constructed in about 1883 of limestone with a curved face and stepped back. An unusual feature is a culvert, a little under 1 m wide and 2 m high, which runs the full length of the wall with a number of outlets into the dock. The culvert was not in use and had been allowed to silt up. The site investigation revealed that the soil behind and below the wall is stiff glacial till.

(b) Design considerations

The Engineer for British Waterways Board invited Fondedile Foundations Ltd to prepare a scheme for stabilizing the dock wall, including filling the culvert with concrete, without interference to the normal commercial activities of the dock. This precluded a conventional ground anchor solution because it was not possible to lower the water in the dock, which was a dangerous procedure anyway considering the unstable condition of the wall. The scheme for stabilizing the wall was required to take several possible loading conditions into account.

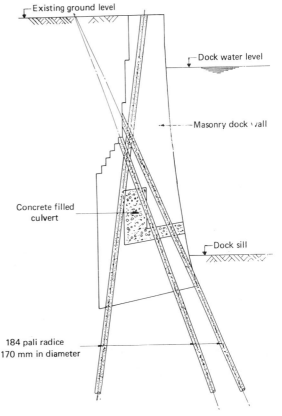

FIGURE 17.7 Arrangement of mini piles used to stabilize Delamere Dock Wall. (*Courtesy of Fondedile Foundations Ltd.*)

The solution proposed, and eventually carried out by Fondedile, was based on *pali radice*, 170 mm in diameter, positioned to resist the resultant of the active forces on the wall in axial compression or tension (Fig. 17.7). The pile lengths varied up to a maximum of 25.5 m and the safe working loads varied up to 450 kN in compression and 160 kN in tension (backward-raking piles). High tensile steel reinforcement was used in the piles; that in the tension piles was provided with corrosion protection. To avoid the sudden change from stabilized to an unstabilized wall the piles on each side of the critical centre section were phased out gradually by reducing the intensity and length.

(c) Construction

To enable the silt to be removed and the concrete placed in the culvert, holes 170 mm in diameter were drilled vertically from the top of the wall into the culvert at intervals of about 10 m. The silt was then removed by air lift, with the aid of divers working in the culvert. After all the silt had been removed, the divers bolted temporary shutters to the face of the wall to seal the culvert outlets. The culvert was then filled with a special concrete mix designed to minimize washout under water; 185 m³ of this special concrete was placed by concrete pump in under 6 h without significant washout or loss.

Because of the unstable nature of the structure the piles were installed from both ends working towards the sensitive centre section. A total of 184 piles was constructed in ten weeks in 1986 using two specially built heavy duty track-mounted drilling rigs. The wall was monitored throughout the contract, but very little additional forward movement was recorded.

17.10.3 The stabilization of a leaning minaret in Iraq

The Al-Hadba Minaret in Mosul, Iraq (Fig. 17.8), was built in about 1200 AD. For a number of reasons, but mainly because of the prevalent

FIGURE 17.8 The Al-Hadba Minaret. (*Courtesy of Fondedile SpA, Italy.*)

northwest winds and extraction of water from a nearby well for many years, the minaret developed a marked lean to the southeast. Overall, it is about 50 m high. It is made up of a stone masonry base, 11 m square and 10 m high, on which a second, slightly smaller, base has been built in brick. The brick trunk is 25 m high and tapers from 5 m diameter at the bottom to 4 m at the top. The dome and balcony have also been constructed of brick.

The displacement of the minaret, measured at 42 m above ground level, was 1.4 m in 1964

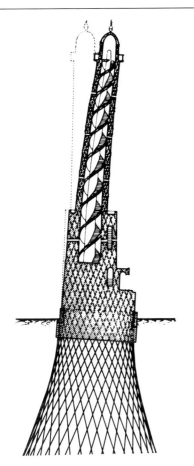

FIGURE 17.9 Scheme for strengthening the Al-Hadba Minaret. (*Courtesy of Fondedile SpA, Italy*).

adapted their system of reticulated *pali radice* (RPR) structures to underpin towers. An RPR structure is a three-dimensional network of mini piles and is normally used to resist landslips and to form *in situ* retaining walls. When used to stabilize a tower the RPR structure takes the form of a frustum of a cone, the top of which is connected to the base of the tower. As a result, the mass of the encased soil and the tower are combined, producing a very stable structure with a low centre of gravity.

The scheme proposed and carried out is illustrated in Fig. 17.9. The subsoil below the minaret is clayey silt. By using *pali radice* of nominal diameter 150 mm with a special expanded base, safe working loads of 300 kN per pile were achieved. Because of the precarious condition of the minaret it was necessary to encase it in very substantial scaffolding without the latter actually touching the structure itself. The scaffolding provided a safeguard against sudden collapse of the minaret, but also provided access for pointing, grouting and stitching the whole structure.

The main works were carried out in about nine months and, when completed in 1981, the final displacement of the minaret was 2.19 m.

17.11 References and further reading

Attwood, S. (1987) Pali Radice: their uses in stabilising existing retaining walls and creating cast-in-situ retaining structures. *Ground Eng.*, **20** (7), 23–7.

Bruce, D. A., Ingle, J. L. and Jones, M. R. (1985) Recent examples of underpinning using mini-piles. *Proc. 2nd Int.Conf. on Structural Faults and Repair*, Engineering Technics Press.

Ellis, I. (1985) Piling for Underpinning. *Symposium on Building Appraisal, Maintenance and Preservation*, at University of Bath, University of Bath, Bath. p. 88–96.

Federation of Piling Specialists (1987) *Specification for the construction of Mini Piles*. Piling Publications Ltd.

Fleming, W. G. K., Weltman, A. J., Randolph, M. F. and Elson, W. K. (1985) *Piling Engineering*. Surrey University Press, London.

Institution of Civil Engineers (1988) *Specification for Piling*. Thomas Telford, London.

Lizzi, F. (1981) *The Static Restoration of Monuments*. Sagep, Genoa.

when readings were first taken. By 1978 it had increased to over 2 m and was accelerating, necessitating immediate remedial action. It was at about this time that Fondedile SpA were invited to submit a proposal for stabilizing the minaret.

Stabilizing a tower is very different from stabilizing a bridge or a retaining wall. Even a tower in good condition can be difficult to underpin because of restricted access and/or the limited working space available. With a leaning tower the safety factor against overturning may be approaching unity, making any form of construction work hazardous and necessitating the maximum degree of expertise and care.

With the above factors in mind Fondedile

Thorburn, S., Hutchison, J. F. (eds) (1985) *Underpinning.* Surrey Univesity Press, London.

Weltman, A. (1981) A Review of Micro Pile Types. *Ground Eng.,* May, p. 43–49.

Stitching

I.W. Ellis

18.1 Introduction

Stitching is a technique whereby existing brick and stone masonry structures can be strengthened to increase their resistance to compressive, shear and tensile forces and to bond together fractured elements. The technique was evolved in about 1952 by Dr Lizzi, Technical Director of the Italian company Fondedile SpA, to strengthen masonry structures damaged during the Second World War. In Italy the system is known as *reticolo cementato*, but since it was introduced into the United Kingdom in the early 1960s it has become more widely known as 'stitching' or 'masonry reinforcement'.

Reticolo cementato or stitching is a method of consolidation carried out by grouting steel reinforcing bars into small holes drilled in a network or other predetermined pattern (Fig. 18.1). The network of steel bars can be likened to the steel in reinforced concrete, thereby changing normal masonry into reinforced masonry.

18.2 Outline of technique

The holes drilled in stitching works are generally of the order of 20–40 mm in diameter and long enough to ensure sufficient overlap of the 'scissors'. Length varies with the thickness of the element to be stitched and the nature of the weakness. The number of holes or bars per unit area will depend upon the condition of the structure and the reason for the strengthening. A rough guide would be three or four holes per square metre of wall, with each hole being about three times the wall thickness in length. Longer holes do not necessarily create a stronger solution. They are more expensive to drill and on most occasions it is better to have a greater number of shorter holes.

The reinforcement should be of the rebar type to provide a good bond between the bar and the grout. On ancient monuments and in a damp environment, e.g. exposed bridge piers, bridge arches and canal tunnels, stainless steel should be used. The diameter of the reinforcement is normally between 12 and 20 mm.

The grout is usually neat cement or pulverized fuel ash (PFA) – cement, with a water:cement ratio of between 1.0 and 1.5. Sand–cement grout should only be used when large voids are known to exist. Epoxy and polyester resin grouts may be used, but they are very expensive and should be employed only when their high strength and penetration properties are required; they are not normally recommended if existing voids in the masonry structure exceed 3% – 5% of its volume. Large voids should always be filled with a material having similar properties to the masonry itself, such as cementitious grout.

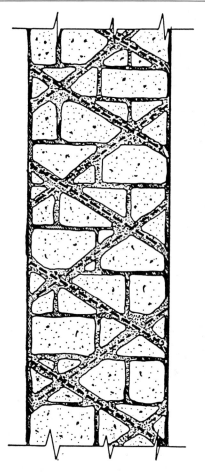

FIGURE 18.1 Scheme of masonry wall stitching. (*Courtesy of Fondedile SpA, Italy.*)

18.3 Applications

Stitching has a wide application in the maintenance of brick and stone masonry structures. The following are a few examples:

1. to strengthen bridge piers cracked as a result of differential settlement or consolidation of rubble fill
2. to strengthen bridge piers underdesigned or subjected to increased loading
3. to consolidate arch rings to reduce deflection
4. to strengthen tunnels weakened by ground movement or settlement

5. to strengthen spandrel walls and wing walls distorted by consolidating fill
6. to strengthen weak masonry behind anchor plates and pattress plates
7. to bond together detached elements of structures including wing walls and abutments, spandrel walls and arch rings, parapet walls and bridge decks.

18.4 Limitations

Structures between 0.5 and 2 m thick can be stitched very successfully. Masonry walls consisting of small pieces of hard irregular stone and less than 500 mm thick are extremely difficult to stitch, but soft brickwork only 350 mm thick can be dealt with quite easily. Massive structures 2 m or more thick do not normally require stitching.

18.5 Plant: selection and operation

18.5.1 Drilling

The plant selected for drilling stitching holes will depend upon the general condition and size of the structure and the hardness of the material to be drilled. The drills are normally hand held and operated by one man.

Small sensitive structures, particularly those constructed of soft stone or brick, are usually drilled using electric rotary drills with diamond coring bits and water flush to cool the bits and remove the cuttings. These drills ensure minimum disturbance to the structure but are slow. Medium-sized structures can successfully be drilled using electric rotary–percussive drills (Fig. 18.2). The main drawback with this type is that the drill bits jam easily when drilling long downward-sloping holes because of the difficulty of removing the cuttings. The vibration from these machines is insufficient to damage the majority of brick and stone masonry structures. Pneumatic rock drills can be used for massive structures, particularly if they are constructed of very hard stone or brick, and where long holes are necessary.

FIGURE 18.2 Drilling stitching holes with a Kango rotary–percussive drill. (*Courtesy of Fondedile Foundations Ltd.*)

Where accuracy in drilling is important, fixed hydraulic rotary drilling rigs are normally used. These rigs, similar to those used to construct mini piles, rotate and feed in short lengths of screw-coupled casing, coring the masonry with tungsten or diamond-tipped crowns and with water flush (Fig. 18.3).

18.5.2 Preparation for grouting

When several rows of holes have been drilled, the reinforcement and grout pipes are inserted into them and a section of the structure is prepared for grouting by flushing out any remaining loose debris with clean water. Each grout pipe is fitted with a valve for monitoring and control. Grouting commences at the lowest level and progresses upwards.

18.5.3 Grouting

The grouting equipment normally consists of a mixer, a storage tank and a pump together with grout lines and control valves. The mixer should be a high speed colloidal type to minimize segregation. After mixing, the grout is normally transferred into a storage tank fitted with a low speed agitator. Overmixing in a high speed mixer will result in a rise in temperature and premature hardening of the grout.

Grout pumps are usually of the double-acting type to ensure a continuous even flow of grout. For very sensitive grouting operations, a hand-operated pump should be used to exercise greater control.

The grouting equipment is normally positioned close to the area to be stitched in order to keep the grout lines as short as possible and, most importantly, so that the grout gang can signal to each other quickly during the operation. Grouting of stitching holes, usually at low pressures of between 1 and 2 atm, is regulated using a three-way valve positioned fairly close to the hole. A pressure gauge is fitted on the line between the

FIGURE 18.3 Drilling a stitching hole 50mm in diameter and 12m long with a small rotary drilling rig. (*Courtesy of Fondedile Foundations Ltd.*)

three-way valve and the grout pipe valves (Fig. 18.4). A return line between the three-way valve and the storage tank will ensure that the grout flow is continuous and that blockages do not occur because of slow grout take.

The pressure gauge adjacent to the three-way valve gives a quick indication when a hole is full and the line should be moved to the next hole. Adjacent grout valves are left open to monitor and control the spread of grout. Large grout takes by one hole should be avoided by switching to adjacent holes. Large volumes of fluid grout, even at minimal pressure, can exert a substantial force. It is therefore safer to allow the grout to harden before continuing, even if it means drilling additional holes.

18.6 Design considerations

Stitching is used to increase resistance to compressive and shear forces and to provide tensile strength. In reinforced concrete, evaluation is possible because its properties are known, but with reinforced masonry evaluation is not practicable because of the possibility of voids and variation in the strength of mortar, brick, stone etc. For this reason stitching is more of an art than a science and should rely more upon experience and examples than upon calculations.

The engineer responsible for preparing a scheme to repair or strengthen a structure will need to know the cause of deterioration, what changes in loading are anticipated, and the general condition of the masonry. He will also want to know whether the cause of deterioration has been rectified (e.g. underpinning to prevent further settlement) and whether any other work has been carried out.

18.7 Specification

Grouting and stitching are not covered by

FIGURE 18.4 Grouting a stitching hole. (*Courtesy of Fondedile Foundations Ltd.*)

standard codes of practice and specifications and it is therefore very important that those entrusted with the task of preparing schemes and carrying out the works are experienced specialists with professional as well as operative skills.

18.8 Testing

It is possible to compare tests carried out on treated and untreated panels of stitched brick and stone masonry. The only problem is that the panels must be fairly large, about 5 m² at least, to be meaningful. The cost of the preparation and testing of such panels is very expensive, and in many cases the money may be better utilized in additional strengthening works.

The efficiency of the grouting operation can be checked by random core drilling, by cutting out 'windows' or by borescope. A water test or grout test is probably more revealing.

18.9 Advantages

One of the main advantages of stitching is that the strength of a structure is improved immediately work commences. Very sensitive structures can be stitched, even those constructed of very hard stone in a weak mortar, if they are repointed and gravity grouted prior to drilling. Because the reinforcing bars are not stressed, the structure is consolidated gradually without alteration to the existing stress pattern.

18.10 Case Histories

18.10.1 Consolidation of a bridge in Northumberland to reduce arch deflection

Underbridge 116 at Longhoughton carries the Newcastle to Berwick railway line over the road from Alnwick. The bridge has one span of about

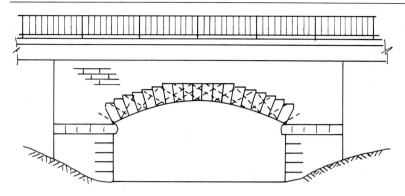

FIGURE 18.5 Longhoughton Railway Bridge: consolidation of masonry arch by *reticolo cementato*. (*Courtesy of Fondedile Foundations Ltd.*)

7.5 m and was built of local sandstone (Fig. 18.5). The surface voussoirs of the arch are about 600 mm thick but the arch ring is probably thinner. Although the bridge was generally in very good condition, deflection of the arch under normal live loading was considered to be excessive. In 1972 British Rail invited Fondedile Foundations Ltd to prepare a scheme to reduce this deflection to within acceptable limits.

From previous experience Fondedile were confident that a *reticolo cementato* solution would reduce deflection. The only difficulty was deciding upon the intensity of bars that should be proposed, bearing in mind the many unkown factors, i.e. the condition of the masonry, the thickness of the arch ring, the thickness of the backing etc. Accordingly, a network of high tensile steel bars 20 mm in diameter and 1.2 m long, allowing four per square metre, was proposed on the understanding that additional bars would be added should tests prove that the deflection was still not acceptable.

Altogether in three weeks 234 holes of 30 mm diameter were drilled, the reinforcement was placed and nearly 5 tonnes of neat cement grout were injected. Scaffolding was erected for access to the arch soffit in two stages, allowing single-line traffic and pedestrians to pass under the bridge at all times. Loading tests carried out by British Rail after the work was completed proved that deflection had been reduced to well within acceptable limits.

18.10.2 Strengthening of a bridge in the Pennines damaged by frost

Heatherycleugh Bridge is a five-arch masonry bridge built at least 130 years ago to carry the road (A689) between Stanhope and Alston over a ravine in Weardale (Fig. 18.6). Constructed of local stone, the bridge rises about 16 m above the Heatherycleugh, which flows through the centre span. It is 6.5 m wide, has five semicircular arches of 5.5 m span each and is about 60 m long overall.

The Pennines are very bleak and Heatherycleugh Bridge, because it is fairly high and crosses a ravine, is particularly exposed to driving rain and icy cold winds. During the severe winter of 1981–2 rain and snow, which had penetrated the weathered pointing of the spandrel walls, piers and arch soffits, froze to a considerable depth. This caused some distortion to the spandrel walls, some of the piers to develop cracks and mortar to be dislodged. The damage was discovered during a routine inspection, when a gap was noticed between the recently surfaced roadway and the parapet walls, indicating outward movement of the spandrel walls. Further inspection disclosed vertical cracks in the ends of the centre piers, considerable deterioration and loss of the mortar generally and dislodgement of some arch stones.

The County Engineer for Durham County Council invited Fondedile Foundations Ltd to prepare a scheme for strengthening the bridge.

FIGURE 18.6 Heatherycleugh Bridge after completion of strengthening works. (*Courtesy of Fondedile Foundations Ltd.*)

This was to include stabilizing the spandrel walls by means of tie-bars, consolidating and tying the centre piers, repointing as required and provision for grouting, arch stitching and any other works that further investigation might reveal to be necessary. The work actually carried out by Fondedile was as follows.

1. Over 400 m² of repointing was applied to arch soffits and spandrel walls;
2. Twenty-one medium tensile tie-bars 30 mm in diameter, each with two 600 mm x 600 mm square pattress plates, were installed. The bars were protected from corrosion by grease and Griflex plastic hose. The pattress plates were hot-dipped galvanized, painted with bitumastic paint and bedded on sand–

cement mortar. Finally, the tie-bars were tensioned and grouted;
3. Fifty-one high tensile steel bolts 25 mm in diameter and 1.5–2.0 m long were installed at 1.5 m intervals through the centre piers. The 150 mm x 150 mm end plates were painted with bitumastic paint and bedded on sand–cement mortar, and finally the bolts were tensioned and grouted with neat cement grout;
4. Approximately 900 high tensile steel bars 16 mm in diameter and 700 mm long were grouted in a network of holes 32 mm in diameter (*reticolo cementato*) to strengthen the arch masonry;
5. Over 270 high tensile steel bars 25 mm in diameter and 2.25 m long were grouted in a

network of holes 32 mm in diameter (*reticolo cementato*) to strengthen the parapet walls damaged by repeated vehicular impact over many years;

6. Approximately 110 tonnes of neat cement grout was injected in items 2–5 above.

The plant used to drill the 75 mm diameter holes for the tie-bars through the spandrel walls and the arch fill was a lightweight hydraulic horizontal drilling rig with a small down-the-hole hammer. All other drilling was carried out by hand-held rotary air drills. The pointing was carried out by hand, and the neat cement grout was mixed and pumped by a combined colloidal grout mixer and pump.

Because of the severe weather during the winter of 1982–3 the work had to be carried out in two visits, totalling five months; there was a period of two months during which work was impossible.

18.11 References and further reading

Cestelli, G. C. (1983) Introductory report. *IABSE Symp. Venice,* International Association for Bridge and Structural Engineering, Zurich, pp. 81–113.

Lizzi, F. (1981) *The Static Restoration of Monuments,* Sagep, Genoa.

Lizzi, F. (1983) Final report. *IABSE Symp. Venice,* pp. 313–20.

Ground anchors

I.W. Ellis

19.1 Introduction

A ground anchor is a structural member which can transmit an applied tensile force to a load-bearing stratum. The main application of anchors in the maintenance of brick and masonry structures is to prevent horizontal movement of retaining walls, bridge abutments etc. Anchors may be inclined or vertical, passive or prestressed, temporary or permanent. Anchors used to restrain existing structures from sliding and/or rotation are normally inclined, prestressed and permanent.

The main component of an anchor is the tendon, one end of which is grouted in the load-bearing stratum whilst the other end is attached to the structure by the anchor head. The lower section of the anchor in Fig. 19.1 is known as the fixed anchor length and the upper section, where the tendon must be debonded, is the free anchor length. Anchors with an expected life of more than two years are known as permanent anchors and normally require the tendon and anchor head to be protected from corrosion. Construction methods vary considerably, influenced by existing soil and site conditions and the plant operated by the anchor contractor. An understanding of

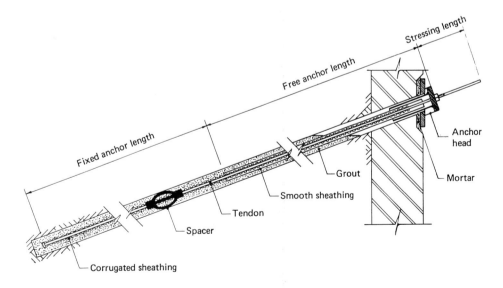

FIGURE 19.1 Typical permanent anchor with double corrosion protection. (*Courtesy of Fondedile Foundations Ltd.*)

FIGURE 19.2 Fondedile multibell anchor. (*Courtesy of Fondedile Foundations Ltd.*)

anchor design principles and construction methods is essential for all engineers responsible for specifying and supervising ground anchors.

19.2 Design considerations

19.2.1 Existing structure

When anchors are proposed to stabilize an existing structure, the initial design consideration should be that of overall stability, i.e. the stability of the anchored ground–structure system. Next, an appraisal of the structure is required including examination of old drawings and/or a site investigation to determine the cause of movement, the

capability of the structure to withstand the proposed concentrated anchor loads and what effect the vertical component of the applied load will have on foundation stresses.

19.2.2 Anchor load capacity

The load from the tendon is transmitted to the soil through the cementitious grout in the fixed anchor length. In granular soils which allow good penetration of grout, the skin friction at the soil–grout interface will be high. However, in soft clays the skin friction will be very low. To increase the load-carrying capacity of anchors in cohesive soils it is now general practice to form several underreams or bells along the fixed anchor length in order to increase the area of shear and provide additional support in end bearing (Fig. 19.2).

The skin friction developed along the grout–soil interface is greatly influenced by the techniques employed by the contractor to construct the borehole. In general, fast drilling and belling methods that produce clean boreholes and well-defined bells are to be preferred.

At the grout–tendon interface care must be exercised to ensure that all grease, soil and other extraneous matter is removed from the steel tendon before grouting. Normally the tendon in the fixed anchor length is encapsulated in a corrugated plastic sheath using resin or cement grout before the tendon is placed in the borehole.

Ground anchors should be designed with an adequate factor of safety against failure of the ground mass, the soil–grout bond, the grout–tendon bond, the tendon and the anchor head. In general, the tendon, anchor head and other factory-produced components should have a minimum safety factor of 2.0. The safety factor against pullout should be at least 2.5 and over 3.0 in soils liable to creep. The safe working load of anchors used to stabilize existing structures is usually in the range 200–600 kN.

19.2.3 Materials

(a) Grout

In the fixed anchor length neat cement grout is normally used for grouting the encapsulated tendon in the borehole and for grouting the tendon in the plastic sheath. Cement grout should have a minimum compressive strength of $40 \, \text{N/mm}^2$ at 28 days and the allowable maximum bleed should be controlled, particularly when grouting the tendon in the sheath.

(b) Tendon

The tendon may consist of prestressing steel in the form of bar or strand, either singly or in groups. Quality control during manufacture is essential and each delivery of steel should be accompanied by a test certificate. Spacers should be designed to ensure that the tendon can be placed centrally in the borehole without contamination of the grout with extraneous matter.

(c) Anchor head

The anchor head normally consists of a bearing plate bedded against the structure and an anchor plate, normal to the line of the anchor, against which the tendon strands or bars are locked off.

Restressable anchor heads may be required if monitoring of the anchors is considered necessary.

19.2.4 Corrosion protection

Anchor tendons are very highly stressed and a small reduction in the section of a strand or bar due to corrosion may result in sudden failure which could have disastrous consequences. It is therefore common, as well as economic sense, to ensure that the tendon and anchor head are adequately protected from corrosion at all times.

By encapsulating the tendon in the fixed anchor length two layers of protection are provided. Care must be taken to ensure that the corrugated plastic sheath will transmit the high anchor stresses without damage.

Corrosion protection of the tendon in the free anchor length is normally provided by factory-applied sheathing. Care must be taken that the sheathing is not damaged in transit, storage and handling. Special measures will be required to ensure adequate protection to the tendon where joints in the sheathing occur and of the anchor head including locking nuts, friction grips and the bare bar or strand on either side of the anchor head (Fig. 19.3).

19.3 Construction

19.3.1 Drilling

In general, ground anchors used to stabilize existing structures are installed at an inclination of

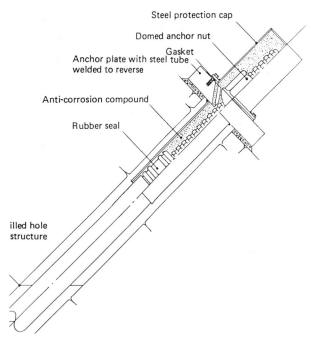

Steel protection cap

Domed anchor nut

Gasket

Anchor plate with steel tube welded to reverse

Anti-corrosion compound

Rubber seal

illed hole structure

FIGURE 19.3 Dywidag Anchor Head with corrosion protection. (*Courtesy of Dywidag Systems (UK) Ltd.*)

between 20° and 45° to the horizontal (Fig. 19.4). When drilling at shallow angles close to ground level particular care is required to minimize disturbance to the surrounding ground if damage to existing services, roads, property etc. is to be avoided.

Ground anchor drilling methods and plant are similar to those used to construct mini piles except that in cohesive soils anchors are usually underreamed. The borehole diameter of anchors is normally between 100 and 200 mm. By constructing a number of bells or underreams the effective diameter of the fixed anchor length can be trebled. The underreams are formed mechanically, either singly or several at once. One of the most successful clay anchors is the Fondedile multibell anchor which is constructed using a belling tool capable of forming up to seven bells simultaneously. All belling tools should be fitted with a device to indicate when the bells have been fully formed.

In most soils maximum borehole deviation is about 1 in 30 but this may be exceeded in soils containing large stones.

19.3.2 Tendon fabrication and handling

Fabrication of the tendon should take place under cover to ensure that the components are kept clean and dry. Care must be exercised to prevent damage to the prestressing steel by impact, corrosion or welding spray, and to the plastic sheathing from rough handling. Spacers must be securely fixed to the tendon to prevent them from being dislodged during emplacement in the borehole but at the same time they must not damage the plastic sheathing.

Heavy tendons will require a crane for lifting and placing in the borehole. Care is needed to ensure proper slinging to avoid excessive bending and proper alignment to avoid damage to the borehole and contamination of the tendon.

19.3.3 Tendon grouting

The cementitious grout may be placed either immediately before or immediately after the tendon is placed. Usually the grout is placed first. The plant used is similar to that used to grout mini piles, i.e. a high speed colloidal mixer, an agitating tank and a grout pump. The grout should always be placed by tremie after first ensuring that all debris has been removed by flushing the borehole with clean water.

Grouting should be carried out as soon as possible after the drilling operation has been completed. The grout should always be brought to the top of the borehole and the bottom of the tremie should be kept below the grout surface at all times. On completion, the grout immediately behind the structure should be washed out to prevent a strut effect. This section should be regrouted after the tendon has been stressed.

FIGURE 19.4 Skid mounted Wirth hydraulic drilling rig with continuous flight auger. (*Courtesy of British Rail.*)

19.3.4 Anchor head

The bearing plate should be bedded against the structure on sand–cement or resin mortar, allowing sufficient time for the mortor to attain a minimum compressive strength of 30 N/mm² before the anchor is stressed. Care should be taken to ensure that the bearing plate will not slide downwards during stressing, e.g. by recessing into the structure, and that the anchor plate is concentric with, and normal to, the tendon.

19.3.5 Stressing and testing

Ground anchor stressing and testing should be carried out by experienced personnel under qualified supervision. Details of the load increments,

extensions, losses etc. must be recorded for all stressing and testing operations (Fig. 19.5).

(a) Stressing

Ground anchors may be stressed when the soil at the grout interface along the fixed anchor length has sufficiently recovered from the effects of drilling and the grout has attained a minimum compressive strength of 30 N/mm². This period may vary from seven days in dense granular soils to 21 days or more in clay soils.

Ideally all the strands or bars in a tendon should be stressed simultaneously using a multi-unit jack. The only difficulty is the weight of the jack, which may require mechanical handling. Mono jacks are lighter and much easier to handle

FIGURE 19.5 Engineer checking extension of Dywidag bar during stressing operation. (*Courtesy of Dywidag Systems (UK) Ltd.*)

but each bar or strand must be stressed in stages if all bars or strands are to achieve the same final prestress. In order to minimize bedding-in losses it is essential to ensure that all bars or strands are completely debonded in the free anchor length before locking off.

All personnel should be aware of the dangers attached to operating stressing equipment. Nobody should stand over a jack in operation or anywhere where they could be injured should either the jack or the tendon malfunction.

(b) Testing

Permanent ground anchors should be proof tested by loading and unloading in increments up to 1.5 times the working load before locking off. Under certain conditions, e.g. for large contracts and when constructing anchors in unknown soils, preliminary suitability tests should be carried out

on the site, but such anchors would not normally form part of the permanent works and would be monitored over several weeks or months to provide information on loss of prestress, creep etc. Anchors that show excessive loss of prestress (i.e. greater than 5%) over a short period of time will require investigation.

19.4 Conclusions

Contractors have been installing ground anchors for many decades and during that time considerable advances have been made, particularly with respect to the use of prestressing steels and corrosion protection. With technological advance comes a greater need for the specialist contractor with a sound knowledge of geotechnical engineering and an experienced and competent workforce.

There is also a greater need for good specification and competent inspection. It is recommended that engineers proposing ground anchor works should become familiar with the new British Standard on ground anchors (BS 8081: 1989) and should read *CIRIA Report 65* on the design and construction of ground anchors (Hanna, 1980). They should also enter into pre-tender consultation with prospective specialist contractors. This should ensure that the engineer provides all the information necessary for the contractor to submit a satisfactory design and tender. It will also help with the engineer's selection of specialist contractors.

19.5 Case Histories

19.5.1 Stabilization of walls of railway cuttings

(a) Introduction

Two sections of retaining wall in the railway cuttings at Gospel Oak and Cricklewood, on the approach to St Pancras Station, have been stabilized with ground anchors to arrest forward and rotational movement. British Rail engineers were initially alerted to the problem when settle-

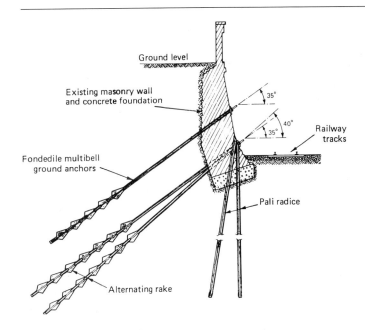

Ground level

Existing masonry wall
and concrete foundation

35°

40°
35°

Railway
tracks

Fondedile multibell
ground anchors

Pali radice

Alternating rake

FIGURE 19.6 General arrangement of
anchors and mini piles used to stabilize the
retaining wall at Gospel Oak. (*Courtesy of
Fondedile Foundations Ltd.*)

ment behind the wall at Gospel Oak was reported. Examination revealed that the centre 60 m of a 150 m section between an overbridge and a tunnel was severely bulged and cracked vertically for the full height of the wall.

Daily monitoring soon established that the wall was still moving and possibly accelerating, indicating that a solution was required very quickly. An investigation to establish why the wall was moving was inconclusive. In recent years some changes had taken place in the area including track lowering during electrification, replacement of track drains, redevelopment of property behind the wall, and possibly a change in the water-table. Any one or a combination of these changes may have caused the wall to move.

At an early stage in the investigation into ways of stabilizing the wall, British Rail consulted well-known contractors who specialized in strengthening operations and within a few days of being approached Fondedile Foundations Ltd produced a scheme for dealing with the critical 60 m length of wall. An order placed almost immediately allowed the contractor to mobilize and commence drilling only 18 days after the wall was first

known to be unstable. When the work started the forward movement of the wall was 10–15 mm a day. The total movement at track level since monitoring commenced was about 100 mm.

(b) Remedial measures

The scheme adopted for stabilizing the 8 m high wall at Gospel Oak is shown in Fig. 19.6. Multibell anchors were selected because London Clay was known to exist in the vicinity. The clay shear strength was confirmed from samples taken in the fixed anchor length of several anchors as the work progressed. The top row anchors had five bells and a safe working load of 375 kN and the bottom row had seven bells and a safe working load of 500 kN. Altogether 82 anchors were installed in the initial 60 m length. The vertical component of the applied anchor prestress was resisted by 96 *pali radice* 170 mm in diameter, approximately 20 m long and having a safe working load of 220 kN.

The bottom row of anchors was installed first and a nominal stress of 20% of the safe working load was applied three days later. This nominal stress was sufficient to slow down the movement

FIGURE 19.7 Retaining wall at Cricklewood after stabilization. (Note the distortion of the gantry supporting overhead conductors due to initial movement of the wall.) (*Courtesy of British Rail.*)

of the wall and within two weeks to stop it almost completely. The underpinning piles were installed before the final stress was applied to the anchors.

Disruption to rail traffic was confined to complete possession of the first track and partial possession of the second track for the 11 week contract period in 1977. Daily monitoring of the wall movement by British Rail engineers established that the total horizontal movement was 375 mm at the top of the wall but less at track level.

(c) Cricklewood wall

British Rail engineers were alerted to movement of the Cricklewood section of wall when a gantry supporting the overhead conductors was found to be distorted (Fig. 19.7). The scheme proposed for Cricklewood was similar to that for Gospel Oak. A total of 184 multibell anchors and 230 *pali*

radice were installed in 11 weeks in 1985 to stabilize 150 m of retaining wall. Better production was achieved on this section of wall because a wider working area was available (Fig. 19.8). This enabled the contractor to employ larger and more powerful drilling rigs to construct the anchors without the necessity for a scaffold platform as was required at Gospel Oak. Additionally, a new drainage channel was provided behind the top of the retaining wall to intercept surface water.

19.5.2 Stabilization of bridge abutment, Reading

Shepherds House Bridge is a three-arch brick bridge carrying the A4 trunk road over the railway line in the Sonning cutting near Reading. When this section of the Great Western Railway was built by Isambard Brunel in about 1840 it was

FIGURE 19.8 Two rigs installing anchors and one rig installing mini piles along the Cricklewood retaining wall. (*Courtesy of Fondedile Foundations Ltd.*)

broad gauge and utilized only the middle arch. It was not until much later that the cutting was widened to take four tracks and the outer arches were used. This necessitated construction of a retaining wall to support the ground between the new track and the north abutment.

The foundation of the north abutment is considerably higher than track level, a fact possibly not appreciated when the retaining wall was built. At a later date both the retaining wall and the abutment were seen to be moving. Monitoring was put in hand and it was quickly concluded that

remedial measures to halt the movement were required urgently. After consultation with Fondedile Foundations Ltd, British Rail decided on a multibell anchor solution (Fig. 19.9).

Within days, work commenced to install 11 anchors in a single row through the abutment foundation. The anchors were 170 mm in diameter and 18 m long overall with four 0.5 m diameter underreams in the 5 m fixed anchor length. They were constructed at 1 m centres with alternate rakes of 20° and 25°. The tendon was made up of six 15.2 mm diameter strands and the

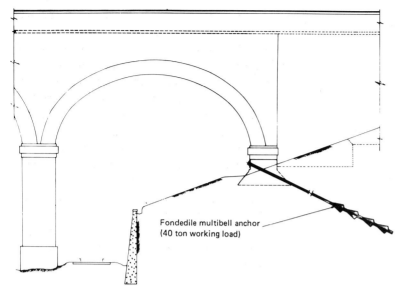

Fondedile multibell anchor
(40 ton working load)

FIGURE 19.9 Eleven anchors were required to stabilize the rotational movement of the abutment of Shepherds House Bridge. (*Courtesy of Fondedile Foundations Ltd.*)

safe working load of the anchors was 400 kN.

Horizontal movement of the abutment up to 10 mm a day was recorded during the beginning of the four-week construction period in 1971. Once stressing commenced the movement slowed down and was soon halted altogether. Six years later, the load in the anchors was checked and found to be substantially unchanged, and so the anchor heads were permanently capped. Other stabilizing measures included a double row, staggered, of concrete piles behind the retaining wall (to counter the formation of local slip circles) and a drainage system (to prevent softening of the clay strata). Sixteen years after the completion of these works, the abutment was observed to be still stable.

19.5.3 Strengthening of quay wall, Barnstaple, Devon

Until recently, during very high tides the River Taw overflowed and flooded some low-lying areas of Barnstaple. To safeguard the property in these areas the South West Water Authority implemented a tidal defence scheme which included raising about 230 m of masonry quay wall (Fig. 19.10).

The wall, which was founded on bedrock, was about 5 m high, with a little over half being exposed above the river bed at low tide. Although the wall was only 500–750 mm thick it has served as a quay wall for at least 100 years with only minor localized deterioration. The wall was to be raised 1.3 m by removing the old parapet wall and replacing it with a reinforced masonry wall bonded to the existing wall. To ensure that the old wall would be capable of withstanding any new stresses imposed upon it Fondedile Foundations Ltd were invited to prepare a scheme for strengthening the deteriorated sections of wall and for improving its overall stability.

The scheme proposed and eventually carried out comprised localized stitching and prestressed rock anchors installed vertically through the wall to improve its resistance to rotation and sliding. The bulged and cracked areas of masonry were strengthened by stitching with networks of 12 mm diameter Staifix stainless steel bars approximately 1.5 m long. Anchors 133 m in diameter and 10 m long overall, of 250 kN safe working load, were installed at 2 m centres through the centre of the wall. The neat cement grout used to grout the anchors also consolidated the masonry, which was particularly important at

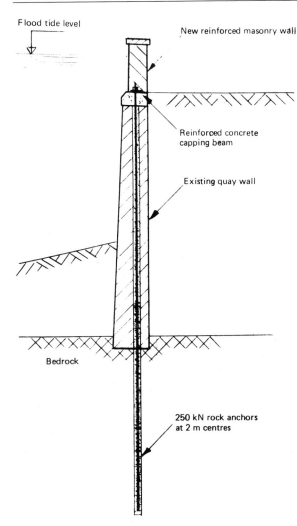

Flood tide level

New reinforced masonry wall

Reinforced concrete capping beam

Existing quay wall

Bedrock

250 kN rock anchors at 2 m centres

FIGURE 19.10 Tidal defence scheme at Barnstaple for strengthening and raising the old quay wall. (*Courtesy of Fondedile Foundations Ltd.*)

the toe of the wall where the bearing pressure was almost doubled after the anchors were tensioned.

The anchor tendons consisted of 26.5 mm Dywidag threadbars with double corrosion protection over the full length and a smooth plastic sheath to debond the free anchor length in the wall. At the top of the wall anchor plates 250 mm × 250 mm × 50 mm were bedded on a continuous reinforced concrete capping beam.

Starter bars for the masonry reinforcement were also set in the capping beam.

The 114 anchors were drilled by skid-mounted hydraulically powered Wirth drilling rigs with down-the-hole hammers. The grout was mixed in a Colcrete double-drum mixer and pumped to each anchor position from a central mixing area by a Colcrete Colmono 10 grout pump. The 32 mm diameter stitching holes were drilled by rotary-percussive hand-held drills. The works were carried out jointly with the Direct Labour Department of the South West Water Authority in 1983/4 and were programmed so that only a short length of wall was exposed to flooding at one time. Overall the contract took 14 weeks to complete.

19.6 References and further reading

Bastable, A. D. (1974) Multibell ground anchors in London Clay, *7th F.I.P. Congress, New York*, Session on Ground Anchors, pp. 33–7.

British Standards Institution. BS 8081: 1989. *Code of Practice for Ground Anchorages.*

Hanna, T. J. (1980) Design and construction of ground anchors. *CIRIA Report 65*, London.

Littlejohn, G. S. and Bruce, D. A. (1977) *Rock Anchors: State-of-the-Art*, Foundation Publications, Brentwood.

Waterproofing

R.O. Hall

20.1 Introduction

Seepage of water through masonry structures is one of the principal causes of their deterioration. It washes out the mortar and, when associated with freeze–thaw cycles, is responsible for spalling and splitting. When the water carries salts (either applied to the road surface for melting ice or picked up as the water percolates through the structure), it may evaporate from the surface leaving behind inorganic crystals which can push the surface off as flakes. The water may be a nuisance when it emerges, by reducing the value of a space under viaducts. It is particularly annoying to pedestrians when it is contaminated with oil and may cause damage to road surfaces. There is therefore considerable advantage in installing waterproofing and drainage.

Waterproofing should be considered whenever other work is being carried out on a structure, and it is frequently worthwhile in its own right. A large number of railway underbridges has been treated, but relatively few road bridges. The latter are less obviously in need of waterproofing where their tarmacadam wearing course, together with the falls due to road gradients and camber, direct a lot of the rainwater to a drainage system. Unfortunately, the road surface is unlikely to be impervious and there will be leakage. There will

inevitably be cracks along the kerb line, next to parapets and around gullies and manholes, as well as cracks caused by traffic and the imperfect reinstatement of trenches dug to reach services. Although a great deal of water may be shed, there will be some leakage and it is likely to be contaminated with salt, and so a waterproofing layer is still necessary. The methods available are described in this chapter and some typical applications to railway bridges are given. The same principles apply to road bridges, although modifications may be necessary to the membrane protection, the drainage and provision for services.

20.2 Existing waterproof layer

The waterproofing systems which were installed when the structures were built, or which have been added during subsequent maintenance, will generally be limited to naturally occurring bitumens or clays. In many very old structures there will be no such layer, and in later ones a tarmacadam wearing course will have been deemed sufficient. If waterproofing exists it will typically consist of tar or asphalt applied to the extrados of the arch. Even if originally effective, this is likely to have failed because it is brittle and subsequent movements in the structure have cracked it.

Puddled clay may also be encountered, either at the level of the arch or higher up in the structure. In either case, tracking the leak and effecting a repair may well be difficult. However, a case can be cited in which an old – and hitherto effective – puddled clay waterproofing layer over a series of arches was damaged by a service authority, who excavated in the road above in an attempt to lay a main where there was, in fact, inadequate depth to do so; although the road surface was reinstated, there were soon complaints of water percolation through the arches into the tenanted spaces below. Repairs to the waterproofing layer were undertaken using a 100 mm layer of 3:1 sand:bentonite, which was mixed and laid dry and then lightly watered to render it impervious before back-filling and re-surfacing. This technique was apparently successful and has obviated further complaints.

20.3 New waterproofing layer

In general terms:

1. A new waterproofing layer may be installed at a high level, collecting water as soon as it has penetrated the wearing course or track ballast;
2. A new waterproofing layer may be installed at a low level, intercepting the water when it has drained down to the level of structural masonry;
3. Occasionally the fill above the barrel of the arch is rendered waterproof;

4. Waterproof treatment to the visible external face is sometimes tried, to improve the appearance. This is not recommended, as any waterproof coating will keep the brick or stone saturated; it could even cause a build-up of water pressure leading to failure of the coating or spalling of the masonry. It must be admitted that it is very tempting to apply such treatment to try to improve the appearance and inconvenience of wet subways or retaining walls, but it is most unlikely to be successful.

With all the methods, a variety of waterproofing systems and materials is available. The method selected will depend on local circumstances and whether other work is being carried out, on the existing system and drainage, the time available, the size of the job, the time of year and the presence of services.

20.3.1 High level systems

The waterproof layer itself may consist of a self-supporting sheet or it may be applied to a solid substrate, usually a concrete deck slab. Generally, a sheet is cheaper and easier to install but will be

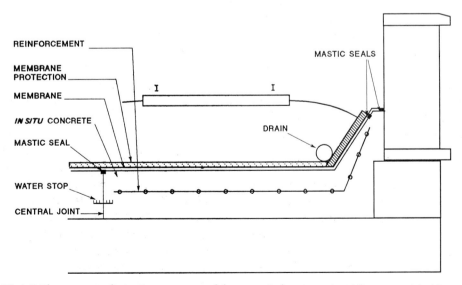

FIGURE 20.1 Bridge waterproofing using a concrete slab – a typical cross-section (diagrammatic). (*Courtesy of British Rail.*)

more vulnerable to mistreatment.

Strengthening a bridge or viaduct is frequently achieved by casting a continuous reinforced concrete slab over it. In addition to its structural advantages, such a slab can be used to carry a new waterproofing layer. The resulting waterproofing system is robust and resistant to damage during subsequent maintenance operations, especially if a well-bonded liquid system is used in conjunction with good quality concrete. However, allowance for services and provision of drains may be a problem.

Figure 20.1 is a composite diagram based on a number of viaducts on the British Rail West Coast Main Line that were treated in this way. Each job was carried out in two halves, one track at a time, under long weekend possessions; this caused no problems apart from necessitating a central joint along the entire length of each viaduct.

Detailing of joints and treatment of edges are particularly important with all bridge waterproofing. The larger flat areas are relatively straightforward, but the labour-intensive construction details made under difficult site conditions and, frequently, poor weather conditions are much more likely to fail. They should be as simple as possible and tolerant of weather and site conditions. It is important to allow for mastic beads at the centre where the two slabs meet and at the edge of the concrete, where shrinkage cracks would otherwise provide a leakage path. Similarly, construction joints across the slab and any expansion joints must be detailed with the greatest care to make them watertight or to allow controlled leakage into an adequate drainage system.

Particularly recommended are liquid-applied waterproofing membranes which give an intimate bond to concrete. They have the advantage that any minor defect will allow leakage through to only the very small volume of concrete in the immediate vicinity. If the concrete is well compacted this will not be important. With many sheet systems the bond is not adequate to prevent water spreading out from the site of any small leak, which thus renders the whole membrane useless. Also, it is usually easier to provide a neat and waterproof edging with a liquid system, e.g. by continuing the membrane a short way up the parapet. In this case, adequate protection of any exposed area against impact and sunlight is needed.

Various systems are available with Agrément Board approval. They may be based on bitumens, polyurethanes or acrylics.

1. Bituminous systems may be applied molten or as solutions or suspensions. They are also available as composite sheets which may be bonded down, frequently with hot bitumen. They are very resistant to water but readily dissolve in petrol or oils. They are not usually very robust and are liable to crack if applied to a substrate which is subject to movement or cracking. Many grades will embrittle in cold weather, and slump in hot weather if applied to verticals. Mastic asphalt is usually satisfactory but is slow to apply and expensive. With other hot applied systems the bond to the concrete is often inadequate, giving the same disadvantages as sheet systems.

2. Polyurethanes are tough rubbers applied as liquids by brush, squeegee or, preferably, spray. They are basically very tough, flexible and tear resistant, but may be heavily filled and extended with inert materials. This greatly reduces not only their cost but unfortunately also their tear strength, and may mean that they shrink and crack in sunlight. With suitable primers they adhere well to concrete and are good at bridging minor cracks. They are very resistant to water, oil and petrol when cured, but are sensitive to water during application. The minimum temperature for application is 5–10 °C, but when cured they have a wide range of service temperature. Systems are available from Sika Inertol Ltd and Permanite.

3. Acrylics are in many ways similar to polyurethanes. Although not as tough as pure polyurethanes they are as good as most commercially

available formulated polyurethanes. They are rather less sensitive to water during application and can be applied at temperatures down to 0 °C. They will bond to many polyvinyl chloride (PVC) water stops and some PVC sheet waterproofing systems (see below). An acrylic membrane is available from Stirling Lloyd.

Alternatively, if a concrete deck is undesirable or too expensive, a flexible sheet system may be used. A single sheet of flexible plastic is laid over the whole of the bridge a short way below the ballast or wearing course and simply brought up a short distance at the sides and continued over the ends. A number of suitable membranes, manufactured as roofing, is available.

To prevent accidental damage, as robust a protection as possible is required. A sand blinding or geotextile is adequate below. The protection above will depend on the likely hazard and may be a rigid board such as glass-reinforced concrete or paving slabs. Alternatively a stabilized sand layer may be adequate for railway ballast, and sand asphalt for road surfaces.

One of the largest applications of this method to date was on Welwyn Viaduct on the British Rail East Coast Main Line. The details are given separately as a case history (Section 20.5) but essentially the structure was in danger of rapid deterioration due to water penetraton of the piers. Owing to its position on a critical two-track main line the periods available for possession were very short. This combination of factors made the selection of a sheeting system inevitable.

The method has a number of advantages and disadvantages:

1. It is very quick to lay;
2. It involves the minimum disturbance to the structure;
3. It is very versatile and is largely independent of the structure; for instance, it has even been used on wooden decks;
4. It is usually the cheapest option;
5. However, it is not very robust; a single leak can make a large area of sheet valueless;
6. Services, if they exist, are a major problem. Putting them in the fill above the sheet means that the sheet must be laid at a depth which necessitates considerable excavation; putting them below the sheet means that the waterproofing is ruined the first time that access is required.

A number of materials is available: either rubbers, such as Hypalon, Butyl or EPDM (ethyline propyline) rubber or plastics such as plasticized PVC. They are similar (or identical) to the materials used for single-ply roofing. The most convenient is PVC which is also available with an attached glass scrim to ensure adequate adhesion to bitumen-based wearing courses. The sheet is formulated to have a life in excess of 30 years on a roof (a much more severe environment than a bridge). It is tough and tear resistant and is not attacked by oils or aqueous solutions, although some brands are not resistant to bitumen. It is easily fabricated on site by solvent or hot air welding and may also be solvent welded to unplasticized uPVC drainage. A range of accessories is available, e.g. pre-coated metal edging strip to which the sheet can be bonded.

An alternative material, used in a very similar way, is the modern analogue of clay puddle. This is dry bentonite sandwiched between a polypropylene textile protective webbing and a biodegradable top layer. It is available in roll form under the trade name Rawmat and is very useful where it is necessary to waterproof large areas rapidly. A good example was at Snow Hill, Birmingham, where a number of businesses occupied the arches under the disused station site. Water penetration was becoming a problem. The bentonite layer was unrolled directly onto the compacted fill. Membrane protection was provided by a 150 mm layer of sand, followed by further fill material as available.

The method has similar advantages and disadvantages to the use of a PVC sheet. In addition, there may be problems such as are found with canals and reservoirs when drying out causes cracking of the clay layer, although the manufacturers claim that this is not a problem.

20.3.2 Waterproofing at low level

Similar methods and materials have been used to waterproof the extrados of the arch, jack arches etc. This practice has the disadvantage that a great deal of fill material must be removed, with the possibility of damaging the spandrels. The method is generally used as a subsidiary to other major jobs such as the addition of a saddle to the arch or maintenance of the spandrels. Essentially there are two possibilities: direct waterproofing of the extrados or waterproofing of a saddle.

A liquid membrane has occasionally been applied directly to the arch, but this is dependent upon finding a reasonably smooth sound surface and on good surface preparation. This is an unlikely combination. Further, any structural movement is likely to cause failure of the membrane. Nevertheless, the method has been used with success on a number of occasions.

A loose laid sheet system is more likely to be successful and is not as vulnerable to accidental damage as the same sheet installed at a higher level – at least not after it has been successfully buried. It is very easy to install. The pre-welded membrane is simply unrolled both ways from the centre of the arch down at least to the level of the springing. The ends of the sheet should rise to form a trough to contain a drain.

Alternatively, if a concrete saddle is being added to an arch it makes sense to waterproof it at the same time. Any of the methods described above may be used, although particular care is required when a spray system is applied to new concrete. With special primers, such systems have been applied to concrete only a few hours old.

20.3.3 Waterproofing the fill

Waterproofing wet patches of arches by impregnation of the fill material with chemical grouts has been tried occasionally, apparently with success. The method is merely a palliative but may be the only possibility. Chemical grouts are produced for the control of groundwater during excavation of shafts and tunnels. Early materials were too hazardous and this restricted their use, but modern materials, which are two-part water-based gels, are non-toxic. They are mixed with water and pumped into the porous ground. After a few minutes the solution sets to an impervious jelly. Some polyurethanes, including materials which foam, are coming onto the market but experience of them is limited as yet.

20.4 Conclusions

It is possible to overcome many problems with brick and stone structures, to reduce maintenance costs and to extend their life by keeping rainwater and chemicals away from them. Rainfall may be intercepted at various heights in the structure using a variety of materials. The choice of method and material will depend on site details and external factors, but waterproofing is always worth considering, especially if it can be installed when other jobs are carried out. Whatever the system used, it must be conscientiously carried out and sufficient time must be allocated to allow it to be done properly; it may be necessary to provide protection so that the work can proceed in adverse weather conditions.

20.5 Case History: waterproofing Welwyn Viaduct

Welwyn Viaduct is a large two-track railway viaduct on British Rail's East Coast Main Line. It has 40 arches, each of about 9 m span, with a maximum height of some 27 m, and was built in 1850 from locally produced bricks which were soft, red and porous. By the 1930s there were signs of deterioration and the structure was extensively refaced with blue engineering brickwork, 230 mm thick.

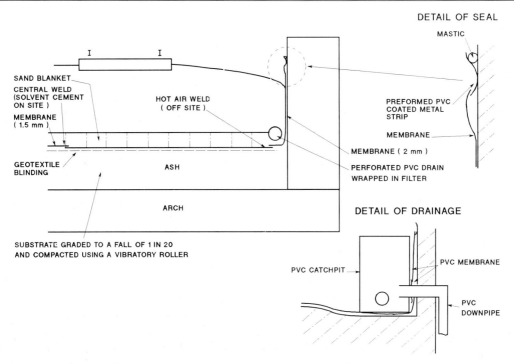

FIGURE 20.2 Welwyn Viaduct: waterproofing using sheet, showing a typical cross-section (diagrammatic). (*Courtesy of British Rail.*)

The original waterproofing and drainage were typical of this type of structure. The waterproofing consisted of a bituminous membrane laid over the extrados and backing brickwork to direct the water into porous collector pits above each pier. These were drained by cast iron pipes built into the centre of the pier and discharging at ground level. Over the years the drainage system had become choked. As it was buried deep inside the structure it was completely inaccessible and impossible to clear. Furthermore, the bitumen membrane had become ineffective, so that water was percolating through the arches and down through the piers.

A particularly serious problem was the effect of water penetration between the original structure and the blue brick cladding. Presumably as a result of frost action, many of the piers had developed vertical cracks of considerable length 200–300 mm from the corner, i.e. roughly the thickness of the cladding. These had been stitched over a number of years, but obviously both piers and arches would have continued to deteriorate without treatment of the fundamental problem of water ingress. It was decided to renew the waterproofing and drainage systems and to pressure grout the cracks between the new and old brickwork.

The waterproofing operation had to be undertaken in conjunction with essential track renewals, a series of 18 h possessions being allowed which included only 6 h for the installation of the waterproofing and drainage. This meant that the only practicable method of waterproofing was to use a loose laid sheet under a shallow depth of ballast.

The method used was to install a prefabricated PVC membrane some 300–400 mm below the level of the sleepers, directing rainfall into new drains laid against the inside faces of the parapet walls; these drains, in turn, would discharge into new external downpipes on the outside faces of

FIGURE 20.3 Welwyn Viaduct waterproofing: laying waterproofing membrane. (*Courtesy of Beton Construction Materials Ltd.*)

the viaduct (Fig. 20.2). Although the viaduct is a grade II listed structure, there was no difficulty in obtaining planning permission for the new external downpipes; the only requirement was that they should be black, so as not to appear obtrusive.

The length of the viaduct (including allowances for extensions off the ends and overlap of sections) is 480 m and the width 8 m. It was decided to deal first with the southbound line in four stages and then follow a similar procedure on the northbound line, so completing the work in eight weekends. Thus in each weekend an area of approximately 120 m × 4 m would be waterproofed, plus the upstand at the edge, and watertight joints would be made with the previous stage(s).

The track and then the old ballast, ash and

other fill were loaded onto wagons standing on the adjoining line; the resulting surface was prepared and compacted to the correct levels and falls. A geotextile (Terrafix 1014R) was laid next. This gives some protection to the waterproof membrane from below; it also provides an easier surface on which to work and is particularly useful in keeping loose dirt out of joints during solvent welding. The geotextile was supplied in rolls 2.2 m × 50 m, which were simply unrolled with overlaps.

The waterproofing membrane itself was then installed. The material used was a plasticized PVC, Wolfin 1B, 1.5 mm thick on the horizontals and 2 mm thick on the verticals, where it was left unprotected. The membrane is supplied in rolls 1.5 m wide by 15 m for the thinner sheet and 1.0 m wide by 10 m for the thicker sheet, but

FIGURE 20.4 Welwyn Viaduct waterproofing: laying membrane protection. (*Courtesy of Beton Construction Materials Ltd.*)

to reduce the number of site joints it had been pre-jointed off site by hot air welding. The large rolls were unrolled on site and solvent welded to adjoining sections. The edge along the parapet was finished by welding the sheet to a PVC-coated aluminium strip which had been fixed to the wall previously (Fig. 20.3).

One of the basic requirements was to divert the drainage to external downpipes. These were fitted in advance and the necessary holes were drilled in the spandrel walls. The end section of the downpipe was fitted through the hole, projecting inwards by some 200 mm, so that it would eventually fit into a catch pit. These pipes had of course to penetrate the membrane, and so when the latter was laid it had to be sealed around each outlet. This was a critical point, but proved to be very simple to effect. A small patch of the

membrane material was cut with an undersized hole for the pipe; this was pushed over the pipe stub and solvent welded to both the pipe and the main membrane. The PVC catch pits were fitted over the downpipe stubs and drains were installed parallel to the spandrel walls. Membrane protection was provided by Lotrack Geomesh (a synthetic netting) covered by Geoweb, which is a plastic honeycomb, 100 mm deep, that was subsequently filled with sand. This provided a base on which plant could run to spread the track ballast (Fig. 20.4).

The viaduct now shows no signs of water penetration. The new drainage system discharges water after rain and seems to be working very satifactorily. The waterproofing, undertaken in 1986 by direct labour, cost a little over £40 per square metre.

PART FIVE

Special types of structure

Tunnels

A.M. Sowden

21.1 Introduction

The purist would probably define a tunnel as a subterranean passage formed without removing the overlying rock or soil. With scant regard for semantics, this chapter will also deal with the type of underground passage which has been constructed by 'cut and cover' methods. It will not deal with those which are so constructed as to be, in effect, long bridges, but will adopt a criterion based on the form of loading and the interaction of the load with the structure. In a tunnel, because of arching effects, the distribution of load onto the lining is a function of the relative flexibility of the lining and the surrounding ground, the timing of the installation of the lining and the nature of the contact at the interface. The loading on a bridge, in contrast, is not dependent upon such interaction. Tunnels carrying drainage and sewage are dealt with more particularly in Chapter 22. Since this book concerns itself with brick and stone masonry, the tunnels on which it concentrates here are those built in the eighteenth and nineteenth centuries to accommodate canals and railways.

Among these there was little consistency in cross-sectional profile, but they were generally of a basic horseshoe shape. The thickness of the lining depended upon the nature of the ground through which the tunnel was built, but it was determined on an empirical basis without the benefit of any scientific knowledge of either soil or rock mechanics. In hard strata, the lining was sometimes limited to the upper part, with arched vaulting sprung directly from the rock side-walls and often intended merely as a canopy to provide protection from falling pieces of rock rather than as a structural support. In softer ground, a completely lined opening was usual, so utilizing to the full the interactive soil/lining effects which help to minimize the bending moments on the lining. The omission of a structural invert or its inadequacy to resist heave or swelling of the tunnel floor has been the cause of many of the troubles subsequently experienced (see Section 21.6.8).

Figures 21.1 and 21.2 show the way in which most tunnels in rock in Britain were excavated and timbered, but Fig. 21.3 is over-optimistic in its portrayal of the accuracy with which the excavation matched the profile required for the extrados of the lining. In practice, rock could not be removed with such precision and a considerable amount of 'overbreak' occurred beyond the required profile. The additional removal of material over the top of the tunnel was intended to improve access to enable the brickwork to be built at the crown and to allow space for the timbering which supported the ground above

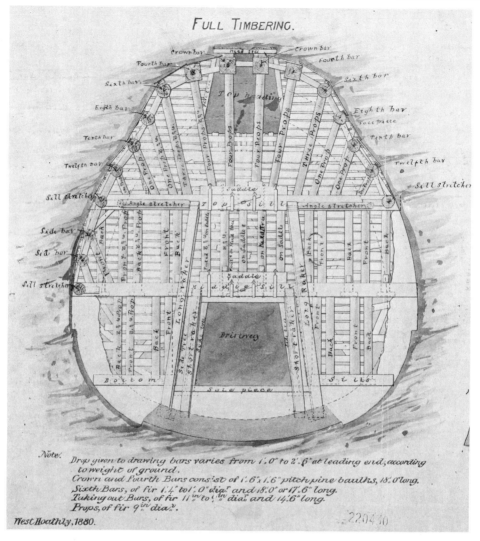

FIGURE 21.1 Typical timbering for the excavation of a tunnel (cross-section). (*Courtesy of British Rail.*)

while this was being done. The drawing (Fig. 21.3) bears a note to the effect that, as the timbering was withdrawn, the resulting spaces were to be 'packed with bricks or other dry material'. In some cases, sleeper walls of masonry were built to support the rock off the back of the tunnel vault (Fig. 21.4d). According to Simms' classic book *Practical Tunnelling* (1844), 'the work was built solid against the earth whenever there was more space than was necessary for the insertion of the intended thickness of the brickwork' or, alternatively, 'vacuities were rammed solid as the work advanced. Whichever of these plans may be adopted, it is of great importance that it should be carefully executed; it should, therefore, in all cases be well attended to .... It should be an invariable rule never to leave a vacuity behind the work.'

Unfortunately, execution often differed markedly from intent; in many instances, tempor-

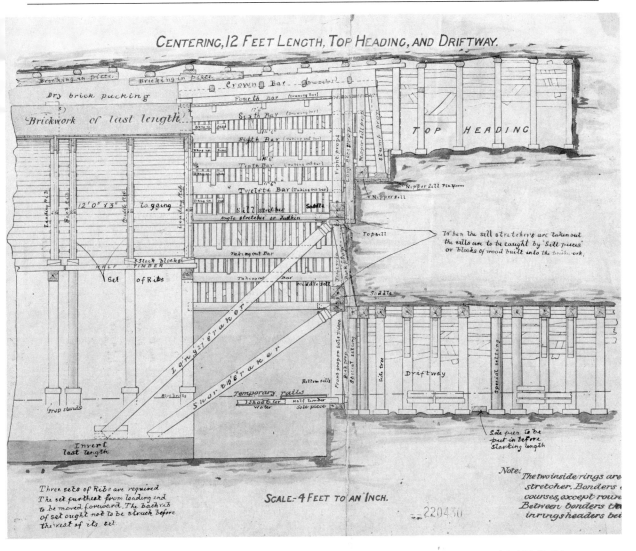

FIGURE 21.2 Typical timbering for the excavation of a tunnel (longitudinal section). (*Courtesy of British Rail.*)

ary timbering was left in to rot away and voids were not even filled, much less packed tight. In other instances, insufficient rock was excavated to accommodate the intended thickness of the lining, which was reduced accordingly and then not even 'built solid' against the surrounding ground (Fig. 21.4c). The adverse effects of such practices are apparent; inward distortion of the lining under load causes outward deflection of adjacent sections and this is resisted by the surrounding

ground only if there is good contact with it. The development of some passive resistance provides the reaction to limit arch deformation and hence inward movement.

In tunnels of any significant length, it was normal to begin by constructing vertical shafts from the ground surface down to the line of the intended tunnel excavation, in order to increase the number of working faces and facilitate the removal of excavated material, so reducing con-

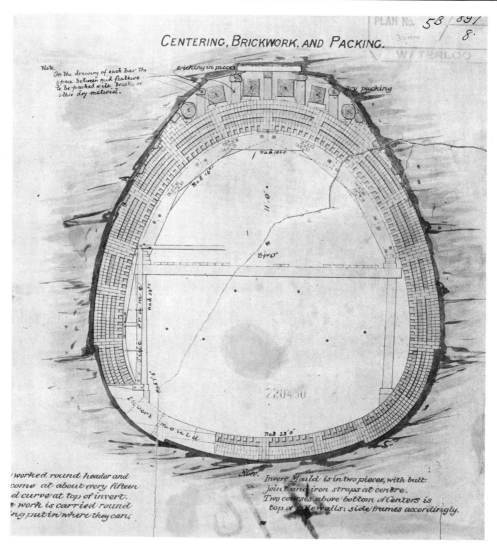

FIGURE 21.3 Typical construction of a brick-lined railway tunnel. (*Courtesy of British Rail.*)

struction time. Often – but not always – the shafts were retained to provide permanent ventilation to the tunnel and they remain a constant liability in terms of condition and stability. In those cases where they were filled in on completion of the construction or where they were sealed at their top and bottom in such a way as to be invisible either from within the tunnel or on the ground above it, their very existence can be

forgotten, leaving an even more potentially hazardous situation.

In 1953, the vault of a tunnel in Manchester caved in under the weight of such a filled and sealed construction shaft, causing the collapse of two houses built over it; five lives were lost. The subsequent inquiry report emphasized the need for maintenance staff to know of the existence of old shafts and recommended that their positions 'should be permanently marked in the tunnels themselves, so that these places can be particularly watched'. That remains good advice, which should be adopted whenever infilling is now undertaken. The initial identification of shafts infilled in the past presents a problem, but it may be overcome by evidence from construction records, tell-tale signs at ground level (often best observed by aerial photography) or radar surveys (see Section 6.4.1(a)).

Furthermore, every construction shaft, whether left open or not, effectively provides a collector drain from any water-bearing strata through which it passes and carries the discharge down to the level of the tunnel. Thus shafts are invariably a source of considerable water ingress into the tunnel, such as may be beyond the capacity of any garland drain provided at the base of the shaft to intercept. If voids are left between the lining of the shaft and the surrounding ground or are caused by the flow of water, a situation can arise where the shaft lining loses much of its peripheral shear support from its interaction with the ground and increasingly the entire weight of the shaft lining is thrust onto the crown of the tunnel, to be carried by the vault at the shaft–tunnel intersection.

21.2 Special aspects of tunnel maintenance

Deterioration of masonry is of a basically similar nature wherever it occurs and typical countermeasures are described elsewhere in this book. However, there are certain features of its incidence and treatment which are particularly significant in the case of tunnels.

1. Tunnel linings had to be built in difficult, cramped, obstructed and usually wet conditions, in poor light, with the result that the quality of the finished work almost inevitably suffered. Even if the work was well supervised, it cannot reasonably be expected that all jointing and bonding would be to a high standard and that the cross-sectional profile would always be geometrically accurate or even that it would be constant throughout the length of a tunnel. If supervision was inadequate, this was likely to be reflected in further shortcomings – bricks laid dry on the extrados or omitted altogether, poor collar joints, timbers left in and voids inadequately filled (see Fig. 21.4) or centring removed before the mortar strength was sufficient to resist distortion of the lining. Furthermore, there can be no assurance that the masonry thickness or the standard of workmanship will be uniform throughout the length of the tunnel;

2. Apart from the portals, only the intrados of the lining (excluding the invert) is reasonably accessible for inspection;

3. Patch repairs, previously undertaken, may obscure the progress of on-going failures, and there is commonly a lack of records of construction or of movement monitoring in the past;

4. Without considerable investigation, the characteristics of the ground immediately surrounding the tunnel, the nature of the contact between it and the lining and the thickness and quality of the lining are all unknown. Even with that information, the strength and stability of the lining cannot be calculated with accuracy;

5. Water, with all the problems associated with it, is almost invariably present; (Fig. 21.5)

6. Inspection and repair are rendered more difficult by the lack of natural light, by problems of access, logistics and possible air pollution;

7. The temporary closure of operational tunnels or the diversion of the traffic they carry is difficult, if not commercially impossible, and so maintenance work may have to be undertaken during short occupations;

8. Most repairs can be effected only from within the tunnel but cannot be undertaken in such a way that they cause a significant permanent reduction in its internal dimensions. There is

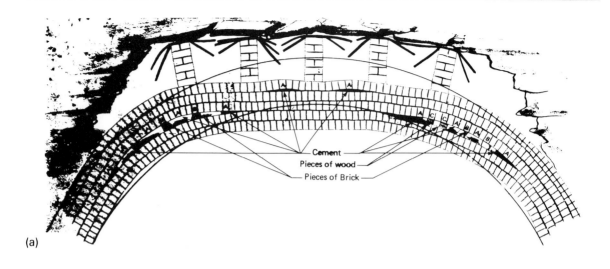

Cement
Pieces of wood
Pieces of Brick

(a)

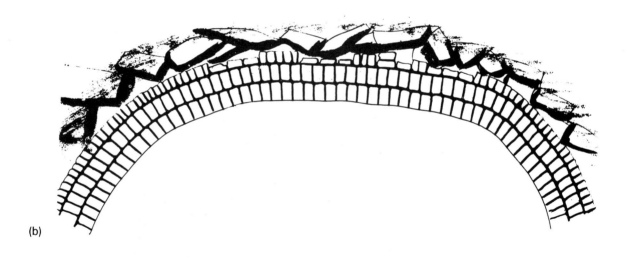

(b)

FIGURE 21.4 Deficiencies in the construction of railway tunnels as revealed by opening out. (*Courtesy of British Rail.*)

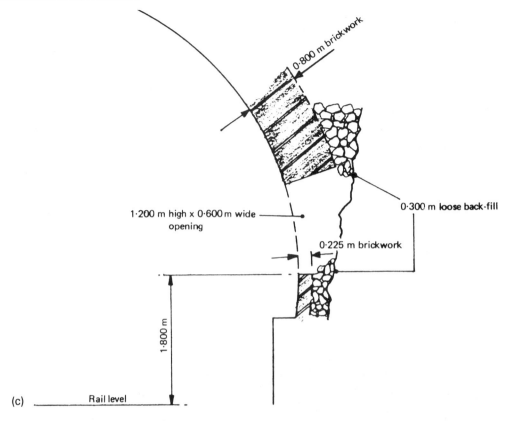

0·800 m brickwork

1·200 m high x 0·600 m wide opening

0·300 m loose back-fill

0·225 m brickwork

1·800 m

Rail level

(c)

(d)

299

FIGURE 21.5 Water percolation into the Severn Tunnel before cementation (injection of neat cement grout) repairs to the brick lining in 1929–30. (*Courtesy of British Rail.*)

only limited scope for such reduction even for temporary works, so that working space is severely restricted.

On the credit side, it should be noted that parts of tunnels away from the portals and open air shafts are often less prone to freezing conditions and work on them is less affected by weather delays than operations in the open. As a result, it may be possible to undertake repairs to those parts when similar work on more exposed structures would be impossible. By programming the labour force to repair work well inside tunnels during the coldest months of the year and taking appropriate precautions with materials, the amount of non-effective time incurred can be minimized.

21.3 The nature of tunnel defects

The tunnels under consideration have been in existence for upwards of 70 years. It may therefore be assumed that gross defects in design and construction will have already manifested themselves and, generally, been dealt with. Current problems therefore derive mainly from the gradual degradation of the lining masonry or from some progressive change in the loading applied to it from the surrounding ground. They rarely occur without forewarning in the form of visible signs of distress.

Defects in the lining arise from

1. degradation and loss of mortar in joints

2. degradation of the lining through the aggressive action of water and the chemicals carried by it
3. freeze–thaw cycles, the effects of which are most marked towards the ends of tunnels
4. ill-advised works of alteration (e.g. lowering of the track formation to increase headroom or the construction of additional refuges) which may remove critical support from the lining or change stress paths within it
5. abrasion and impact damage from traffic in canal tunnels, and track loadings that break up drains and/or aggravate the softening process of the formation in railway tunnels

Except in the case of the shallowest cut-and-cover tunnels, the amount of overburden is such as to distribute the live loading from traffic passing

FIGURE 21.6 Icicle formation in a tunnel. (*Courtesy of British Rail.*)

above them to the extent that it usually has negligible effect on the lining, as the stresses arch around the tunnel.

In the surrounding ground, changes may increase the pressure exerted on the lining or reduce the support afforded to it. Water may cause swelling and/or softening of clays and marls, disintegration of gypsum, dissolution of limestones and loss of shear strength in rock joints and bedding planes; it may carry away fine particles, causing voids or removing support. Where the orginal construction has left cavities or where they subsequently develop, the surrounding rock may undergo disintegration, ravelling out and progressive caving into the voids; this sort of action can cause pressures to build up that are beyond the capacity of the lining to resist, especially if it is not in full perimeter contact with the surrounding ground. Softening of the strata on which side-walls are founded will allow them to settle. If the invert is absent or inadequate and the ground under the tunnel becomes saturated and weakened, there may be inadequate resistance to possible inward movement of the side-walls. Ground may 'flow' around under the side-walls and up into the base of the tunnel. Swelling of certain strata in the presence of water can also lead to heaving of the floor.

Tunnels are likely to be damaged by major earth movements affecting the strata through which they pass. Adjacent activities, such as mining and quarrying, may have adverse effects.

Problems associated with tunnel portals are similar to those involving the face rings, spandrel and wing walls of bridges, but they are more likely to be due to water and inadequate drainage than to the effects of traffic.

It is apparent that water is the principal source of deterioration of tunnel structures. In addition, water entering a tunnel, unless collected and led away in a controlled manner, may damage service equipment and installations. If it freezes, ice formations can short-circuit electrical conductors and even give rise to physical hazards to passing traffic (Fig. 21.6); when the thaw comes, ice

FIGURE 21.7 Tunnel inspection from purpose-built gantry. (*Courtesy of British Rail.*)

falling from the lining and from shafts may cause injury to personnel or damage to passing traffic and fixed equipment or it may pile up to such a depth on running lines as to give rise to a risk of derailment. Ice formation in shafts is exacerbated by cold air draughts, drawn down by the passage of traffic, and its removal constitutes a special problem.

21.4 The identification of defects

Tunnel maintenance must take into account both the lining and the surrounding ground, as well as the interaction between them. Initially it must do so subject to the constraint of having direct access only to the intrados and the portals.

As with all masonry structures, the first line of defence is a disciplined system of regular inspection to detect open joints, spalling, cracks, hollow patches, bulging and distortion, serious water percolation or ineffective drainage (see Chapter 5). Bearing in mind that lack of natural light imposes limitations on the effectiveness of any form of superficial examination, principal or detailed inspections should be carried out at rather more frequent intervals than in the case of other structures – say biennially at least, and annually if there are any doubts about condition. Such inspections should comprise a close visual examination supplemented by sounding (tapping the intrados with hammers) to detect loose bricks, detached rendering or ring separation at the first, and possibly the second, collar joint.

In order to achieve a sufficiently close examination, there is need for some form of mobile access platform, even if this is no more than a scaffold tower mounted on a trolley or barge. To ensure comprehensive coverage in an occupation of limited duration, there must be sufficient examiners each with the task of covering a part of the tunnel's periphery; there is thus a prima facie case for a purpose-built mobile gantry (Fig. 21.7) but the economic justification for such a piece of equipment will depend upon there being an adequate workload to afford it reasonable utilization. An inspection platform on an articulated and telescopic arm, mounted on a rail wagon or canal barge, is an alternative possibility.

If potentially important details are not to be overlooked during the visual examination, it is essential to have good general lighting supplemented by a spot light facility by which a high level of illumination can be directed onto areas of special concern. It is preferable to vary the viewpoint during examination, possibly by occasionally moving through the tunnel in the opposite direction, as some defects are more easily identified when seen from a particular angle.

As with any structural examination, observed defects must be accurately recorded in such a manner and in such detail as to allow proper comparison with previous and subsequent observations. In the usually damp and dirty environment inside a tunnel, an accurate and legible manuscript record is difficult to achieve; a commentary dictated into a tape recorder is generally more expedient. There is scope for photographic and, increasingly, video recording. Suspect areas of masonry, considered worthy of further investigation, should be marked for easy identification later.

The examination of tunnel shafts inevitably involves the use of special access equipment in the form of some kind of personnel hoist and demands a particular concern for safety, in compliance with the Construction (Lifting Operations) Regulations, S.I. 1581. It may be necessary to regard the shaft as a confined space (see Section 5.9.4) and to consider the effects of pressure surges generated by traffic in the tunnel below.

21.5 Appraisal and diagnosis

Routine examinations of the type described in Section 21.4 serve to identify potentially critical features and concentrate attention on them; these then need to be monitored and investigated in greater depth, in order to appraise the significance of the observed defects and diagnose the cause.

It is usually worthwhile to undertake a preliminary 'desk' study, in an attempt to make good deficiencies in existing records, by research into the history of the original construction and of any subsequent problems or repairs, by consulting geological records and maps, by comparing old Ordnance Survey maps with recent aerial survey photographs and by reviewing mining records etc. (see Section 1.7).

The next requirement is to confirm or establish the physical details of the geometry and integrity of the tunnel lining, the characteristics of the surrounding strata and the nature of the interaction between them. Various non-destructive methods of investigation are available, the more effective of them being described in Chapter 6; all need to be undertaken by specialists and most will record anomalies in the structure and/or strata, but meaningful interpretation of the results is often unreliable at present and invariably depends upon specialist expertise. Radar seems to hold the greatest promise for successful development, particularly with respect to the identification and delineation of voids and the location of 'blind' shafts.

More detail and greater reliability is achievable

through direct observation as a result of drilling holes, taking core samples, cutting 'windows' or using endoscopes, but such techniques yield only localized information, with the result that the building up of a comprehensive picture may become a time-consuming and costly operation, which must be reserved for specially suspect areas.

Surface deterioration is readily observed during routine examinations, but indications of more serious defects such as distortion and bulging are less easily spotted in their early stages. Irregularities of shape, even if they are observed, may be of long standing, possibly from the date of construction (see Section 21.2) and stable. In such circumstances they are of little consequence. However, if they are 'live', they may well be the first indications of a potentially serious situation. The vital need is therefore to detect and monitor progressive development of distortion; this must rely on recording variations of tunnel shape, by accurate profiling at intervals of time. Conventional surveying methods of precise levelling and/or convergence measurement are inadequately comprehensive and unduly laborious for this purpose. Recently developed techniques of profile measurement and recording have recourse to semi-automated systems, using photography or lasers, coupled to computerized data processing equipment (see Section 6.4.2(e)). Provided that sufficient care is taken to ensure repeatability of readings, by correlating them to fixed markers in the face of the masonry, changes of shape at successive intervals of time can readily be detected. The computer equipment facilitates the analysis of the resulting mass of data and identifies changes exceeding specified limits; it thus provides a mathematical sieve which focuses attention immediately onto potential problem areas.

Such attention may call for more precise and locally concentrated monitoring of movement, the assessment of stresses (using the types of strain gauge described in Chapter 6), establishing water levels and the permeability of surrounding strata, analysing percolating water and tracing its source, and the sampling and testing of the lining masonry and of the ground surrounding it. The objective must be to build up as comprehensive an understanding as possible of the structural behaviour and incipient failure mechanisms, to allow effective remedial measures to be devised.

21.6 Types of repair

21.6.1 Control of percolating water

Few tunnels escape problems due directly or indirectly to the ingress of water. The digging of the tunnel will effectively have formed a drain towards which any water in the ground above and around it will naturally flow. It is possible, by grouting up all flow paths in, and/or immediately behind, the lining, to prevent percolation, at least in the short term and at low pressures, but usually the water excluded from one section of a tunnel will be driven to find an outlet through an adjacent (possibly previously dry) part beyond the grouted area. If the vault alone is sealed, water may find its way under the side-walls and soften up the tunnel floor. Rarely can the grouting of the entire periphery and length of a tunnel be contemplated and if it were undertaken there would be a risk of the build-up of excessive hydrostatic pressure.

Generally, therefore, the aim should be to control percolating water rather than to attempt to exclude it from the inside of the tunnel. The use of sheeting (normally corrugated) to line parts of the intrados, in order to deflect flows away from vital equipment or to insulate water from freezing draughts, is a dubious practice, despite the fact that it has been widely adopted. The fixings and metal sheeting are prone to rapid deterioration, to the extent that they are unable to resist the aerodynamic forces generated by traffic for long; the sheeting obscures those parts of the masonry which it covers and so inhibits proper examination; thermal linings, depending upon the material used, might constitute a fire hazard, either from flammability or by emitting toxic fumes. (However, closed cell polyethylene plastic panels have been used effectively and without ill effects in unlined rock tunnels in Scandinavia and Canada.) Trials of surface-applied coatings (10 mm thick) using microsilica–cement sprayed mortar and polymer-modified cement rendering have proved effective in preventing (at least initially) the ingress of water through the areas so treated, but they do not, of course, obviate saturation of the masonry and they may be susceptible to frost action in the longer term.

The recommended objective should be to reduce the spread of points of inflow of water by grouting behind the lining and then re-drilling weep holes at points where the inflow can conveniently be collected and led down to drains at floor level. The drains must be of adequate size and fall and kept clear of obstructions, to ensure that the water is carried away to a suitable discharge point. In old tunnels existing drainage arrangements are often neglected and possibly crushed; it is highly likely that they will need overhaul, if not renewal.

There remains a risk that the concentration of flow through a limited number of weep holes may encourage the washing out of fine material, giving rise to cavities, but this may be overcome by the judicious use of filter material. To this end, an extractable drainage unit has been developed, based on the premise that a hole of adequate diameter (40 mm) can be formed economically using conventional hammer drills. A nominal 18 mm gas barrel, about 200 mm long, is mortared into this hole. A polyethylene tube (of 12 mm bore) is inserted into the gas barrel, extending as far as required into the hole; at its back end, a length of porous polyethylene, close fitted within the tube, provides a drainage medium which allows water to pass without loss of soil particles. Alternatively, a slotted poly vinyl chloride pipe, covered with a filter sleeve, can be sealed into a somewhat larger cased hole to serve the same purpose. In either case, it is possible to extract the filter components for cleaning or replacement. By the use of suitable pipe fittings, a

further length of plastic tube, insulated if necessary, can connect the drainage unit to an outfall drain.

Individual air shafts may be effectively treated against water percolation by grouting voids, fissures and permeable ground around them, as described in Section 15.3.2. Injection holes may be drilled through the shaft lining (as in the case described in Section 21.9) or, if the condition of the lining might make this a difficult or hazardous operation, drilled down around the outside of the shaft from ground level or up from within the tunnel. Alternatively, the shaft may be encircled with a grout curtain, installed through a ring of boreholes drilled from ground level.

21.6.2 Renewal of deteriorated mortar

Loss of mortar from joints can destroy the integrity of the masonry and allow individual units to drop out; special attention must therefore be paid to its restoration. Manual methods of raking out perished mortar and of repointing are slow and labour intensive and they may well fail to achieve the depth of rehabilitation required. Preparation by high pressure jetting followed by mechanical (pressure) pointing (see Chapter 14) is more effective and, being considerably faster, more economical; its somewhat unattractive finish gives little cause for concern in a tunnel location.

For deeper joint restoration, it is necessary to have recourse to internal grouting (see Section 15.3.1), but, as described in Section 10.4.2, satisfactory penetration and reliable dispersion of the grout is not easily achieved. To enhance the likelihood of success, it is desirable to seal both the rear and front faces of the lining to contain grout flow; this implies that, initially, voids in the annulus around the extrados should be grouted and the intrados repointed.

21.6.3 Like-for-like renewal of inner ring(s)

In common with any other masonry structure, if

FIGURE 21.8 Tunnel strengthening using steel sections specially curved to match the distorted profile. (*Courtesy of British Rail.*)

surface spalling is diagnosed as indicative of distortion or overstress under load, it should provoke early investigation into what is amiss and remedial measures should be taken accordingly. More commonly, in tunnels it is due to atmospheric or chemical action, in which case it becomes a cause for concern only when it approaches a condition in which it may hazard the integrity of an arch ring or a face wythe – when the loss of material from the surface extends to, say, about half the thickness of the face units.

Re-facing on a like-for-like basis has been, and to an extent still is, a widely used repair technique. Isolated patches may be replaced using simple centring but they need to be properly bonded into the existing masonry. This is a slow and labour-intensive technique and, for the reasons explained in Section 10.4.4, there is only limited scope for its effective application.

21.6.4 Underlining

Where clearances permit, deficiencies in the lining of a tunnel may be made good and further deterioration prevented by the provision of a semi-structural lining to the intrados, using the methods described in Section 10.4.5 (Fig. 21.8).

Railways on the Continent of Europe have made extensive use of the technique of reinforcing tunnel linings with sprayed concrete (see, for example, Alias, 1983; Eraud, 1980, 1984; Spang, 1971, 1973, 1976). Past reluctance to emulate them in the United Kingdom has probably been due to apprehension over potential problems of adhesion, rebound and percolating water. It is clear that successful application depends critically upon the proper cleaning of the surface of the existing masonry, the incorporation of drainage channels at its interface with the new concrete and the provision and adequate anchoring of suitable reinforcement (see Chapter 16). It is equally necessary to make provision for the material which rebounds during the spraying operation, so as to prevent it from fouling track ballast or drains.

Although thinner sprayed coatings are possible (see Section 16.5.3), a thickness of at least 100 mm, with mesh reinforcement, is normally preferred for tunnels. In order to maintain adequate clearances it may be necessary to cut the intrados back but, in the case of brickwork, probably by no more than the defective inner ring. Rather than being used for localized patch repairs, a sprayed coating should extend over the full profile of the vault.

In circumstances where the masonry arch is sprung from exposed rock side-walls and the latter show signs of loosening or weathering, partial re-lining in masonry or concrete (conventionally placed behind shutters or sprayed on) would suffice as protective facework to the rock and underpinning to the existing masonry.

21.6.5 Grouting behind the lining

The injection of grout to fill voids at the interface between the lining and the surrounding ground (annular grouting) is intended to prevent the loosening of the rock or to allow the development of composite action between the rock and the masonry; it may have the added advantages of rehabilitating the joints in the outer rings of the masonry, of strengthening the rock by filling fissures in it and of providing a degree of water-proofing to the lining.

The practice is to drill holes through the lining at 1.5–2 m centres, to a staggered pattern; grout is then injected through the lowest holes and allowed to set before proceeding to the next line of holes above, and so on, working progressively up to the crown. Allowing each lift to take at least its initial set guards against excessive local loading on the lining, while working symmetrically upwards on each side of the tunnel avoids unbalanced loading. The grout for the crown section should be stiffer than at lower levels, in order to keep its face at as steep an angle as possible and so reduce the risk of trapping air pockets against the rough surface of the rock. As an additional precaution against inadequate filling and in order to contain the spread of grout to a specific length of the tunnel (such as can be completed within the available occupation time), stop ends can be formed in the annular void at appropriate longitudinal spacing using a specially designed (foamed) grout. In any case, re-drilling or coring is necessary to check the efficiency of the grouting operation, and where significant voids remain a second stage of grouting, with a more fluid mix, should be undertaken.

The grout used comprises cement, sand and pulverized fuel ash (PFA) in various proportions and with a range of water:cement ratios depending upon the properties sought. Proportions of sand to cement may vary from 1:1 to 5:1 depending on strength requirements; pumping difficulties at the higher sand contents can be resolved by the use of PFA and/or bentonite.

Where the only requirement is the filling of sizeable voids with a low strength material, this may be achieved by the use of a cheap foamed grout; alternatively a lightweight aggregate may be blown in by compressed air and subsequently a normal grout be injected.

Effective grouting requires expertise in the specification of the mix and injection procedure, skill in implementation and care in monitoring the consistency with which the stipulated results are achieved in practice. Furthermore, it must always be recognized that the water used in the drilling operations may weaken both the masonry and the surrounding ground, while the injection pressures apply additional loading of a relatively 'shock' nature to the as yet unstrengthened lining.

21.6.6 Stitching and rock bolting

Stitching, as a technique to restore coherency across fractures, ring separation and the like (see Chapter 18), occasionally finds application in tunnels.

Rock bolts, taken through the lining of a tunnel and anchored into the surrounding ground, serve (in association with annular grouting) both to reinforce the rock mass and to associate the lining (including any underlining) with a ring of rock so that they act together as a composite structure in resisting loads. Rock bolts can be used to provide temporary support in place of centring, as well as permanent improvement in the stability of a tunnel. They may be tensioned (active) or untensioned (passive) reinforcement.

Fully tensioned reinforcement, installed radially before significant loosening of the ground has occurred, will produce a zone of strengthened rock, but this can only be effective in competent rock with clean discontinuities, where stresses induced normal to the applied pressure increase the shear resistance of the discontinuities. If the

rock quality is poor, and particularly if clay-coated joints are present, this effect may be difficult to achieve. Nevertheless, tensioned reinforcement offers resistance to ravelling, by limiting the rotation of blocks and by dowel action which develops an additional interactive support pressure as the lining distorts.

Untensioned fully grouted bolts rapidly build up resistance to movement, but movement must occur before the strength of the bolts can be mobilized. There will be a tendency for high local stresses to build up in the bolts adjacent to discontinuities. In contrast, with ungrouted bolts the strain is distributed throughout the bar, giving a less stiff support which can sustain greater movement before failure. In practice, the benefits of both systems can be obtained by tensioning followed by grouting, but with the free length of the bolt (i.e. between end anchorages) debonded; the debonding material also provides a degree of protection against corrosion.

Annular grouting is a necessary preliminary to a system of tensioned rock bolting, to ensure good contact between lining and rock and to avoid distorting the masonry. Bolts are normally spaced at 1–2 m centres and must be anchored in competent rock, well beyond any potentially unstable blocks. Large bar diameters (20–32 mm) are essential for dowel action, with the advantage of a long corrosion life. The drilled holes should have diameters 5–10 mm larger than the nominal bolt size. Resin anchorages are preferred to those of cement mortar; they require a shorter bond length to mobilize a given load, they can be installed without difficulty even in upwardly inclined holes and they can be stressed shortly after installation. The usual anchoring technique involves the use of a capsule which is pushed to the end of the drilled hole; the bolt is inserted and rotated through the capsule to ensure mixing of the constituents of the resin. At the other (exposed) end of the bolt, a face plate at least 150 mm square is needed to distribute the load on the lining and it should be as nearly as possible normal to the bolt axis. Since bolt orientation must have regard to the direction

FIGURE 21.9 Restricted working space, even with rail tracks singled. (*Courtesy of British Rail.*)

of the stratification and alignment of discontinuities in the rock, it may be necessary to provide a concrete pad between the face plate and the tunnel intrados. Cement mortar is beneficial for grouting the free length, since it can be injected under low pressure for a sufficient length of time to ensure efficient grouting of any remaining fissures and cavities as well as the annular space around the bolt. Tensioning should be undertaken as soon as possible after installation and before secondary grouting and should be to a load compatible with the crushing strength of the lining.

Untensioned rock bolts should be fully grouted using cement mortar. Face plates and corrosion protection are necessary.

21.6.7 More radical remedies to defective vaults

The deterioration of the lining or the build-up of pressures on it may have progressed so far or so rapidly that there is no feasible alternative to partial or complete reconstruction. Rather than reconstruct on a like-for-like substitution basis, current practice favours the use of prefabricated sections to reinforce or replace the existing masonry. Such segments may be of precast reinforced concrete or spheroidal graphite cast iron. The graphitic element in the latter confers greater tensile strength and ductility, allowing the adoption of thinner and lighter units. In the case of a normal twin-track railway tunnel (of some 4 m radius), it is normal to be able to achieve sufficient strength with cast iron units of no more than 200 mm overall thickness, and possibly of as little as 150 mm; by cutting away only two rings of brickwork, some support to the surrounding ground can be retained during the installation of units which will not encroach on existing clearances.

The critical aspects of the reconstruction operation concern the method of ensuring stability during the cutting away of the old lining and the method of erecting the new units. The work usually has to be undertaken in short sections, using some form of protective shield supported by heavy centring; a mechanical erector is needed to place the units and hold them in position until a ring is self-supporting. Even with traffic restricted to a single line, working space is very limited and there are considerable problems of logistics and safety (Fig. 21.9). The organization of such an operation becomes complex and details are beyond the scope of this book.

The ultimate solutions, in the case of a tunnel which is beyond economical repair, lie in opening it out to form a cutting, if that is feasible, or in diverting the service it carries, either through a new bore or around the topographical feature which originally gave rise to the construction of the tunnel.

The options discussed in this last section are particularly difficult, costly and disruptive to commerical operations. It is apparent that they should be avoided by ensuring that appropriate repair work is undertaken in good time. The effects of neglect of a tunnel lining can all too rapidly overwhelm the capacity to implement economic repairs.

21.6.8 Underpinning and inverting

Occasionally trouble is experienced because of inadequate foundations to the tunnel side-walls; frequently the weakness or total absence of an invert is to blame for serious defects. Lack of adequate vertical support to the side-walls may allow the tunnel to settle, causing fractures and reduced operating clearances; this calls for some form of underpinning. Lack of horizontal resistance in the floor of the tunnel enables the side-walls to move inwards under externally applied loading and the ground between them to heave – a situation which may first reveal itself through variations in track levels in the case of a railway and by boats grounding in the case of canals. This indicates a need for appropriate strutting between the bases of the side-walls. A substantial reinforced concrete invert, keyed into or carried under the side-walls, is a solution commonly adopted, since it can provide the strutting effect as well as improved distribution of vertical loading.

In the interests of maintaining structural stability, both conventional underpinning and inverting need to be carried out in short sections (of not more than about 2 m width, measured in the direction of the axis of the tunnel). Except in the case of a canal tunnel, it may be possible to construct a new invert without complete closure but by withdrawing traffic from one half at a time and arranging reversible working over the other half; each section of invert can then be cast *in situ* in two separate stages, albeit with a construction joint on the tunnel centre line, where provision has to be made for bending stresses.

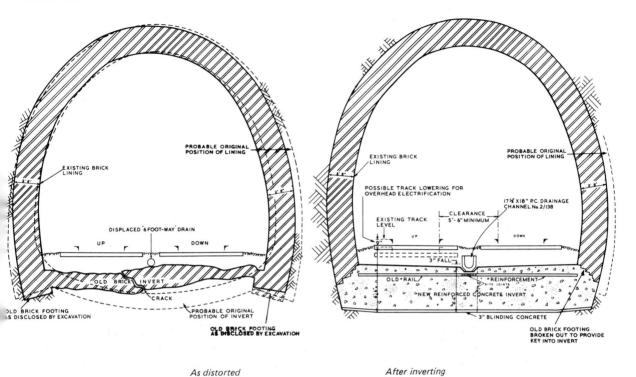

As distorted   After inverting

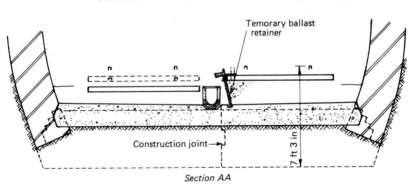

Section AA

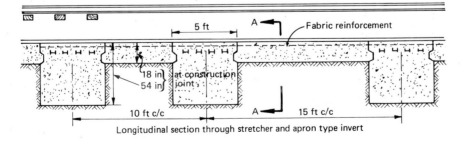

Longitudinal section through stretcher and apron type invert

FIGURE 21.10 Re-inverting of Gillingham Railway Tunnel. (*Courtesy of British Rail.*)

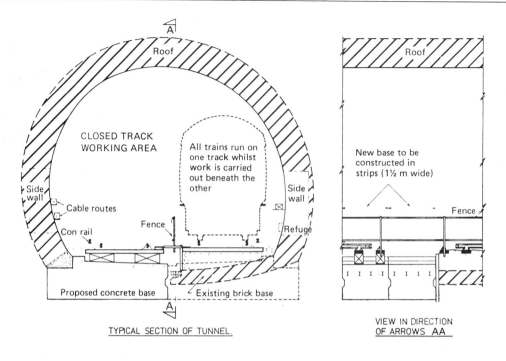

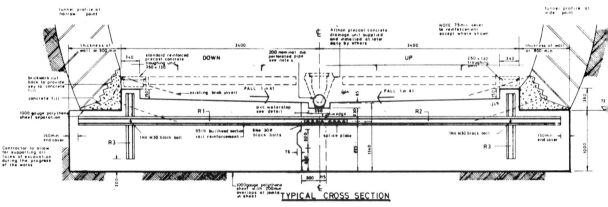

FIGURE 21.11 Re-inverting of Southampton Railway Tunnel. (*Courtesy of British Rail.*)

Alternatively, closure may be kept to a minimum by arranging for the invert to be formed of a series of precast units extending over almost the full width of the tunnel, with reinforcement protruding from each end and linked into *in situ* cast concrete underpinning of the side-walls; in this case, a single reversible line of railway is best located temporarily in the centre of the tunnel's cross-section. Examples of some solutions adopted are shown in outline in Figs. 21.10–21.13. A case involving a variety of tunnel stabilization techniques (jet-grouted column supports, mini pile underpinning, slab grouting, stitching and spreader beams), together with the systems used to monitor their effect, is described by Riddell (1986).

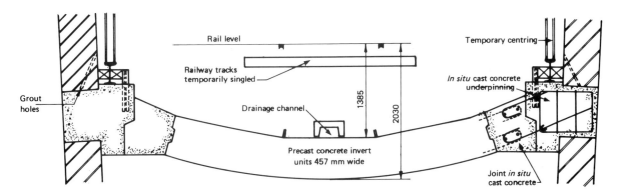

FIGURE 21.12 Inverting of Arley Railway Tunnel. (*Courtesy of British Rail.*)

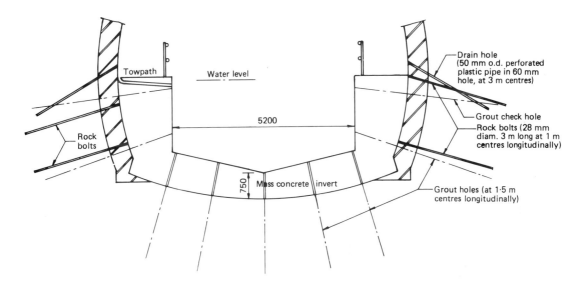

FIGURE 21.13 Re-inverting of Netherton Canal Tunnel. (*Courtesy of British Waterways Board.*)

21.7 Scope for further development

Until relatively recently, the understanding of the behaviour of old tunnels was largely superficial and the remedial measures adopted to deal with defects were empirical. Developments in the last decade have gone some way towards changing the approach, but there is still scope for progress in various areas.

There is a need, for example, for improved methods of non-destructive investigation of the geometrical and physical properties of the lining and of the stresses in it, together with ways of identifying the existence and extent of hidden voids and cavities, without having to drill through or cut windows in the masonry.

It would also assist the precision monitoring of convergence and similar progressive movement if existing rapid semi-automated systems of profile measurement could be developed to achieve the greater degree of accuracy which such monitoring requires and to allow the measurements to be taken more conveniently, without interruption to traffic.

There would be advantages in evolving a rationale for strengthening tunnels in rock by associating the characteristics of the existing lining with those of the rock mass and, by using annular grouting and rock bolting (and possibly also sprayed concrete), to develop the interaction between them and so to achieve a composite structure of greater load-bearing capacity than any element of it. This must involve a knowledge and understanding of both lining and rock mass properties to the point where a well-founded decision can be taken as to the most appropriate combination of remedial measures for any given circumstances. Methods of classification are needed to reflect the properties of the lining and the rock mass, so as to avoid recourse to rigorous and complicated design analysis.

Some studies along these lines have already been undertaken by James Williamson and Partners, Consulting Engineers of Glasgow, in 1981. An apparently similar system is described by Abbruzzese and Poma (1986). As a measure of one of the parameters, rock quality, it has been suggested that an index developed by Barton might be used; this describes rock in terms of rock mass quality Q, the derivation of which is given by Barton *et al.* (1974). Although intended (and widely accepted) as an indication of the tunnelling quality of rock for new excavations, Q appears to offer a convenient starting point as a measure of the quality of the rock surrounding existing tunnels, although the degree of support needed may be affected by the changes with time which will probably have taken place.

Such an approach is partly empirical and it needs validation and refinement on the basis of experience and performance. In furtherance of this aim, it is necessary to accumulate relevant data; to that end, it is recommended that any expert investigation of the characteristics of rock around an existing tunnel should include an assessment of its rock mass quality index and the efficacy of any repairs should be related to this.

21.8 Case History: bulge repairs to Bramhope Tunnel

21.8.1 Defects

Bramhope (railway) Tunnel, near Leeds, was constructed in 1845–9 and was lined with brick or sandstone masonry 18–24 in thick. Bulges at haunch level had been under observation for many years but were considered stable. Drilling had revealed no cavity behind the lining nor any indication that the bulging was due to water pressure. From about 1978 it had become apparent that one particular bulge on the west side, some 6 m long, was developing and by November 1979 it was approximately 300 mm out of profile. In mid-1980 it was deemed necessary to single the railway to allow the erection of substantial steel portal frame supports (using 24 in × 7½ in steel joists, mostly at 1.25 m centres) to safeguard the stability of the tunnel.

21.8.2 Investigations

Core samples in the area of the bulge showed the surrounding strata to be an argillaceous feldspathic coarse Namurian sandstone, containing vein mineralization, with some voids between it and the lining. The veins consisted of marcasite and pyrites (both iron sulphides), which had changed with time, by oxidation and hydration, into sulphates; this chemical mineral change caused a volumetric expansion of 100% so expanding the sandstone and exerting pressure on the lining, in an area where the lining was only 18 in thick. Core sampling elsewhere in the tunnel did not reveal the same degree of fracturing or metallic intrusions.

Acoustic emission monitoring at the bulge showed little activity apart from that which was train influenced. Pull-out tests on 20 mm rebars anchored into the rock indicated that they were capable of carrying loads of 10.5 tonnes with no visible evidence of yielding, but acoustic emission readings suggested distress in the rock strata when

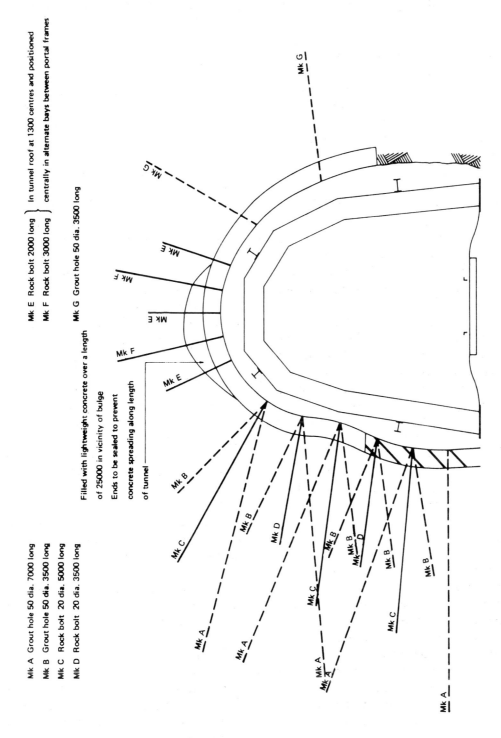

Mk A Grout hole 50 dia. 7000 long
Mk B Grout hole 50 dia. 3500 long
Mk C Rock bolt 20 dia. 5000 long
Mk D Rock bolt 20 dia. 3500 long

Filled with lightweight concrete over a length
of 25000 in vicinity of bulge
Ends to be sealed to prevent
concrete spreading along length
of tunnel

Mk E Rock bolt 2000 long ⎫ In tunnel roof at 1300 centres and positioned
Mk F Rock bolt 3000 long ⎬ centrally in alternate bays between portal frames
Mk G Grout hole 50 dia. 3500 long

FIGURE 21.14 Bramhope Tunnel: arrangement of temporary support frames and repairs (by grouting, stitching and rock bolting). (*Courtesy of British Rail.*)

315

the applied load exceeded 8.4 tonnes. The water entering the tunnel was analysed and found to be non-aggressive.

21.8.3 Repairs

A grout curtain was installed at each end of the length to be repaired, voids between the rock and lining being filled with quick-setting grout (Pozament BBC 79).

On the west side, voids were grouted using a PFA–cement mix. The holes (Fig. 21.14, marks C and D) were then re-drilled (at approximately 1.25 m horizontal spacing) using hand-held rotary–percussive drills; high yield rebars 20 mm in diameter were inserted and anchored with Selfix resin cartridges (bond length 450 mm). At the intrados end, the bars were threaded and provided with pattress plates, 100 mm square and 20 mm thick. Each bolt was tightened to a load of about 2 tonnes and grouted in. Further holes (Fig. 21.14, marks A and B) were drilled and 20 mm rebars were grouted in, untensioned, as stitching bars, using neat sulphate-resisting cement grout under a pressure of 0.35 N/mm^2. All rock bolts, pattress plates and stitching bars were sheradized.

On the east side, which was free of intrusions and much less fractured, two rows of holes (mark G) were grouted in two stages.

In the roof, after stabilization of both walls, a void above the lining was grouted using a 7:1 PFA:cement mix, giving a compressive strength of 4 N/mm^2 at 28 days. Holes were re-drilled and rock bolts were installed, tightened and grouted in, as on the west side (marks E and F). The above repairs were carried out by Colcrete Ltd.

Finally, the lining in the bulged area was cut away in panels not exceeding 1 m wide. The fractured nature of the sandstone was then evident, as were the intrusions of green-coloured marcasite hydrosulphates. The rock was trimmed back to permit re-lining in Class A blue engineering brickwork, 450 mm (four rings) thick, in English bond, using sulphate-resisting cement

mortar. The steel frames were removed and the railway was restored to double track in June 1981.

21.9 Case History: repairs to a ventilation shaft, Llangyfelach Tunnel

21.9.1 The problem

Constructed in 1913–17, Llangyfelach Tunnel, near Swansea, was one of the last railway tunnels in Great Britain to be lined in brickwork. One of its construction/air shafts had become particularly subject to the ingress of water and to the deterioration of its brick lining. The shaft in question is of 3 m diameter and some 42 m deep. It passes through Upper Pennant Measures, with about 7 m of glacial gravel overlying 3 m of fissured/shattered sandstone, below which are shales and coal measures and a band of impermeable clay. The gravel allows surface water to penetrate the sandstone, whose fissures permit random lateral flows; overcutting of the shales during construction left an almost continuous annular void, up to 300 mm wide, over the lower half of the shaft, with the result that water could readily flow down to the tunnel. At the sandstone–shale interface, water emerged from the rock and entered the shaft in a number of strong jets. At this level the brickwork was in particularly poor condition, as it was also adjacent to three old and blocked drainage garlands and where timber collars had been used to support the brickwork during construction of the shaft.

21.9.2 The nature of the repair work

The repairs undertaken were intended to prevent water ingress by the injection of grout behind the shaft lining in order to render the gravel and sandstone impervious, to fill the annular void and also to replace defective brickwork and make good mortar joints. The work was undertaken from a platform suspended within the shaft. An interesting feature was the provision of an inflatable rubber ring around the periphery of the platform to close the gap between it and the lining and so prevent material from falling onto the tracks below; thus all operations other than the repositioning of the platform could be undertaken without interference with railway traffic.

21.9.3 Grouting

Initial grouting used a 5:1 PFA:cement mix with a thixotropic additive; this was successful in filling the annular void but had limited effect in the upper portion of the shaft. Reducing the spacing between grouting points, increasing the length of the holes drilled into the surrounding ground and using a thinner mix (10:1 PFA:cement) gave no marked improvement: neither did the use of a neat cement mix (0.40 water:cement ratio). It appeared that, although large fissures were being filled, fines in the gravel were preventing grout penetration of the interparticle voids. However, by drilling the holes at an upward inclination of 20° and using water flushing, sufficient fines were

expelled to achieve an effective uptake of cementitious grout over the upper portion of the shaft, although the walls remained damp.

Resin grout (Borden MQ5) was then injected behind the lining over most of the shaft depth to seal minor leaks and a significant reduction in dampness was achieved.

Demonstrably, a relatively inexpensive PFA–cement grout had served to fill fissures in the rock and the large voids between the brickwork and the rock; it was less effective for penetrating gravel, although some improvement was achieved by water flushing. The more expensive resin grout successfully completed the sealing operation.

21.9.4 Brickwork repairs

Defective brickwork was replaced. Decayed mortar was removed using water jetting (at 2000 lbf/in^2). The brickwork joints were then pressure pointed using a 3:1:1½ sand:PFA:cement mix of mortar with a detergent plasticizer and a sodium silicate accelerator.

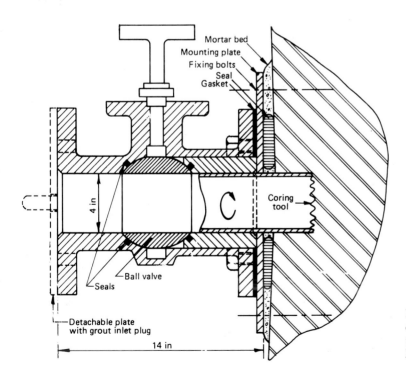

FIGURE 21.15 Valve assembly for coring the brickwork of a subaqueous tunnel. (*Courtesy of Port Authority of New York and New Jersey, USA.*)

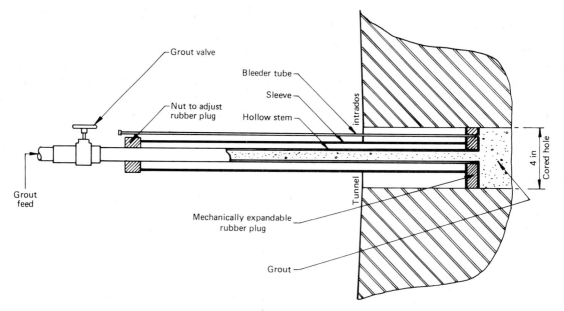

FIGURE 21.16 Mechanical packer for grouting masonry (valve assembly omitted for clarity). (*Courtesy of Port Authority of New York and New Jersey, USA.*)

21.9.5 Costs

The above work was undertaken in 1983, by Thyssen (Great Britain) Ltd, at a total cost of about £142 000.

21.10 Case History: coring a subaqueous structure

This section is based on US Patent No. 4625813 of December 2, 1986.

In subaqueous tunnels the opening of holes through the lining presents obvious difficulties because of the risk of an inrush of water and silt under the pressure of a significant hydrostatic head.

In order to investigate the condition of the brick lining of a rail tunnel under the Hudson River in New York and, if found necessary, to improve its structural integrity by grouting, the Port Authority of New York and New Jersey recently patented in the USA a sophisticated full

port valve assembly, the essential features of which are shown in Fig. 21.15. With the ball valve open, the barrel of a coring device can be inserted through the cylindrical bore of the assembly to cut a core from the brickwork and retrieve it for examination and evaluation. At any stage the coring device can be withdrawn and the valve closed to prevent any inflow of water or other material or to permit, for example, the resumption of traffic through the tunnel. A grouting device can be connected to the valve assembly, after which the ball valve is opened to allow a grouting compound to be injected through the assembly into the brickwork. Alternatively, a mechanical packer (Fig. 21.16) can be inserted through the open valve assembly into the cored hole. This packer comprises a hollow stem on which is mounted an expandable rubber seal which is adjustable to fit closely to the inside of the cored hole, so as to prevent any leakage. Grout is fed in through the packer stem, while a bleeder tube allows for the removal of any air and water trapped in the hole.

21.11 A cautionary tale

This section is based on a *Times Law Report* of February 27, 1861.

In a legal action, in 1861, between the South Eastern Railway Company and a firm of contractors who had been employed to construct a railway from Tunbridge Wells to Robertsbridge, it was claimed that the builders 'did not construct the walls or arches of two tunnels of the thickness of four rings, as required by the contract, but of the thickness of one ring only, with the exception of some parts of the Strawberry Hill tunnel, in which the contractors put two rings; loose bricks and rubbish, without cement or mortar being used . . . in place of the other two or three rings'.

An engineer who had been called to inspect the tunnels some five years after their completion stated that he had had holes cut through the sides and crowns and 'in the Grove tunnel he found only one half brick setting on cement instead of four. The second, third and fourth rings were stacked in dry, which made the work more injurious than beneficial . . . . The openings showed that there were spaces behind the brickwork which ought to have been filled in or backed up with earth. The Strawberry Hill tunnel proved, when inspected, to be better work. In some places there was a thickness of 9 inches of brick and cement, but in no case could they find more than 9 inches.'

Another witness said that about 100 holes cut in each tunnel had revealed that 'there were large spaces without backing and that between the upper surface of the crown of the tunnel and the rock above, the space was so large that he believed he could have made his way from one end of the tunnel to the other'.

Mr Barlow gave evidence that he was the principal engineer of the line and that he had appointed a Mr Richardson as resident engineer to supervise the work. At the time, he himself was 'much engaged upon other works and devoted as much time as he could to the work in question, paying a visit about once a fortnight to the tunnels . . . . It was difficult at all times to see the work in the tunnels owing to the want of light and it was still more difficult to detect bad work, as the contractors always took care when they saw the engineer coming that the work should appear quite right. Richardson had always reported to him that the work was correct and that the brickwork was not allowed to be less than that specified for in the contract; he had applied to the railway company for more inspectors to be appointed but they had refused the application.' Richardson had subsequently been dismissed and did not give evidence.

The contractor's defence was to the effect that 'the engineer had seen all the work as it progressed without making a complaint; if the engineer had failed in his duty, then it was very hard that the defendant should be called upon to pay in consequence'.

The jury found in favour of the railway company, awarding them damages of £3500 and adding that 'the litigation might have been avoided if the company's engineer had attended to his duty'.

The South Eastern Railway had to construct, in both tunnels, two additional rings of brickwork, at a cost of £4700, reducing their width by 18 in and their height by 9 in. As a result, it was not possible to operate standard rolling stock on the Tonbridge to Hastings line for many years until 1986, when the tracks through Strawberry Hill Tunnel were singled, while those in Grove Tunnel were mounted directly on a concrete slab foundation and a speed restriction of 25 m.p.h. was imposed on trains.

21.12 References and further reading

Abbruzzese, L. and Poma, A. (1986) Gallerie ferroviarie in esercizio. *Proc. Int. Congr. on Large Underground Openings*, ATTI, ITA/SIG, Florence, June, Paper A.1.

Alias, J. (1983) Réparation des Ouvrages Souterrains à la SNCF par Béton Projeté. *Travaux*, May, pp. 68–75.

Association Française des Travaux en Souterrain: Recommandations sur les méthodes de diagnostic pour les tunnels revêtus. *Tunnels et Ouvrages Souterrains* No 44. AFTES, Paris.

Recommandations sur les travaux d'entretien et de réparation. *Tunnels et Ouvrages Souterrains* No. 58, AFTES, Paris.

Recommandations sur les injections. *Tunnels et Ouvrages Souterrains* No 81. AFTES, Paris.

Recommandations sur les reparations d'étanchéité. *Tunnels et Ouvrages Souterrains* No 82. AFTES, Paris.

Barton, N., Lien, R. and Lunde, J. (1974) Engineering Classification of Rock Masses for the Design of Tunnel Support. *Rock Mech.* **6**, pp. 183–236.

Campion, F. E. (1950) *Part Reconstruction of Bo-Peep Tunnel at St Leonards-on-Sea*, Railway Paper 40, Institution of Civil Engineers, London.

Cantrell, A. H. (1952) *Some Major Problems in Railway Civil Engineering Maintenance*, Railway Paper 47, Institution of Civil Engineers, London (covers inverting of tunnel at Hastings, Warrior Square).

Carpmael, R. (1932) Cementation in the Severn Tunnel. *Proc. Inst. Civ. Eng., Part 2*, **234**, Paper 4862.

Chambron, E. and Picquand, J.-L. (1979) Rénovation des tunnels ferroviaires: méthods d'Auscultation et de Confortment. *Annales de l'Institut Technique du Bâtiment et des Travaux Publics*, No. 373, June.

Dutens, C. and Lienart, J.-M. (1983) Evolution of supervision and maintenance methods for underground structures of the Paris Transport Authority (RATP). *French Railway Review*, 1(3), 261–76.

Eraud, J. (1980, 1981) Rénovation des Tunnels SNCF. Technqiues de réparations. *Revue Générale des Chemins de Fer*, **99**, June, 357–68, September, 477–97, and December, 681–70 1980; **100**, March, 145–68, May, 289–314, July and October, 623–44, 1981.

Eraud, J. (1984) La politique de la SNCF en matière de surveillance, d'entretien et de rénovation des tunnels. *Travaux*, February. pp. 28–38.

Haider, G. and Richards, L. R. (1987) Netherton Canal Tunnel: recent repairs of failed sections. *Proc. Inst. Civ. Eng., Part 1*, **82**, Paper 9159.

Hoek, E. and Brown, E. T. (1980) *Underground excavations in rock*. Institute of Mining and Metallurgy, London.

Holmes, G. C. (1982) Bramhope Tunnel: bulge repairs. *J. Per. Way Inst.*, **100**(3), 211–22.

King, C. W. (1951) *Arley Tunnel: Remedial Works following Subsidence*, Railway Paper 41, Institution of Civil Engineers, London.

Martin, D. (1983) Shield in drive to renovate 180-year-old canal tunnel (Blisworth Tunnel). *Tunnels and Tunnelling*, November, pp. 35–8.

Ministère des Transports, France (1979) *Instruction Technique pour la Surveillance et l'Entretien des Ouvrages d'Art*, Fascicule 40, *Tunnels, Tranchées Couvertes, Galeries de Protection*, Direction des Routes et de la Circulation Routière, Paris.

Riddell, J. B. (1986) Strengthening works in High Street Tunnel, Glasgow. *Proc. Inst. Civ. Eng., Part 1*, **80**, Paper 8907.

Robarts, H. E. (1946) *Tunnel Maintenance*, Railway Paper 23, Institution of Civil Engineers.

Simms, F. W. (1844) *Practical Tunnelling*, pp. 102, 123, Troughton and Simms, London.

Spang, J. (1971) Design, construction and maintenance of tunnels of German State Railways. *Tunnels and Tunnelling*, September, 341–8.

Spang, J. (1973) Extensive guniting and grouting repairs completed in Reilerhals Tunnel. *Tunnels and Tunnelling*, January, 35–9.

Spang, J. (1976) Bauarbeiten an eisenbahntunneln der Deutschen Bundesbahn. *Eisenbahningenieur*, Nos 8, 307–12, and 9, 364–71.

Tanner, P. W. and Burton, R. A. H. (1963) Installation of a reinforced concrete invert in the Gillingham (Dorset) Tunnel on the Southern Region of British Railways. *Proc. Inst. Civ. Eng.*, **24**, Paper 6656.

Thompson, N. (1986) The renovation of Ashford Tunnel, Monmouthshire and Brecon Canal. *Proc. ICE/Univ. of Manchester Conf. on Infrastructure Renovation and Waste Control*, pril 1986, Manstock, Manchester.

Wyllie, D. C. (1988) Railroad tunnel modification and rehabilitation in North America. *Proc. 5th Inter. Symp. on Tunnelling '88* organized by the Inst. of Mining and Metallurgy. London, April. pp. 427–437.

Sewers and culverts

G.F. Read

The following definitions have been assumed for the purposes of this chapter, notwithstanding the fact that both types of underground structure present similar problems as far as repair and maintenance are concerned. Reference in the subsequent text to sewers includes culverts.

SEWER: An underground conduit or duct formed of pipes or other construction and used for the conveyance of surface, subsoil or foul water.

CULVERT: A large pipe or enclosed channel used for conveying a watercourse or stream below formation level.

22.1 General

Intense media coverage of the continuing number of sewer collapses with all the attendant (if unmeasured) social, environmental and traffic disruption costs they bring, their causes and the escalating costs of repair has highlighted an inter-national problem. Although the United Kingdom was the first country to install large-scale sewerage systems, the Victorian civil engineers were not backward in exporting their expertise and consequently the current problems of sewer dereliction are not just confined to that county. Sewers are part of our national industrial heritage and rehabilitation of many of them is long overdue. There are about 250 000 km of public sewer in the United Kingdom and the present-day cost of replacing the network by conventional means amounts to probably £70 000 million. The repair and maintenance of an asset of this magnitude should accordingly be of paramount importance, particularly when it is remembered that 95% of the population are connected to the network, which in fact is the highest 'linkage' in the world.

Many of these assets are old, some dating back to the last century, and the need for investment in this infrastructure is evident in view of the continuing number of collapses and blockages – 5000 a year costing some £20 million to rectify. The distribution of age, size and materials of the public sewerage network in England and Wales is given in Table 22.1.

From Table 22.1 it can be assumed that some 10 000 km of the total public sewer network is in brick construction but this does not include the considerable mileage of culverted watercourses etc.

Many of our cities are located on deep alluvial plains within river valleys; as they developed, particularly rapidly during the Industrial Revolution, extensive culverting of minor rivers and

TABLE 22.1 Sewers of England and Wales: Age and Material

Age	Brick (%)	Clay (%)	Concrete (%)	Other (%)	All (%)
Pre-1914	3 (95)	19 (26)	0.1	0.3	22
1914–45	0.7 (100)	23 (24)	1.4 (71)	0.5	26
Post-1945	0.2 (100)	36 (18)	13 (69)	3	52
All	4	78	14	4	100 (30)

Figures in parentheses are percentages of sewers 300 mm in diameter or larger.
Some 16% of brick sewers are judged to be unsound by present-day standards.

streams became commonplace and such watercourses were often spanned by the early buildings. This led to a massive network of private culverts whose condition today is generally perhaps much worse than the adjacent public sewerage network constructed at probably the same period. Nevertheless in the repairs and maintenance of such private culverts similar techniques to those adopted for public sewers having similar construction are utilized. Figures, even broadly estimated, are not available for the extent of this culvert network but in the case of Manchester it has been assessed that within a five mile radius of St Ann's Square alone, right in the heart of the city, there are nearly 100 miles of culverted river! As cities grew and the demand for more housing convenient to the mills and factories grew, the rivers and streams became convenient waste collectors and, thus despoiled, were later culverted and conveniently forgotten. New sewers were constructed to remove surface water more efficiently, yet in most instances the culverted rivers survive and continue to flow beneath houses, factories, railways, canals and highways; they still remain the responsibility generally of the property owners above, who frequently completely disregard their existence until some structural failure becomes manifest.

22.2 The problem

Engineering works do not last for ever and regular maintenance works are of paramount importance if collapses, with the resultant risk to life and limb, are to be avoided. Unfortunately in the past sewers and culverts have not always attracted the necessary maintenance funding.

It is generally accepted that the older parts of the sewerage network are in poor structural condition, their hydraulic performance is often inadequate and they are having to carry much greater live loads than those for which they were designed. Consequently, substantial civil engineering works are undertaken each year to deal with sewers which have serious structural defects or where lack of capacity is causing river or stream pollution or flooding – often with associated public health risks. The immediate cost of these works is high, currently around £190 million per annum, but in addition there are indirect or 'social costs' arising out of the consequential impact on the community – perhaps varying between three and even ten times, in the most critical places, the civil engineering costs. There are now indications that the Government will expect the industry to move towards a cost–benefit approach and this will mean the inclusion of such social costs along with benefits. The author of this chapter is involved in a research programme in the Department of Civil and Structural Engineering at the University of Manchester Institute of Science and Technology which is aimed at developing easy-to-use models for the prediction of such indirect costs associated with different sewerage rehabilitation techniques.

Age is not the only feature in classifying sewerage dereliction but construction before the middle of the nineteenth century will in general

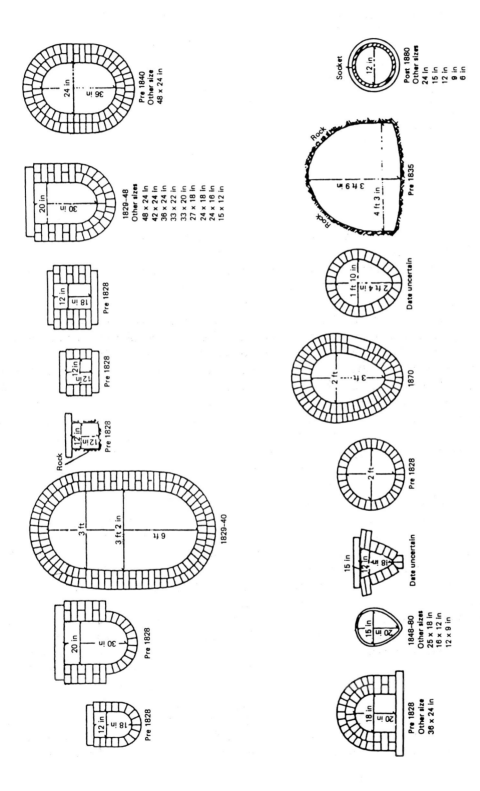

FIGURE 22.1 Early forms of sewer construction. (*Courtesy of City of Manchester.*)

indicate a very serious risk of dereliction, much higher than for the younger parts of the network. Other significant factors include the type of road in which the sewer or culvert is situated, subsoil conditions and the care taken in the original bedding and back-filling. There is no substitute for a proper engineering survey leading to a clearly identified need. On average over 15% of the national sewerage network is over 100 years old but when the situation in the older industrial cities such as Manchester is considered, where the comparable figure is 38% (570 miles), the ever-present potentially serious situation which generally exists under the older parts of such towns and cities is appreciated. Manchester, in fact, possesses the oldest extensive sewerage system in England; early forms of construction are shown in Fig. 22.1. This drawing is a copy of one made in 1896 to accompany the City Engineer's Report to the Rivers Committee entitled 'Manchester Main Drainage and Works'. Recent investigations suggest that some of the construction dates shown require amendment, usually to earlier dates. It seems that many of these early forms of sewer construction are similar to the format utilized under other towns and cities as they subsequently developed in the nineteenth century.

22.3 Original construction methods

Until about 1850 the construction of sewers was generally of brickwork in lime mortar, often combined with natural stone inverts and soffits.

Today the cross-sections of such sewers are typically non-uniform throughout their length and frequent variations in size and gross distortion of shape are generally apparent. Brickwork coursing is uneven and erratic and the bonding is irregular. Missing bricks are frequent and sometimes the areas of missing brickwork are extensive. The bricks themselves are of poor quality and irregular in shape by present-day standards. Tests on bricks taken from such sewers indicate high water absorption (approximately 19% of dry weight) and a relatively low crushing strength

(around 20 N/mm²). The lime mortar used for construction has softened and been eroded leaving the brickwork even more open jointed than was envisaged by the designers. Stone flag tops have fractured or become displaced, and even where the sewer itself has not collapsed, voids can often be found outside the sewer which invariably accelerate the first stages of a subsequent collapse.

The structural performance of all brick sewers is dependent upon their being maintained so that the structure is in a constant and uniform state of compression. The shape and construction of these sewers is generally such that their compressive ring strength is more than adequate to resist the loadings they have to take and normally they transfer the load to the surrounding ground. Any deterioration of the fabric or reduction in ground support is therefore significant and will result in overall weakening of the structure.

Structural condition apart, it is interesting that with the early sewers the size appears to have been determined by construction methods rather than by hydraulic consideration, and the result tends to be that many local sewers have spare hydraulic capacity when analysed by modern-day methods.

A major problem with the existing network is often access. Manholes as we know them today are virtually non-existent. In Manchester, access to the sewers during the construction was by brick-lined circular shafts, generally 900 mm in internal diameter, constructed in 225 mm brickwork laid radially or occasionally in stretcher bond, the shafts being capped with natural stone flags at depths of 1–1.5 m below ground level; the road was reinstated without any provision for permanent access to the system or any record as to location.

The old brick sewers exhibit wandering alignment and varying gradients. Sometimes the variations are extreme. Connections and access shafts are often badly made and have caused local structural failure at the point where they enter the main sewer.

The conditions encountered in the older sewers considerably increase the health and safety risks borne by those working in them today. In addition to the physical danger of working adjacent to potentially unsound structures, the large quantities of debris and silt in the sewers increase the likelihood of septicity and toxic gases. The large number of disused connections, the long lengths of derelict connecting sewers and the open-jointed brickwork and associated voids all provide ideal conditions and breeding grounds for rodents.

Nevertheless, the efforts of the early nineteenth century sewer builders should not be belittled since many of their brick sewers have lasted between 150 and 200 years with the minimum of maintenance, but clearly, by modern standards of quality of materials and construction techniques, the early brick sewers represent a relatively 'primitive' form of civil engineering. From around 1870 onwards, standards of construction generally improved dramatically to levels which can be appreciated today as being based on sound engineering principles.

Sewers in clayware pipes began to appear again about the middle of the nineteenth century. The word 'again' is used because there is some evidence that such sewers were known to the Romans but

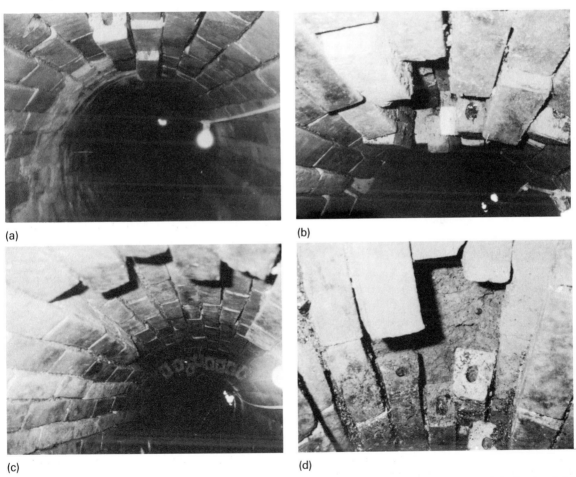

(a)

(b)

(c)

(d)

FIGURE 22.2 Typical collapse mechanism sequence of a double-ring brick sewer: (a) mortar eroded allowing brick movement; (b) inner ring squat following brick loss and joint closure; (c) progressive collapse of inner ring; (d) support retained only by ground arching. (*Courtesy of Sewer Services Ltd.*)

the art seems to have disappeared with their civilization. The use of clayware pipes continued to increase in the latter part of the nineteenth century, leaving brick construction for the larger size of conduit. Inroads into the latter form of construction continued in the early part of the twentieth century with the introduction of concrete pipes and concrete segmental construction although the latter technique often involved an inner lining of engineering brickwork, particularly where the sewer was carrying a strong effluent or was required to be particularly abrasion resistant.

22.4 Structural considerations

Pippard and Baker (1938) established the important role that jointing material plays in the construction and behaviour of arches and the use of cement mortar was shown to increase the loading capacity by factors of up to 4.0 compared with lime mortar.

One of the common factors related to the deterioration of old brick sewers is the gradual breakdown of the lime mortar. It was also quite common in the Victorian era for the engineers to leave out jointing of the brickwork in the sides of the sewer near the invert, endeavouring thereby to arrange that it would also act as a land drain; but they apparently disregarded the problems this would bring during times of peak flow when the sewage would escape from the conduit into the surrounding ground, draining back into the sewer when the flow reduced but bringing with it a quantity of ground from outside. The effect of cavitation of this type is a typical cause of sewer collapse following many years of the flow's escaping into the adjacent ground.

The excavation techniques which were used in the construction of the early brick sewers may also have been instrumental in the formation of voids outside the permanent works. These could include

1. timber left in which subsequently decays
2. slips and ground movement not properly consolidated

3. sub-drains not removed or filled
4. voids between permanent and temporary works not adequately filled and compacted

Once soil migration has taken place, allowing the formation of voids and zones of soil weakness immediately adjacent to the sewer, then the loads on it are not distributed uniformly and localized stress in the brickwork will occur. The sewer then becomes particularly sensitive to slight disturbance, perhaps resulting from heavy traffic vibration or surcharging of the sewer in storm conditions. A typical collapse mechanism may well develop as indicated in Fig. 22.2. This shows a case of inadequate side support causing joints in the crown and invert brickwork to open widely and lose mortar rapidly until the brickwork begins to fall away. The infiltration of groundwater or the exfiltration of sewage through these open joints will accelerate the rate of deterioration. Figure 22.3 indicates the stages by which renovation, using the rendering–grouting technique, might be undertaken.

22.5 Assessing structural condition

Over 90% of the public sewerage network is non-man-entry, i.e. it has an internal diameter of less than 900 mm, and it is only during the last decade, with the development of closed circuit television, that we have been able to examine in detail the inside of this non-man-entry part of the network. Prior to this, an internal survey had to rely on visual inspection from a manhole, using lights and strategically placed mirrors, which obviously had its limitations to say the least! Equipment is still not available for the extent of cavitation outside the barrel of the sewer to be accurately determined but development currently under way should eventually enable this to be achieved.

Two basic methods are therefore available at present for a structural inspection:

1. remote inspection – normally using closed circuit television equipment

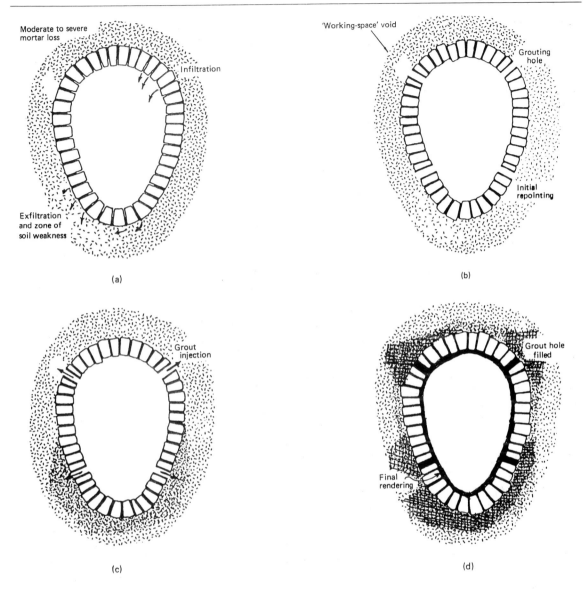

FIGURE 22.3 Typical collapse mechanism and stages of renovation: (a) existing conditions; (b) first stage works; (c) grouting in progress; (d) completed works. (*Courtesy of Sewer Services Ltd.*)

2. direct inspection – by walking through the sewer

The choice is governed mainly by sewer size, although cost and safety are important considerations.

The assessment of the structural condition of sewers is as much an art as a science and has of necessity to rely heavily on the informed judgement of civil and structural engineers experienced in this particular field. Research suggests that collapse of a sewer is normally the result of a

327

complex interaction of various mechanisms with a significant element of chance in the timing of final failure – it is not possible in fact to predict accurately when or how a sewer will fail. However, the Water Research Centre's *Sewerage Rehabilitation Manual* suggests levels of deterioration which imply different levels of collapse risk.

22.6 Options

In sewerage generally, the actual conduit only represents some 20% of the initial cost – excavation, support and reinstatement account for the remainder – so that the largest part of the infrastructure asset is 'the existing hole in the ground'. Therefore if we can put a new pipe inside the old sewer or renovate it without further excavation, we can theoretically save up to 80% of the traditional cost. Consequently, in recent years, sewer renovation has been increasingly advocated as a means of coming to terms with the immense problem now faced by the water industry generally of dealing with an often outdated sewerage network. Nevertheless, it should be appreciated that renovation still involves considerable resources in time, manpower and funding. Renovation must be considered as an option but one should never lose sight of the potentially hazardous nature of some of the renovation work and the long-term implications of renovated sewers. Nevertheless, if it is economically practical, renovation should be faced at the appropriate time – if not, the stage will be reached when the more expensive replacement will be the only solution.

22.7 Access

Many of the early sewers had either insufficient or substandard manholes by present-day health and safety standards, and this often necessitates a manhole or shaft construction programme as a prerequisite to either renovation or even in certain cases repair. This two-phased approach has the advantage that partially blocked sewers can often be cleared and their condition examined in more detail before a rehabilitation scheme is designed. Generally the shafts are constructed using precast reinforced concrete bolted segments similar to those utilized for new sewers in tunnel. Construction is generally carried out by underpinning each 600 mm depth of ring, and where appropriate the internal shaft diameter is selected to enable the shaft to accommodate a shield should it ultimately be found necessary to replace the sewer in tunnel. If the ground is loose water-bearing material, the shafts can be sunk by the caisson technique with the segments being built at the top of the shaft while excavation is carried out at the base. The built-up shaft lining sinks under its own weight or with the addition of kentledge.

In certain instances precast concrete manhole rings can be utilized but they do not have the same flexibility for accommodating the adjacent mains or cables etc. likely to be encountered.

Old manholes generally require upgrading to present-day standards of safe access.

22.8 Repair and renovation (man-entry sewers)

22.8.1 Stabilization techniques

The rehabilitation of brick and masonry sewers and culverts is not new and many local authorities have regularly carried out extensive repointing programmes to deal with the deterioration of the original mortar along with the replacement of missing or defective areas of brickwork. Rather than repointing old brickwork, rendering or 'larrying' over the defective surface using cement mortar developed as a common practice – perhaps this was not unrelated to the demise of the specialist bricklayer who at one time was a significant member of every sewer maintenance gang. Additionally, it was quite common practice to apply a 'grano' screed to the invert, often at the same time, all of which was labour intensive and led no doubt to the development of pre-shot gunite segments or precast concrete segments and in due course to the thin shell linings of today. There are nevertheless still cases where partial re-lining only is considered appropriate and pressure pointing of the remaining brickwork is done using equipment specifically designed for the purpose. Alternatively, where design loadings permit the adoption of this approach, it has been found that the provision of an impermeable front barrier, formed by the use of high strength polymer-modified cementitious mortar rendering, to the internal surface of an open-jointed brick sewer provides an ideal situation for external grouting to the exterior of the barrel.

Some specialist contractors are developing spray-on coatings involving the use of epoxies, polyesters and polyurethanes. All systems have two component materials which are either pump batched, mixed and delivered under pressure to a spray head or reach the head through two separate hoses and are then mixed under pressure when combined in the head, the spray head either being winched or driven along the sewer.

A smooth resilient coating of the required thickness, usually 10–20 mm, is applied at a spraying rate of 40–60 m/h. A major advantage of spray-applied coatings is that they deal with lateral connections by leaving a smooth transition of material 'feathering' up into the lateral. Spray-on coatings suffer from the fact that generally the materials currently being used are not water tolerant; hence the sewer to be treated must be virtually dry. Further development work is progressing with regard to this technique.

22.8.2 Lining techniques

It became clear to the chapter author in the early stages of the dramatic sewer collapse era in central Manchester during the late 1970s that some form of internal lining or permanent formwork, to stabilize the old sewers quickly before collapse took place, needed urgent examination in an endeavour to preserve parts of the network which had not deteriorated too far. Hence much of the initial experimental work in this field was carried out in the City Engineer's Department, utilizing new materials which were becoming available to the construction industry. This laboratory and site assessment formed much of the early development work in the UK.

Lining consists of the insertion of pipes or segments into a sewer to provide improvements in structural and hydraulic capacity and chemical resistance, the annulus between the lining and the sewer structure then being grouted to ensure satisfactory performance. A feature common to all techniques is the reduction of the internal surface roughness of the existing sewer. Thus, although there is in general a loss in cross-sectional area associated with renovation work, this is usually counteracted to a greater or lesser extent by the improved flow characteristics.

In the Water Research Centre's *Sewerage Rehabilitation Manual* linings are categorized into three types:

Type I   utilizes the structural capacity of the existing sewer and requires bonds between lining and grout and grout and existing structure;

Type II   is designed as a flexible pipe requiring no bond between lining and grout or grout and existing structure and assigning no long-term strength to the existing sewer;

Type III   provides permanent formwork for the grout and the lining cannot be assumed to contribute to the long-term strength of the sewer.

It is suggested that non-man-entry linings are designed as Type II structures as an adequate bond to the existing sewer cannot be guaranteed. Man-entry sewer linings may be designed as Types I, II or III. Type II linings are generally required if corrosion is apparent because a bond cannot be confidently assumed. Consideration of thin-walled Type III linings is only appropriate if the existing sewer is not fractured or visibly deformed

FIGURE 22.4 Early Victorian brick sewer prior to re-lining. (*Courtesy of Tunneline Ltd.*)

and the groundwater level is below the sewer.

The manual recommends detailed procedures with a view to a 50 year design life for all three types of lining utilizing the results of the Water Research Centre's research and monitoring programme. The suggested simplified design methods include a safety factor of 2.0 against the design loads in comparison with the safety factor of 1.25 generally incorporated in 'new construction' design even with the associated problems of bedding, back-filling and compaction.

(a) Engineering brickwork

As a logical extension of the localized repair of deteriorated brickwork the stage may be reached where complete re-lining using Class A engineering brickwork in cement mortar is selected in preference to a pre-formed segmental lining in

view of the nature of the old sewer, particularly if it has excessively sharp deviations, rapid changes of gradient or varying cross-sectional dimensions (Fig. 22.4). The lining is appropriately tied to the existing structure and enables a regular profile to be restored to the original sewer; with this repair method the annulus and all voids within and immediately external to the original structure must be grouted to ensure monolithic action.

Generally the bricks are laid conventionally but in some of the experimental work in Manchester use was made of a brick type in which the largest face was pressed; the bricks were laid with this face presented to the flow, thereby reducing the thickness of the lining.

(b) Precast gunite segments

The use of precast gunite segments is one of the

oldest and most established sewer renovation techniques. The segments, usually grade 40 (40 N/mm²) sprayed concrete with a maximum aggregate size of 10 mm, are manufactured on the surface by spraying onto the outside of a mould supporting reinforcement mesh (normally 6 mm mesh) sufficient for handling and jointing purposes. In order to provide sufficient cover for the steel, the sections are necessarily thick, typically 40 mm. In the sewer, joints are made with *in situ* gunite using specially designed nozzles for the confined working space.

Gunite is also suitable for providing *in situ* linings to old sewers: it is sprayed onto and through a heavier mat of reinforcement fixed to, and following, the internal profile of the conduit. This arrangement overcomes the jointing problems referred to but nevertheless is a dusty operation that is difficult to supervise, although the lining can be designed as a reinforced lining rather than the precast type where the units have to be relatively light for handling. Other uses of this treatment are to protect the sewer from chemical attack and abrasion as well as the long-term reduction of infiltration.

(c) Precast concrete segments

During the 1960s in particular a number of larger brick sewers were strengthened and improved by the insertion of precast concrete segmental linings with projecting reinforcement incorporated in the *in situ* joints. They were usually some 3–4 in thick and accordingly were difficult to handle but have now generally been superseded by the newer materials such as glass-reinforced cement (GRC) or glass-reinforced plastic (GRP) which are relatively lightweight.

(d) *In situ* pumped concrete lining

An established concrete lining technique for tunnels of larger diameter is currently being developed by Tunneline Ltd for the rehabilitation of man-entry brick sewers. Manchester has recently used this patented system to repair a

partially collapsed 1.2 m diameter brick stormwater sewer which was built about 1880, the lining being formed by placing reinforced concrete between a steel circular shutter and the deteriorating tunnel wall to form a new 1.05 m smooth bore conduit. It is claimed to be a cheaper method than the more common lining techniques.

Hoops of 10 mm diameter steel are placed at 100 mm intervals along the old sewer, after pressure cleaning the brickwork, and are secured to six longitudinal lacer bars 8 mm in diameter equally spaced around the circumference of the conduit. The shutter design is based on standard flanged bolted segmental tunnel linings. The 12 m long tubular form, made from 3 mm steel plates, is built up using six flanged rings 2 m long, each comprising four segmental panels fixed together with special hammered-in bolts to all segment and ring joints. Each panel has two integral screw jacks pushed out onto the existing tunnel wall.

Crown segment panels incorporate a flanged port-hole for injecting the 30 N/mm² concrete mix, which, in the Manchester scheme, is being compared with a mix including a proportion of synthetic fibres in fibrillated form (i.e. bundles of interconnected strands) to serve as a crack inhibitor. The concrete with a 100 mm slump is delivered through a steel pipe 100 mm in diameter to the 75 mm wide annulus between the shutter and the brickwork via a standard concrete pump. Vibration is not necessary, as the pressure on the concrete provides the required compaction for a smooth finish.

(e) Glass-reinforced cement

GRC is one of the 'newer' materials for sewerage renovation and was developed by the Building Research Establishment in conjunction with the glass fibre manufacturers, Messrs Pilkingtons, as an alternative to asbestos products. It is a composite material of high strength (70–80 N/mm²) generally consisting of the following:

Fine sand (parts by weight) 1

Ordinary Portland Cement (OPC) (parts by
weight) 2

Water:cement ratio 0.3

Chopped glass fibre 5%

In an automated process GRC is normally sprayed
onto a flat surface, vacuum dewatered and then
transferred to a formwork. The units are usually
struck from the formwork after one day and cured
under moist conditions for 7–28 days. The
finished units are generally about 12 mm thick
and supplied as crown and invert sections (Fig.
22.5), although in the larger sewers three or more
segments per ring are more common. It is therefore necessary to joint both circumferentially and
longitudinally and the segments, which normally

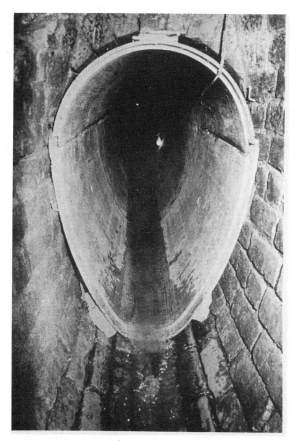

FIGURE 22.5 Glass-reinforced cement lining. (*Courtesy of
Cel-line Ltd.*)

incorporate a lap joint, are bolted to give a good
grout seal. Drilling the bolt holes *in situ* is a
simple and quick operation.

A suitable fixing would be one which has a
nylon sleeve and acts in a similar way to a rawl-
bolt, so that on being tightened it will expand and
draw the GRC laps together. In addition to the
mechanical anchorage provided between the linings and the grout there may be further advantage
if the fixings penetrate the brickwork and give
extra stability. Appropriate manufacturing techniques result in a roughened back to provide a
good grout key.

(f) Glass-reinforced plastic

GRP is a composite material with a high
strength:weight ratio. It consists of a polyester
thermosetting resin reinforced with glass fibres.
Sand is used within the structural wall to increase
the thickness and stiffness of the material and to
remove the necessity for temporary or permanent
stiffeners during annulus grouting.

The liner units are formed onto the outside of a
suitably shaped mandrel. A resin-rich layer known
as the 'gel layer' is applied, which gives a smooth
bore and provides chemical resistance. This is
followed by wound-on bands of glass rovings and,
as necessary, unidirectional woven tape. The glass
rovings pass through a resin bath immediately
before winding and carry a controlled quantity of
resin, sufficient to bond the structure. Layers of
graded sand are introduced into the wall thickness
at suitable intervals, a final layer of coarse sand
being applied to the outer surface to provide a
good grout key. After initial curing, by means of
infra-red lamps, the mandrel is removed and the
liner units are cured in a gas-fired oven and are
then available for use. The units can be formed
with either collar or socket joints to any shape or
length up to 6 m and are generally manufactured
as complete integral units without horizontal
joints.

(g) Polyester resin concrete

Resin 'concrete' is really a misnomer. It is in fact a

FIGURE 22.6 Polyester resin concrete lining. (*Courtesy of City of Manchester.*)

thermosetting polyester resin similar to GRP but in this case with concrete type aggregate used to bulk-out the matrix. It therefore behaves as a plastic rather than a rigid concrete type material. Lining units made of this material are thick and heavy, similar in handling to precast gunite segments (Fig. 22.6). They are manufactured by a closed moulding process, compaction and consolidation being achieved by vibration with resin cure controlled by heating.

(h) Grouting

Grouting in sewer renovation covers two main operations:

1. grouting of the space between the lining and the existing sewer, referred to as 'annulus grouting'

2. grouting of voids external to the existing sewer, referred to as 'void grouting'

Annulus grouting is an intrinsic part of the renovation procedure and, as a result of the design procedures developed by the Water Research Centre and described in the *Sewerage Rehabilitation Manual*, there is now a good understanding of composite renovated structures, making allowance for both the structural strength of the lining and the importance of the grout–lining interface. Grouts used in renovation work for annulus grouting are generally OPC and pulverized fuel ash mixtures (1:3 by mass with a maximum free water:solids ratio of 0.40). In addition a long-chain polymer additive is often specified to inhibit the absorption of free water. The annulus gap is normally a minimum of 25 mm and the grout is forced through the grout holes in the lining immediately following the erection of a number of units.

Void grouting of old sewers is also a most important part of renovation since voids are common, their frequency depending to a large extent on the type of the surrounding ground and the standards of the original construction. It is particularly important if pressure pointing is being used, for on its own the latter treatment may be little more than cosmetic. The main difficulty to be overcome is often the physical problem of drilling through the composite structure from within the confines of the existing sewer, particularly when these are only just above the accepted man-entry dimensions.

For void grouting, the material is usually an OPC and pulverized fuel ash mixture (1:7 by mass).

22.9 Repair and renovation (non-man-entry sewers)

As indicated earlier, working within a sewer of uncertain structural integrity is always hazardous. The risks are further increased if movement is physically restricted by the extremely confined working conditions frequently encountered in sewer renovation schemes. When the trend in our society has been towards reducing hazards at work and improving working conditions, the situation presented by renovation work from within the smaller sewers could be regarded as a retrograde step, to say the least!

In the author's view, manual methods of renovation should not be contemplated where the finished vertical dimension would be less than 900 mm. The following remotely controlled techniques are available for dealing with non-man-entry sewers.

(a) Slip lining

Slip lining generally involves drawing or jacking butt-welded polyethylene pipes through the existing sewer, the annulus being filled with specially developed low density foam grout.

Polyethylene is a thermoplastic material produced from ethylene gas in either high pressure or low pressure processes. The material in pellet form is melted, compressed and extruded through a circular die, similar to the manufacture of clayware pipes. For sewer linings it is usual to cut pipe lengths from the extruded pipe, jointing normally being carried out on site by butt fusion welding on the surface or at the bottom of existing manholes. Alternatively, the ends of the pipe sections are threaded and the separate sections are screwed together within the existing manhole. With the surface welding technique, quite a considerable length of lead-in trench is required which is normally not acceptable in urban areas.

(b) *Insituform* lining

In this method of renovation a flexible lining which follows the existing shape of the sewer in detail is inserted. A thermosetting polyester resin is utilized which is maintained in position in the sewer by a woven polyester needle felt bag or sock until cured. The outer surface is finished with a thin waterproof polyurethane membrane.

The needle felt bag is manufactured to the appropriate dimensions in the factory. The resin–catalyst mixture is fed into the bag to impregnate the felt thoroughly. The felt bag is then transported to site in a refrigerated van where it is 'inverted' into the sewer using a pressure head of water. Once in position the lining is cured by heating the water. This produces a structural lining that closely fits the existing sewer including its imperfections.

(c) Remaking lateral connections

In both the previously mentioned techniques there remains the problem of remaking the lateral connections, and although various development projects are under way involving the use of remote-controlled equipment etc., currently the connections are generally dealt with by excavation from the surface.

22.10 The future

The brick-built sewers and culverts in the United Kingdom represent a remarkable tribute to the skills, knowledge and qualities of workmanship of the nineteenth century civil engineers and contractors. Generally these structures have withstood the effects of time extremely well and their present often dilapidated condition is due primarily to the unavoidable decay of the orginal lime mortar, significant changes in superimposed loading conditions and a lack of planned maintenance. This chapter is not intended to cover all the techniques in use or under development. There are others in addition to those mentioned. Suffice it to say that the industry is generally responding to the need for cheaper and safer forms of sewer rehabilitation techniques which involve minimum environmental and social costs.

It is still in fact very much a case of 'horses for courses' as far as sewerage rehabilitation generally is concerned. Each defective sewer should be treated individually and a solution should be determined without predisposition towards any one method. Progress will be made by developing modern technology for innovative repair and renovation aimed at reducing the hazards and laborious nature of conventional methods. It is interesting that manufacturers consider that they still have to reach the peak of the learning curve in effecting cost-effective systems for sewerage rehabilitation. This is good news which could suggest that, as the technology improves, so sewer renovation and replacement costs will fall. It is encouraging to all concerned, particularly in view of the magnitude of the problem and the current level of funding.

22.11 References and further reading

Fluid Engineering Centre (1984) *Planning, Construction, Maintenance and Operation of Sewerage Systems.* Proc. Int. Conf. at Reading in September, 1984.

Institute of Water Engineers and Scientists. (1983) Proc. of Symposium on *Deterioration of Underground Assets.* London, December 1983. I.W.E.S.

Pippard, A. J. and Baker, J. F. (1938) *The Analysis of Engineering Structures.* Arnold & Co., Maidenhead, Berks.

Read, G. F. (1985) The brick sewers of Manchester. *Civil Engineering,* Nov./Dec., 1985.

Read, G. F. (1986) *The Development and Rehabilitation of Manchester's Sewerage System.* Inst. of Civil Engineers, N.W. Association Centenary Conference at U.M.I.S.T., April, 1986.

Read, G. F. (In press) *Sewers: Rehabilitation and New Construction.* Edward Arnold, London.

Read, G. F. & Vicridge, I. G. (1988) *The Indirect Costs of Sewerage Rehabilitation.* NO-DIG '88 U.K. N.W. Regional Seminar at U.M.I.S.T., June, 1988.

Underwood, B. D. and Rees, C. W. (1983). *Brick Sewer Renovation.* Inst. of Public Health Engineers Symposium on Practical Aspects of Sewer Renovation at U.M.I.S.T., September, 1983.

Water Authorities Association: Sewers and Water Mains Committee (1986) *Sewerage Rehabilitation Materials Specifications.* Information and Guidance Note Series. Water Research Centre, Swindon.

Water Research Centre. (1986) *Sewerage Rehabilitation Manual.* (Second Edition). Water Research Centre and Water Authorities Association.

Chapter twenty-three

Structures in maritime locations

A.D.M. Bellis

23.1 Types of structure considered

23.1.1 Port structures

(a) Dock walls and quays

Most docks in the United Kingdom and the majority of ports abroad were constructed before the First World War. Whenever possible the water areas were enclosed with solid retaining walls of masonry, brickwork or a combination of both. In some cases the walls were hewn from solid rock and/or the retaining wall was founded on the rock. In other instances walls were founded on piles or large footings. Between the wars and after the Second World War much less dock building has taken place and of that most has been of concrete, steel or a combination of the two.

(b) Lock entrances

Access to a system of impounded docks is through a lock or locks with walls of solid stone masonry, brick, brick and stone or brick-faced construction and inverts of similar materials.

(c) Dry docks

Very many dry docks are of similar construction to entrance locks but frequently they have stepped sides and altars.

(d) Slipways

Although a diminishing facility, some slipways still exist and some incorporate brick and stone in their construction.

(e) Jetties and piers

Jetties within dock systems may be of the same construction as the dock walls and will thus present the same maintenance problems. Jetties in open harbours may be for ship berthing or as 'lead-in' jetties to lock entrances or communication passages. Being in more open water and with shipping moving at high speeds, they can suffer both from tidal and wave action and from severe ship contact. Some piers come into the category of breakwaters.

(f) Wharves

Outside dock systems the term wharves applies to berthing facilities in open harbours or waterways.

As with some jetties, deterioration from natural conditions and from contact by vessels berthing at higher than design speeds will be accelerated. The stability of retaining wall structures may also be affected by scour from the watercourse or tidal flows.

(g) Paving

From the early use of stone flags and setts for paving port areas, through a long period of replacement by concrete or bituminous surfacings, treatment has now come full circle in many places where the use of brick or block paving is becoming widespread and fashionable. With the increase in wheel loads of port vehicles and equipment and the use and stacking of containers there is some justification for the use of block paving for ease of re-levelling and modification. Maintenance of these surfaces is now a significant commitment, although the blockwork is usually laid dry and presents its own individual maintenance requirement.

23.1.2 Harbour structures

Some harbours will contain the structures described above, but many are purely areas of safe anchorage or refuge. Where this funcion is achieved artificially it will be by the provision of breakwaters.

(a) Breakwaters

Breakwaters can range from very small structures giving protection to fishing and recreation areas to those standing in 30 m of water that provide shelter for the largest bulk-carrying ships. Between these extremes lie the 'piers' at river and harbour entrances. Although present-day deep water breakwaters can be of masonry to the extent of being stone rubble, the material is not shaped or jointed. Their maintenance can be a very considerable problem and details cannot be included here. In general, the difficulties are mainly those of accessibility by plant large enough to handle the displaced blocks or units and to introduce new ones. The choice of units used in armouring should always be within the capacity of an available floating crane – with sufficient outreach to plumb any part of the structure – or of a mobile crane operating from a roadway running on top of the breakwater. A choice of units large enough to be effective armouring but meeting these maintenance criteria poses a problem for the designer.

Many older breakwaters are constructed of sliced blockwork or ashlar stone similar to dock walls. Problems then are much the same as for dock walls but access to those parts exposed to tides and waves can be difficult.

(b) Sea defences

Sea walls in the form of promenades and sloping revetments, where these are of masonry, are usually exposed at low tide and therefore access for repairs in the dry can be gained for limited periods.

(c) Lighthouses

Many lighthouses are of masonry construction and some are incorporated in piers or breakwaters. Most are subject to extreme maritime conditions. Access for maintenance by labour, equipment and materials in isolated situations is a particular problem and has become a specialist field of work.

23.2 Types of maintenance

23.2.1 Introduction

The maintenance of brick and stonework in a maritime context is characterized by the fact that substantial sections of the structures in quesion are totally, partially or intermittently submerged beneath tidal waters. This produces its own initial problems of inspection and survey and in turn often determines the way in which maintenance can be carried out.

23.2.2 Preventive maintenance

It is normally not practicable with static structures to carry out preventive maintenance in the accepted sense of renewing a part in advance of its deterioration to the point where it would be likely to fail in a short time. Some preventive or limiting measures are possible, however, and these are outlined in Section 23.8 below.

23.2.3 Programmed maintenance

In order to carry out a programme of maintenance there is need for a planned system of inspection. As the major portion of a maritime structure is completely or periodically under water this presents a special difficulty. Ways of carrying out inspections are described in Section 23.6 and Chapter 5.

With programmed maintenance, however, deterioration due to the typical causes outlined in the following sections concerning breakdowns, failures and damage can be remedied before it worsens. It is all too easy to ignore underwater degradation on the philosophy of 'out of sight, out of mind' and no competent engineer should be so lulled into a sense of false security. A good system of inspections and programming, with follow-up checks, is the best way of minimizing what may so easily be overlooked. It also enables a realistic figure to be included in the owner's maintenance and revenue budget and helps to reduce the chance of unexpected high expenditure.

23.2.4 Breakdown failure

Whilst many structures are periodically subject to attack from the elements, water flow and subsidence, maritime structures are peculiar in so far as they are constantly subject to the effects of moving water, such as currents, river and estuary flows and tides, as well as to storm conditions in exposed harbours, sea defences and some port works. Masonry retaining walls forming docks can be subject to rapid variations of water levels at their faces with a higher water-table and superimposed loads behind; 'draw-down' can then lead to the risk of wall movement. There is a constant attrition of immersed and semi-immersed structures, leading usually to gradual deterioration but occasionally to major failure. Therefore, apart from surface examinations for breakdown, the alignment of quay walls, jetties and the like should be checked from time to time to watch for outward, downward or rotational movements which might indicate impending large-scale failure.

23.2.5 Damage repair

Another peculiarity of maritime structures, and port works in particular, is that the traffic using them has not only large mass and potentially high energy but is restrained in its horizontal movements only by the skill of its handlers. Vessels of up to (say) 50 000 dwt (dead weight tonnage) are normally manoeuvring at speeds up to 2 knots (1 m/s) in port areas. They will expect to berth against masonry quays or wharves or, as in the case of a vessel negotiating a lock, will grind along the walls.

Damage will occur to the structure even at acceptably low speeds, but on too many occasions ships badly controlled or affected by high winds, engine failure or other imposed conditions will impact a structure with very considerable energy. The resulting damage will also be considerable.

Although very large ships would normally berth at steel or reinforced concrete structures, they could accidentally come into contact with masonry in piers and training works.

All these factors bring about a high maintenance commitment to a maritime undertaking's assets.

23.3 Materials

23.3.1 Concrete

Because of the high cost of the original materials, especially stone, in repairs, there is an attraction in using concrete. Not only can it be precast or cast *in situ*, but it can fill awkwardly shaped gaps where the replacement of masonry would involve removing sound materials in order to insert new, so increasing the extent and the cost of the work. In many instances there is no practicable alternative to using concrete or 'artificial stone', particularly for underwater work, but its use as a cheap automatic substitute for masonry should be examined closely and critically. For example, concrete used as a replacement for good granite in dock and wharf copings is generally inadequate. It chips and splits too easily under 'attack' from shipping and cannot easily be reinforced against this. It lacks durability and early hardness, which cannot necessarily be gained by using expensive materials. Adequate curing time is not possible for much *in situ* work.

23.3.2 Salvaged materials

It may be worthwhile salvaging granite and similar dense stone from less used or redundant sections of a port for repairs to new damage.

23.3.3 Other stones

Softer stones such as sandstone and some limestones, probably used below cope level, are not normally replaceable. Patch repair by brickwork may be adequate.

23.3.4 Brickwork

Engineering bricks will have been used in some original constructions. Complete dock walls and lock inverts – excluding copes, sills, quoins etc. – are sometimes faced in bricks and these have given good service, only deteriorating under mechanical damage.

23.3.5 Guniting

Where surfaces of stone structures are worn away to an insufficient depth to justify replacement of masonry blocks or by brickwork, a gun application of a cement mix can often give an adequately strong finish. Preparation of the surface, cleaning by grit blasting and dry conditions are important and this work can only be done above water level. Quick setting and hardening additives can be incorporated in the mix if the area is to be submerged within the tidal or flooding cycle.

23.3.6 Grouting

Retaining wall structures can develop cracks and splits and a remedial treatment in some instances is to seal the exit of the fissures and pressure grout behind. The same treatment can be used to tighten and stabilize loose blockwork.

23.3.7 Epoxy cements

Various epoxy cements are available for small-scale repairs requiring high early strength and durability, particularly in situations below water but in a temporarily dry working environment. These are relatively easy to handle in restricted spaces and cure rapidly. Care must be taken in selecting the appropriate quality and in establishing the right working conditions and preparations for their use. They are relatively expensive.

23.4 Deterioration

Deterioration will arise from some or all of the following.

23.4.1 Natural causes

Direct wave action, scour from the effect of currents and river movement and draw-down effects are natural causes of deterioration. The penetration of waves behind revetments can wash out the backing material, causing collapse of the blockwork face. Sea spray and sand attack occurs in the 'spray' zone, and salts or other chemicals in the water, and occasionally marine borers, can attack jointing and in certain cases the stone itself. This is usually a slow process in the United Kingdom but can be more rapid in warmer waters.

23.4.2 Mechanical damage

By far the most extensive mechanical damage to port and harbour facilities comes from shipping. Regular callers such as ferries, which nowadays may be very large, will always be responsible for a steady stream of small defects and every now and then for spectacular damage. The very dramatic and extensive damages usually come from the commerical vessels which are less frequent visitors. In both cases, damage is likely to occur with regularity to lead-in jetties, lock walls and knuckles, quay walls, copes and berthing faces. Damage to masonry also occurs in dry docks but there it is an occupational hazard to some extent, though none the less a repair problem. Portable (hand-held) fenders used by berthing staff, if made of hardwood or similar material, can cause damage when inserted between moving ships and dock or harbour walls. More discipline is required in this activity and the port-owning authority should specify what is acceptable.

Mechanical damage to inverts of passages and locks and gate sills and quoins can be caused by shipping, by abnormal semi-buoyant objects jamming in watertight meeting faces and occasionally by chains and wires used to operate gates and caissons.

Mooring ropes can cause grooving, scoring and chasing in copings and decks, especially at joints. Eventually these lead to hazards as ropes can jam or spring, and so they need regular remedial attention.

The stability of quay walls and wharves can be imperilled by the imposition of loads from surface equipment in excess of those allowed for in the design. Old port facilities on which modern heavy lifting capacity mobile and semi-mobile plant is used without restraint are at risk. Operators should always be required to obtain the engineer's

approval for the acceptability and placing of such equipment.

23.5 Inspection

Much deterioration and damage occurs below permanent water level and between high and low water levels, making not only repair but also inspection difficult. Inspection falls into two categories.

1. Programmed inspections are required to check that structures are in a safe condition and to estimate the extent of wear and tear and any consequent maintenance work;
2. Detailed and specific inspection of damage is necessary in order to judge the residual strength of the structure, to gauge the total extent of the damage and to estimate the cost of the repair. The last mentioned may be necessary as a basis for a claim on the party responsible for the damage or their insurance company. Care should be taken not to rush, even under pressure, calculations of the extent of the damage, as first impressions may not indicate the full extent of failure. It is often worth insisting that the cost of remedial works be met on completion and not settled on an estimate in advance, however long it may take for the damage to be made good.

23.6 Possession and access

Although some possession of the area to be maintained is necessary, even for inspections, it is of course essential to have adequate and uninterrupted possession to carry out the work. It is self-evident that most damages caused by, say, shipping are in locations where there is a constant movement or berthing of vessels, so making possession a problem. That is particularly so where damage has occurred in a narrow passage or in an entrance lock. Even when there are not

FIGURE 23.1 Habitat for dock gate sill repairs. (*Courtesy of Associated British Ports.*)

FIGURE 23.2 Habitat being slung into position (with ballast on top). (*Courtesy of Associated British Ports.*)

problems of ship movements, there is still the difficulty of working below water or in limited time periods in the 'dry' between high and low tide levels. These difficulties will be compounded by adverse weather conditions in exposed situations and it is surprising how water can become rough in windy conditions in what are normally regarded as sheltered locations. Certain weather conditions may cause tide levels to be above or below those predicted and this can adversely affect planned access to tidal zone areas of work.

These problems lead to many inventive solutions both for carrying out inspections and work under water and for increasing the available possession time whilst minimizing the disruption to shipping or other facility users.

For many decades now divers have been used for underwater inspections and maintenance work. Self-contained breathing equipment, with its independence and mobility, has greatly reduced the calls on the traditional helmeted and surface-air-demand diver. Surface air supply, however, can still be very useful for prolonged and continuous activities. A physical connection between divers and the surface support squad is very important in maritime work for safety and communication purposes, especially in the hazardous situations so often encountered.

An increasing variety of useful underwater equipment makes the scope of the work that can be executed by divers much greater. This includes underwater cutting, welding and drilling gear, including thermic lances. Divers are limited in the physical work they can do, however, in a state of semi-buoyancy and in frequently bad or non-existent visibility. They are also expensive.

Ideally, the best way of effecting most repairs is by working in dry and free air conditions and with normal visibility. This can be accomplished by the use of limpet dams, caissons and diving bells. Diving bells, however, need surface air supply and also a means of suspension (lifting and lowering) which can be an impediment to normal traffic movement.

Repairs to quay or lock walls can frequently be carried out using limpet dams without excessive intrusions into the waterway. If they are to be left in position while shipping operations continue, however, they need to be well protected by fenders and lighting and shipping must be fully advised of their presence. No personnel should be allowed in limpet dams while shipping is moving nearby as it takes only a slight movement of the dam to break the seal and cause the chamber to flood.

If there is a repair to be done on the invert of a lock, dock or passage, then an open-topped caisson can be utilized. Like a diving bell, such a caisson will often be a navigational obstruction and it needs a lifting craft or crane in attendance.

A device that has come into its own in more recent years, made practicable by the availability of scuba equipment, is the underwater habitat. This is rather like a diving bell in concept, in that it can be lowered onto the site of the work; thereafter it is quite a different tool, though still requiring an air supply. It is normally made specially for each location and designed to fit as accurately as possible to the profile against which it is to be sealed. After positioning over the site of the work, it is sealed by scuba divers using such caulking materials as rope, oakum or rubber and pumped dry. The divers inside it can then remove their breathing equipment and work in almost dry conditions, with the area for repair being fully visible and accessible. Shipping can move normally with the habitat remaining in position. There is no unreasonable time limit on divers working within it and no air pressure problems at the depths concerned. Compressed air equipment – drills, saws, pumps etc – and hand-held tools can be used in the habitat in the same way as in free air. Figure 23.1 shows a habitat made to effect repairs to the stone sill of a lock gate, while Fig. 23.2 shows it being lowered into position.

23.7 Maintenance procedures

Repairs to damage or failures will need to be carried out to a time scale which accords with the severity and location. Even if the damage is not in

FIGURE 23.3 Sprayed concrete repairs to masonry dock wall. (*Courtesy of Merseyside Development Corporation.*)

a critical location, it may in some circumstances be necessary either to effect temporary repairs or to make good the extremities of the damaged section so as to avoid further deterioration. The sooner the permanent repair is completed the better, of course, as this will help to avoid a spread of the damage or the build-up of silt or marine growth, all of which will increase the extent and cost of the final repair (Fig. 23.3).

Repairs to revetments and embankments which have suffered damage in the tidal and wave zones can take the conventional forms of replacement like for like, by *in situ* concrete or by grouting smaller stones into the voids. However, where large areas are affected, there is now a variety of small interlocking blocks which are hand placed to the gradient required and held in position by geotextile tendons laid through the blocks and down the slope, with anchors at top and bottom.

This type of facing is flexible enough to accommodate small irregularities in the backing material but, with the right choice of block, strong enough to withstand wave action. It also seals the face against water intrusion into the backing to reduce the risk of washout. Other geotextile materials are useful for earth reinforcement or membranes within the body of embankments.

Storms may cause dramatic damage, not only to outer faces but also to the backing of underlying materials. The whole composition of the structure will need to be reinstated in the repair but the specification may require up-grading to prevent future damage under similar conditions.

In the case of gradual deterioration either by natural wear and tear or by continual but small-scale erosion by ships and boats, moorings, fenders or vehicles, a programme of maintenance works will be necessary. This is work that lends itself to being undertaken by a direct labour force rather than by a contractor as far as above-water repairs are concerned. Below water, repairs and inspections by divers may well pose a dilemma in terms of direct or contract labour. If the maritime undertaking is large enough, then the employment of staff divers who can be kept fully occupied is likely to be the more cost-effective system. These men become familiar with the structures that have to be inspected and they know from experience what defects are likely to be encountered and the best methods of repair. It is necessary to make sure that divers are kept up to date in diving techniques and underwater equipment. It is very desirable that engineers or experienced technicians should be able to dive and give professional assessments of underwater defects. However, in most undertakings it is difficult to train such people and to engage them in sufficient diving to ensure that they remain proficient.

Current diving regulations are, justifiably, very stringent – although based more on combating the dangers of deep-sea (North Sea) diving than on maritime work adjacent to land – and it is expensive to employ full-time diving crews. It may therefore be advisable to keep a minimum-sized diving squad in an undertaking's employ and to supplement them with contractors for larger repair works.

23.8 Prevention

There is scope for examining some simple measures for minimizing deterioration and damage.

Where mechanical damage from shipping is experienced, the obvious method of prevention is the introduction of permanent fenders. This can be an expensive provision which, in turn, produces a maintenance commitment of its own. Whether rubber or timber is used in any of the fixed, hanging or floating configurations, the cost of maintenance will be significant and, once provided, the ship owner will always expect it to be there and in good order. The cheapest and most

readily available fendering material is the large vehicle tyre, and if there is no danger of a detached tyre fouling underwater equipment or installations then they are nearly ideal. Scrap tyres have a multitude of uses and, as has been frequently said, 'if motor tyres had not been invented for vehicles, they would certainly have been invented for all these other purposes!'

With the advent of larger ships, especially ships large in beam width, many dock entrances have their side clearances so reduced that there is no room for permanent fenders to be installed. Some ships, such as ferries, are equipped with 'fender' beltings which can cause serious damage to structures. If owners of these ships cannot be persuaded to dispense with beltings they should at least be asked to design them with deep flat rubbing faces to spread the load or, better still, to fit them internally, i.e. as horizontal stiffened ribs around the inside of the vessel at the level at which she is likely to come into regular contact with berthing installations; these do not stop scratches on the ship's hull but they help absorb berthing forces and avoid excessive denting.

In some locations continuous steel plates superimposed on stonework at coping level can reduce damage, though there is a danger that ruptured plates may produce more hazardous conditions if they are not noticed and repaired quickly.

Increases in ships' draughts can cause problems of bed scour because propellers are very close to dock and channel bottoms. Near to structures this scour can lead to undermining and collapse. Regular bed surveys therefore need to be carried out in order to detect any excessive bed movement before it affects the adjoining structures. Ports which have siltation problems will undertake regular hydrographic sounding surveys to check upon available depths for shipping within docks, at their entrances and in any approach channels. However, such surveys will not necessarily show in sufficient detail all those locations where structures may be at risk because of scour or over-dredging. A check made with a simple lead line or other physical means of direct measurement, in

FIGURE 23.4 Repairs to a retaining wall within the tidal zone of the River Severn. (*Courtesy of British Rail.*)

what will inevitably be a fairly small area, is not an expensive exercise. Equally the placing of suitable dense stone on the bed of the waterway can be a comparatively inexpensive way of reducing or eliminating scour.

Deterioration due to chemical or biological attack is difficult to combat, given that there is little or no control over the cause. The choice of repair materials is important and the vulnerability of existing materials that have succumbed to attack should give some indication of what to avoid.

23.9 Manpower

The maintenance of structures is an unromantic engineering occupation but one where much money is spent and where cost effectiveness is vital. It is not an unskilled activity and it is one where familiarity is all-important. Good staff, with a knowledge of the structures and their use, are very necessary. Inspection and supervision are equally important, as both the initial deterioration and the repair work are often obscured or difficult to measure. If the undertaking is large enough, specialist gangs, with skilled craftsmen, divers and supervisors, are justified for this maintenance commitment.

23.10 Case Histories: repairs to retaining walls affected by tidal action

Water had progressively removed mortar from the joints and filling from behind a retaining wall within the tidal zone of the River Severn (Fig. 23.4). The filling had become replaced with silt and each successive tide produced a draw-down on the wall. Normal pressure pointing was undertaken above tide level, with a quick-set additive for the area below it, to seal the face. Pipes were mortared into holes drilled through the wall. Water flush was applied to wash out the silt and cement–pulverized fuel ash grout was injected.

In the case of a sea wall in South Devon, gales and tidal effects had stripped all sand from part of the foreshore and eroded holes in the sandstone on which the masonry was founded. Some of these holes were plugged with concrete-filled sandbags and others with grout bags (geotextile fabric bags filled *in situ* with pumped cement–sand grout). Sprayed concrete, with a quick-setting silicate additive, was then applied as a protective layer at beach level.

23.11 References and further reading

British Standards Institution. BS 6349. *Code of practice for Maritime Structures* Part 1 (1984) General criteria. Part 2 (1988) Design of quay walls, jetties and dolphins. Part 3 (1988) Design of dry docks, locks, slipways and ship building berths, ship lifts and dock and lock gates.

Construction Industry Research and Information Association (In press) *The maintenance and rehabilitation of old waterfront walls*. E. & F. N. Spon, London.

Institution of Civil Engineers. *Conf. Proc. on Flexible Armoured Revetments Incoporating Geotextiles*, March 1984. Thomas Telford, London.

Conf. Proc. on Port Engineering and Operation, March 1985. Thomas Telford, London.

Conf. Proc. on Maritime and Offshore Structure Maintenance, February 1986. Thomas Telford, London.

Institution of Structural Engineers (1943) *Report on Retaining Walls*.

Permanent International Association of Navigational Congresses (PIANC) (Geneva):

Port Maintenance Handbook. (Supplement to Bulletin No. 50, 1985).

Final Report of the International Commission for the study of locks (supplement to Bulletin 55, 1987).

Guidelines for the Design and Construction of Flexible Revetments Incorporating Geotextiles for Inland Waterways, (supplement to Bulletin 57, 1987).

Bramley, M. E., Bray, R. N., Hook, B. J. and Tatham, P. F. B. (1989) *Maintenance and rehabilitation of old waterfront walls*. Report of meeting of Maritime Engineering Group at Inst. of Civil Engineers in October, 1988. Proc. Inst. Civil Engineers, 86, 845–8.

Dry stone walls

C.J.F.P. Jones

24.1 Introduction

Dry stone walls are defined as stone walls over 1.2 m high constructed without mortar; they may be up to 15 m in height and vary in length from a few metres to over a kilometre. A recent census undertaken by the Department of Transport and the Highway Authorities in the United Kingdom has shown that retaining walls are more numerous than previously recognized (Table 24.1), and 50% of these walls are of dry stone construction (Department of Transport, 1987). The maintenance problems of these structures are becoming more apparent and in some districts the maintenance costs of dry stone walls has exceeded the costs required for bridge maintenance (Hayward, 1983).

24.2 Form and structure

Most dry stone walls date from the Industrial Revolution which produced an unprecedented boom in the construction industry. Wall construction was necessitated by the demand for improved transport and the need to build roads and railways on side-long ground typically represented by the dales of Yorkshire and the valleys of Wales and Scotland. In these areas stone was plentiful and today the dry stone wall is a common feature of the landscape. In many cases the walls retain land above the highway as well as acting as a 'burr' wall on the lower slope retaining the carriageway itself (Figure 24.1). Thus, for every kilometre of highway up to 2 km of dry stone retaining wall may exist.

The walls frequently take the form illustrated in Fig. 24.2. They were built either as facing to cuts in a vertical or near vertical unstable or friable material or as 'burr' retaining walls. In many cases walls have been constructed as integral parts of adjacent structures, some of which have subsequently been removed.

The walls were usually constructed directly onto the topsoil without proper foundations. Only the front face of the wall consists of finished material. The stones used for the face were of medium to large size, carefully graded and placed by hand to fit; behind these, small flat stones were used, laid in a horizontal plane grading back from the face (Fig. 24.3). The structures are porous and

TABLE 24.1 Length of retaining walls in the United Kingdom

Owner	Minimum length (km)
English Counties	1254
Metropolitan Counties	2045
Wales	982
Scotland	1117
TOTAL	5398

After Department of Transport (1987).

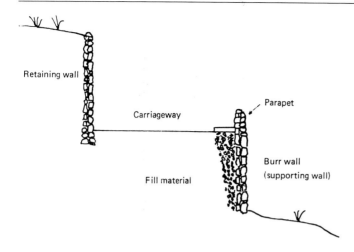

FIGURE 24.1 Dry stone walls supporting a highway and retaining land above the highway.

essentially flexible and would have been built in layers using relatively unskilled labour and little construction plant; in particular, no modern compaction equipment would have been used. The nearest modern approach to this form of construction is a partly reinforced earth structure, as shown in Fig. 24.4, in which layers of fabric reinforcement are integrated with the facing stone. As with any reinforced earth structure, that in Fig. 24.4 would be constructed in layers with minimum compaction. Research by Smith (1977) has suggested that the stability of such a structure could be guaranteed provided that the thickness:

height ratio (*B:H*) is greater than 3:10. Many dry stone walls do not meet these criteria and their longevity and overall stability is frequently difficult to explain. Few dry stone walls meet accepted design criteria for earth-retaining structures.

Where walls pass over underground streams or drainage ditches, the need for competent bridging was usually recognized by the use of especially large blocks near the base of the wall, well matched for a close fit. Elsewhere, different sections of the same wall may vary greatly in stone grading and quality of matching. This suggests

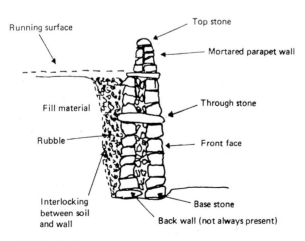

FIGURE 24.2 Form of dry stone wall. (*After Emmerson, 1988.*)

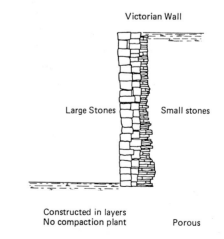

FIGURE 24.3 Construction technique for dry stone walls.

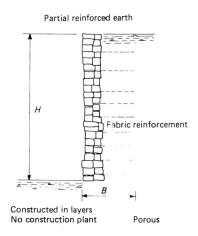

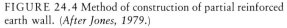

FIGURE 24.4 Method of construction of partial reinforced earth wall. (*After Jones, 1979.*)

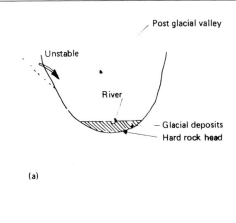

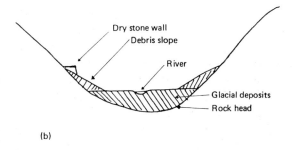

FIGURE 24.5 Position of many dry stone walls in a post glacial valley: (a) stage 1, post glacial valley; (b) stage 2, final valley formation.

that different construction gangs could have been used on different sections.

Many dry stone walls have been raised to a higher level at a later date. These are fairly easily recognized by a change in construction material, the start of which may be identified by a narrow ledge along the line of the original level which may have grass growing on it. The reason why the height has been raised is not always obvious but may have been to improve highway alignment.

The material forming the walls can be very variable. The stone forming the walls is frequently poor with the better quality material apparently being used for building construction. Changes in material quality along the wall are frequent and previous maintenance work is usually identifiable by the use of different quality or differently shaped stones. Material quality, in the form of the stones forming a wall, has a marked influence on durability and the need for maintenance and repair.

24.3 Geology

Many old dry stone walls were built to support highways on side-long ground in valleys formed by glacial action. When first formed the valleys may have been steep sided; with time, the slopes

may weather causing material to be deposited near the valley floor (Fig. 24.5).

Many walls appear to have been constructed on the debris slope and their position may identify the change in slope between the relatively loose debris and the exposed rock head. In these conditions, precipitation falling onto the higher ground finds its way to the valley floor either as surface run-off or as subsoil flow near ground level. As the water reaches the bottom of the scarp slope it is able to percolate through the debris material into the alluvial–glacial deposits in the basin of the valley. This flow, if strong enough, may slowly remove fine material from the composition of the fill and carry it out through the front face of the wall. Evidence of this has been seen in borehole logs where the fill in the lower sections

TABLE 24.2 Factors influencing the deterioration of dry stone walls

Natural	Man-made
Form of construction	Traffic loading and vibration
Percolation and removal of fines	Accidental damage
Material quality	Vandalism
Soil creep	Burst water mains
Storm or flood	Excavation at the toe
Vegetation	New construction (adjacent)
Animals	Removal of support
Old age	De-icing salts
Geological conditions	

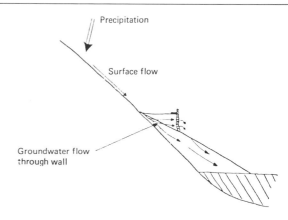

FIGURE 24.6 Groundwater flow relative to a dry stone wall.

of the wall contains significantly less fine material than that near the top (Emmerson, 1988). A second effect resulting from this potential removal of fine material from within the body of the structure will be to cause settlement of the fill. The coarse fill is free to rearrange itself, possibly spreading horizontally at the same time. This action may be partly responsible for the characteristic bulging exhibited by many dry stone walls (Fig. 24.6).

FIGURE 24.7 Typical failure of a dry stone wall. (*Courtesy of C.J.F.P. Jones.*)

352

24.4 Factors influencing wall deterioration

There is a range of factors which influence dry stone wall behaviour and subsequent deterioration. It is possible to identify two groups, classified as either natural or man-made. Table 24.2 is not exhaustive but contains many of the common causes of deterioration. The key elements are seen as the form and nature of the construction, the presence of water and traffic loading. The most usual failure method is a slip failure triggered by the collapse of the front face of the wall (Fig. 24.7). The latter occurs when the bulging face becomes unstable. Prediction as to when this will occur is a matter of experience and instinct.

24.4.1 Natural causes

Many of the natural causes influencing the behaviour and deterioration of dry stone structures are mentioned in Sections 24.2 and 24.3 (see also Fig. 24.7).

(a) Form of construction

The absence of mortar between irregularly shaped and sized stones forming a structure produces large voids in the walls and also results in many stones being in point contact with adjacent blocks. Point contact pressure can be high and may cause crushing and splitting of stonework. This contributes to the weathering process inherent in the walls. The openness of the joints encourages weathering of the surface of the stones and further reduces the area in contact with adjacent stone work. This facilitates additional crushing and the process continues slowly over a period of time, gradually eliminating the interlocking structure of the stone blocks and de-stabilizing the wall. This is a long-term process.

(b) Percolation of water

Although difficult to quantify, the void space in a dry stone wall may constitute 10% or more of the total structure. The voids permit percolation of moisture through the structure but also permit the movement of dissolved salts to the surface of the wall. Percolation also causes leaching of fine material from the fill which may lead to subsidence of the back-fill. Chemicals deposited above the wall are free to pass through the structure and chemical weathering of the rock within the wall is possible.

(c) Back-fill

Little information is available on the behaviour of back-fills used behind dry stone walls or the nature of their composition. It is reasonable to assume that the minimal compaction received at the time of construction would have been improved with time and the passage of traffic. The gradually increasing volumes of traffic from the turn of the century and the increasing axle loadings may have served to eliminate inconsistencies in the back-fill. It is probable that when first constructed large settlements were encountered.

Many back-fills are known to be susceptible to moisture; burst water mains or drainage mistakenly directed into the back-fill have on occasion resulted in an increase in lateral earth pressures leading to collapse. The effect of water movements within the back-fill over many years may also have an effect.

(d) Material deterioration

The quality of the original stone used in the wall is critical to durability. It is apparent that many dry stone walls were built with inferior material and it may be concluded that the best quality stone was retained for use in public buildings and houses. The effect of the use of poor materials is excessive weathering, leading in turn to honeycombing and substantial deterioration of weakly cemented stone. The effects of pollution can be identified as accelerating deterioration, as can the presence of vegetation, water passing through the structure and the action of frost. In badly weathered walls the void ratio may be as high as 50%–60% of the wall face.

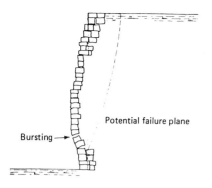

FIGURE 24.8 Mode of failure of a dry stone wall.

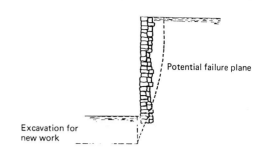

FIGURE 24.9 Failure of a wall resulting from excavation beneath the toe.

(e) Storms and floods

Dry stone walls adjacent to rivers or with their foundations in watercourses are subjected to the usual problems of scour. Because of the low quality of construction which is frequent with this form of structure, the effects of floods may be severe.

(f) Vegetation

The permeable nature of dry stone walls means that no water pressure builds up behind the face of the structure. However, the walls are not dry; indeed, there is considerable evidence of high water flows through most of the walls and much of the stonework is permanently wet or saturated. The presence of water attracts vegetation, either in the form of grass or moss or as small trees. The latter, if not removed, cause displacement of stones from the face and eventually local collapse of the structure. The presence of trees or bushes growing out of the face of a dry stone wall is an indication of poor maintenance.

(g) Animals

Dry stone walls may attract wildlife. Normally these animals are benign and their presence is a sign of relative stability. Occasionally burrows may have an effect on local stability.

(h) Frost

Many dry stone walls are in upland areas and are constantly exposed to severe weather conditions, especially in winter. This, combined with the open nature of the joists between stones, means that moisture within the wall may be frozen to substantial depths behind the face. The effect of water freezing in the structure behind the face may be to force the face outwards, creating larger voids in the wall; in time the effects of cyclic freezing and thawing may contribute to the forward movement of the wall face (bulging) to the point of collapse. The incidence of wall failure is greatest during the spring (Fig. 24.8).

24.4.2 Man-made causes of deterioration

(a) Loading

Most dry stone walls were constructed during the nineteenth century when highway traffic was largely horse drawn. Since that date, there has been a large increase in vehicular traffic and highway loading. The increase in numbers of heavy vehicles has been dramatic, particularly during the last two decades, and this increase has been accompanied by an increase in overall vehicle weight (Chatterjee, 1984).

The direct effect of increased loading is to increase lateral soil pressures acting on the retaining wall. Qualitative and quantitative assessment of these lateral pressures is difficult, particularly as the dry stone walls present little apparent

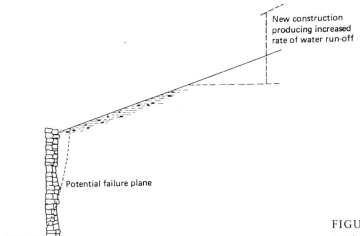

New construction producing increased rate of water run-off

Potential failure plane

FIGURE 24.10 Failure of a wall induced by new construction and changed soil moisture conditions.

resistance to such increase in pressure. A secondary effect of traffic loading is vibration but, as with direct loading, assessment of this effect is difficult.

(b) Impact

Dry stone walls are susceptible to direct impact either by vehicles passing behind the structure or to vehicles travelling in front of the wall. Because of the nature of the construction, vehicular impact normally results in local collapse.

(c) Burst water mains

The action of water on the dry stone wall has been described in Sections 24.4.1(b) and (c) and the presence of water is frequently identified with failure. The standard position for water mains is in the highway and as a result many mains are positioned close to dry stone walls. The presence of a burst water main in a highway supported by a dry stone wall may cause the back-fill to become saturated and soften. Failure under these conditions is common.

(d) New construction

New buildings constructed either in front of a dry stone wall or on a hillside above a wall may change conditions and cause either distress or failure.

The lack of foundations to dry stone walls is frequently not appreciated by present-day contractors and builders. Excavation in front of the wall for the foundation of a new structure or the construction of a car park or road may result in undermining of wall foundations and sudden collapse (Fig. 24.9). Similarly, new buildings constructed on a hillside supported by a dry stone wall at a lower elevation may lead to problems, particularly if the new construction produces an increase in the rate of water run-off. The consequent increased saturation of the wall back-fill may result in collapse (Figs. 24.10 and 24.11).

(e) De-icing salts

The use of de-icing salts in winter has recently been identified as a contributing factor in the deterioration of dry stone walls. The part most prone to deterioration is the base of the wall at the back of the footpath.

(f) Mining subsidence

Dry stone walls are flexible and are usually able to absorb the ground strains produced during mining subsidence. However, the presence of a fault may

FIGURE 24.11 Failure of a dry stone wall caused by new construction producing increased run-off and additional earth pressure. (*Courtesy of C.J.F.P. Jones.*)

concentrate mining movements and in these conditions dry stone walls may be damaged. The damage is usually confined to the effects of compressive ground strains (see Chapter 9).

24.5 Identification of a problem

There is little recorded knowledge concerning the action and behaviour of dry stone walls; the identification of problems and the need for maintenance is often dependent upon local engineering experience.

The number of collapses annually is increasing and in parts of the United Kingdom they have become very frequent. The assessment of dry stone walls consists of regular visual inspections and comparison with adjacent structures. Qualitative judgements are difficult since conditions will vary greatly with quality of stone used, age, subsoil conditions, geometry, weathering factors and local expectations.

Where past movement or the condition of the structure raises doubts concerning stability, regular monitoring should be introduced. Decisions relating to structural safety and conditions often depend upon engineering instinct, although simple visual aids such as tell-tales can be useful to determine whether the structure is moving or in a state of equilibrium. Bulges in dry stone walls are common and the wall may remain in this condition for many years; alternatively, collapse may be imminent. Some visual signs may indicate movement or a change in conditions; particular attention should be taken of

1. newly cracked stonework

2. a recognizable increase in bulging
3. the presence of localized wet areas
4. lush vegetation, localized vegetation
5. horizontal cracks in the carriageway behind the wall
6. settlement behind the wall
7. burst water mains
8. new construction adjacent to the wall

In most cases deterioration may be slow but progressive. The action of de-icing salts on walls may be identified by deteriorating stonework concentrated in a band often less than 1 m from the ground.

Dry stone walls which gain support from adjacent structures present a particular problem. In normal circumstances these are stable, but removal of support as a result of change of use or demolition may result in distress and the need for expensive remedial work to restore stability. Identification of this potential problem is dependent upon prior knowledge of any proposals for demolition or redevelopment.

24.6 Maintenance and repair techniques

The maintenance of dry stone walls may be divided into two groups: maintenance prior to collapse and repair following failure. Normally repair is preceded by action to remove any identifiable cause for deterioration or distress. In some cases this may be possible, as with the repair

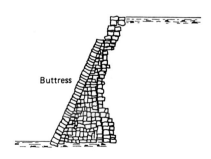

FIGURE 24.12 Construction of a buttress to support a wall.

of a burst water main. Elsewhere the cause of the problem may remain, in which case remedial measures appropriate to the conditions are applicable.

24.6.1 Prior to failure

(a) Mortar

The use of mortar either on the face or between the stones to ameliorate deteriorating conditions is common; the effect is frequently palliative and may even lead to worse conditions. Widespread use of mortar, particularly that applied as a rendering to the face, blocks drainage paths and may hasten collapse.

(b) Stone replacement

The replacement of individual weathered stones is a standard technique for small repairs and is usually successful.

(c) Pressure grouting

Pressure grouting of badly honeycombed walls is a successful repair technique. This method has the effect of transforming a wall and the back-fill into a gravity structure and the retaining wall can no longer be truly classified as a dry stone wall. The technique has two essential elements:

1. Weep pipes must be provided through the wall to ensure wall drainage;
2. Grouting must be undertaken at very low pressures, as the use of too high a pressure may result in collapse of the wall. The use of vacuum grouting may be applicable. Care must be taken not to fill adjacent drainage facilities with grout.

(d) Buttresses

Where conditions permit and land is available the construction of buttresses may prove successful in stabilizing a wall. Typically these may be used when support in the form of adjacent structures has recently been removed (Fig. 24.12).

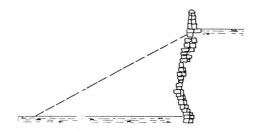

FIGURE 24.13 Use of an earth bund to eliminate potential wall failure.

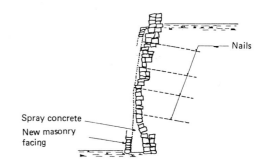

FIGURE 24.15 Soil nailing repair of dry stone wall. (Weep pipes must be provided.)

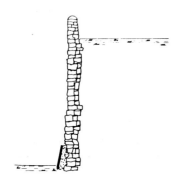

FIGURE 24.14 Use of 'spats' to repair base of wall.

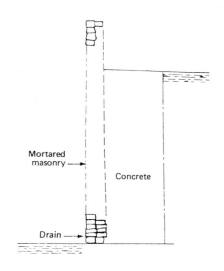

FIGURE 24.16 Reconstructioon of dry stone wall using masonry-faced mass concrete gravity structure.

(e) Earth bunds

The construction of a permanent earth buttress adjacent to the wall provides a permanent resolution of the dry stone wall problem. The wall is effectively removed. This solution may be acceptable to adjacent landowners provided that the resultant slope is usable for agricultural purposes (Fig. 24.13).

(f) Spats

Deterioration at the base of a wall, similar to that resulting from the effects of de-icing salts, may be repaired using 'spats' formed from paving slabs back-filled with concrete (Fig. 24.14).

(g) Soil nailing

A new form of repair for dry stone walls is the use of soil nailing. This method has been described by Bruce and Jewell (1986 and 1987) and the first application of the technique in Britain was for the repair of a dry stone wall in danger of collapse (Fig. 24.15). As with pressure or vacuum grouting, weep pipes must be constructed to alleviate any build-up of water pressure.

24.6.2 After failure

(a) Reconstruction

The most common form of repair following collapse is the construction of a gravity wall faced

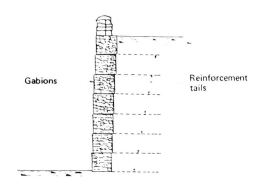

FIGURE 24.17 Reconstruction using tailed gabions.

with masonry (Fig. 24.16). The design and form of a gravity wall follows conventional retaining wall practice and is not based upon dry stone wall techniques.

(b) Gabions

Where land in front of the wall is available, reconstruction using gabions may be effective. The advantage of this form of remedial work is essentially that of economy, whilst the resultant structure is compatible with the dry stone wall in being flexible and porous. The use of gabions to protect walls from scour is particularly advantageous.

The use of tailed gabions, in the form of reinforced soil, provides greater economy and is arguably a close reconstruction of the original structure but with the advantage of increased stability and less reliance on the integrity of the stonework forming the face of the wall (Fig. 24.17).

24.7 Legal considerations

Dry stone walls present particular legal problems. Ownership of a wall is often in dispute and frequently denied. With ownership goes responsibility for repair and maintenance. Walls supporting highways are normally in the owner-ship of the Highway Authority or the Department of Transport.

Retaining walls supporting the land above the highway are usually in different ownership from that of the highway. The wall may have been built as part of the highway but ownership and the responsibility for maintenance may have been transferred together with maintenance funds to adjacent landowners. With the passage of time the land may have changed hands and with it responsibility for maintenance and repair of any walls. However, this responsibility may not have been indicated in any sale and the sums provided to the original owner for maintenance may not have been transferred.

The presence of retaining walls on a property is not usually identified on the deeds and the potential problem of dry stone wall maintenance is seldom addressed during the conveyance of property. The cost of repairs of some walls may be very high and the consequence of failure of a wall may be beyond the means of the owner. Where it can be demonstrated that the retaining wall was built for highway purposes, it is possible to make a case for the Highway Authority to be responsible for maintenance.

Identification of the original purpose of the wall may not be simple, as records are usually deficient. However, walls which have a uniform appearance and which have been built as a single element but which support the land of a number of adjacent owners are likely to have been built as part of the highway. Walls which are contained within the land of a single owner or walls which change appearance or form between adjacent properties may normally be considered to be the responsibility of the landowner.

Some dry stone walls are supported by adjacent structures. In these cases the owner of the wall has the right of support. Should the adjacent structure be removed or demolished the owner may claim support for the dry stone wall. However, this support may only be claimed whilst the structure is in place. Accordingly the rights of the wall owner have to be established prior to demolition.

24.8 References and further reading

Bruce, D. A. and Jewell, R. A. (1986–87) Soil nailing: application and current prcatice. *Ground Eng.*, **19**(8), 10–16; **20**(1), 21–33.

Chatterjee, S. (1984) Assessment of old bridges. *Inst. Highways and Transportation, National Workshop on Bridge Construction and Maintenance, Leamington Spa, April.*

Cooper, M. R. (1986). Deflections and failure modes in dry-stone retaining walls. *Ground Eng.*, **19**(8), 28–33.

Department of Transport (1987) *The Assessment of Highway Bridges and Structures. Bridge Census and Sample Survey,* HMSO, London.

Emmerson, G. R. (1988) Factors affecting the stability of dry stone retaining walls. B. Eng. Dissertation, University of Newcastle upon Tyne.

Flower, I. W. and Roberts, D. H. (1987) Anchored repairs to dry stone masonry retaining walls on a major trunk road. *Proc. Int. Conf. on Structural Faults and Repair*, London, July, Engineering Technics Press.

Hayward, D. (1983) Tumbling walls confound West Yorkshire. *New Civil Engineer*, August, pp. 26–7.

Jones, C. J. F. P. (1979) Current practice in designing earth retaining structures. *Ground Eng.*, **12**(6), 40–5, September.

Smith, A. K. C. S. (1977) Experimental and computational investigation of model reinforced earth retaining walls. Ph.D. Thesis, Cambridge University.

Chapter twenty-five

Structures of historic interest

R. L. Mills

25.1 Introduction

The task of carrying out work on historic structures can be summarized as one of extending their life whilst causing the minimum of disturbance to the original construction.

Engineers tend to adopt an analytical approach to the solution of problems with which they are confronted but this approach does not provide the most sympathetic or appropriate treatment of our historic structures. A better understanding can be obtained by considering such monuments as undergoing an on-going load test, requiring the exercise of skill in assessing the past performance and in the examination of any inadequacies which may have appeared.

An invaluable aid in this assessment is an accurate monitoring system, which need not be expensive to install; in addition to the financial savings which normally ensue, there can be very significant reductions in the type of work which leads to destruction or disturbance of the original fabric.

25.2 Ancient monuments - scheduling

The first relevant legislation was the Ancient Monument Protection Act of 1882. This entrusted to the Commissioners of Works (whose responsibilities are now vested in the Secretary of State for the Environment) the protection and preservation of ancient monuments. The Schedule to the Act listed those Monuments to which the Act applied and these, with subsequent additions, form the published list of Monuments of National Importance.

Reference to the 1882 Schedule gave rise to the term 'scheduling', by which the listing of monuments is commonly known. Apart from the Schedule, the whole of the 1882 Act was subsequently repealed and the Ancient Monuments Acts now in use are as follows:

1. Ancient Monument Protection Act 1882 (Schedule only)
2. Ancient Monuments Consolidation and Amendment Act 1913 (as amended)
3. Ancient Monuments Act 1931 (as amended)
4. Historic Buildings and Ancient Monuments Act 1953
5. Field Monuments Act 1972
6. Ancient Monuments and Archaeological Areas Act 1979

Ancient Monuments Memorandum 102 (Department of the Environment, 1972) outlines numerous Acts and Statutes affecting such monuments. The majority of these contain clauses known as 'saving clauses', which safeguard the provisions of the Ancient Monuments Acts and ensure that they are not overridden.

25.3 Summary of powers and duties conferred by the Acts

1. The Secretary of State shall appoint an ancient Monuments Board as an advisory body;
2. He shall draw up and publish a list of monuments whose preservation is considered to be in the national interest;
3. Before including a monument in the list he shall serve notice of his intention on the owner and occupier in the case of buildings;
4. Owners (and occupiers) served with a notice must give three months notice of any works affecting the monument. Failure to do so renders them liable to prosecution. Other persons damaging a monument may be prosecuted for damage if the monument has been included in an Order in Council;
5. The Secretary of State may use compulsory powers of preservation if a monument is in danger of destruction. If it is liable to fall into decay through neglect he may make a Guardianship Order. He may also draw up a preservation scheme to control all development within its vicinity (preservation schemes are no longer used, except for the one scheme already in existence);
6. Use of compulsory powers entitles those injuriously affected to compensation;
7. The Secretary of State shall appoint Inspectors of Monuments;
8. The Secretary of State or the Ancient Monuments Board may authorize entry onto land to inspect a scheduled monument and they may give advice to owners free of charge apart from out-of-pocket expenses. The Secretary of State shall give advice with reference to any monument protected by a Preservation Order;
9. The Secretary of State may make grants to owners of monuments on condition that the proposed work meets his approval and may supervise any work carried out;
10. The ancient monument architect or engineer is to ensure that these conditions are met;
11. The Secretary of State may make annual payments, known as acknowledgement payments, to owners of field monuments who agree to leave them uncultivated;
12. The Secretary of State may become the guardian of a monument and assume full responsibility for its control, repair, maintenance and management. He has the right of access for himself, his inspectors, agents or workmen, for these purposes;
13. The Secretary of State may carry out excavations on land in his ownership or guardianship and on other sites with the owner's permission;
14. The Secretary of State may make regulations to cover monuments in his ownership or guardianship, including regulations governing public access;
15. Local authorities may also become owners or guardians of monuments in their areas and assume responsibility for them;
16. Local authorities may make grants to owners, provided that the work is approved by the Ancient Monuments Board or the Secretary of State;
17. The Secretary of State may receive voluntary contributions towards the repair of ancient monuments.

25.4 Listing

Under the ancient monuments legislation, the Secretary of State has a duty to schedule buildings and structures above or below ground whose preservation is of national importance because of their historic, architectural, traditional, artistic or archaeological interest. Two types of structure cannot be scheduled – occupied dwelling houses and buildings in ecclesiastical use. If of suitable quality for preservation, both of these can be included in the list of buildings of special architectural or historic interest under section 54 of the Town and Country Planning Act of 1971.

The great majority of Scheduled Monuments are archaeological sites, ruins or structures for which there is no present-day use. Listed buildings are mostly occupied. There is an area of overlap, however, e.g. barns, bridges and guildhalls and, to some extent, industrial structures. Lines of demarcation cannot readily be determined because of the differences in the relevant criteria and the methods of selection. Scheduling has the effect that the Secretary of State becomes directly responsible for the protection of the monument and for ensuring that its treatment, repair or use is compatible with its preservation as a monument. Scheduling is accordingly selective, and identification and selection take time. However, the protection of listed buildings, as an extension of planning control, is primarily the responsibility of local authorities. Every building which reaches a certain standard of architectural or historic interest should be included and listing can be relatively quick. Some structures are scheduled as ancient monuments and listed as historic buildings. As can be appreciated, listing is much more general and comprehensive than scheduling. The duplication between the two may appear confusing, but in practice scheduling takes precedence if a structure is both listed and scheduled.

The lists of buildings and structures of special archaeological or historic interest are intended to constitute a heritage register covering the entire country. All buildings built before 1700 which survive in anything like their original condition and most built between 1700 and 1840 are listed, though selection is necessary. Buildings of character and quality built between 1840 and 1914 and, more recently, between 1914 and 1939 can also be listed.

The items listed are classified in grades of relative importance, as follows:

Grade I   Items of exceptional architectural or historic interest (about 2% of the total)

Grade II*   Particularly important items of more than special interest (about 4% of the total)

Grade II   Items of special interest which warrant every effort being made to preserve them

The lists are available for inspection at local council offices. Buildings and structures included in them may not be demolished, altered or extended without 'listed building consent' from the local planning authority; penalties for non-compliance can be heavy. The criteria for listing ignore the state of repair of the item, the costs of repairing it or its unsuitability for modern needs. However, these factors can be taken into consideration as part of the case for listed building consent.

Local authorities have the power to require a listed asset to be maintained and, to prevent an owner from allowing it to fall into decay by neglect, may serve on him a repairs notice specifying what work needs to be done. It will invariably be a requirement that any alteration, repair or strengthening of a listed structure be carried out in a manner sympathetic to, and using materials in keeping with, the original.

25.5 Remedial works

When considering alternative schemes for remedial works, engineers will sometimes propose demolition and rebuilding as a solution to the structural inadequacies which they have encountered. Conservationists will normally reject such an approach, which, even in those cases where the materials could be re-used, will result in the loss of the historical significance of the orginal structure; furthermore, the character and patina of age cannot be re-captured. This reasoning applies equally to partial rebuilding. It follows, therefore, that proposals to take down a historic structure and erect a facsimile in a different material (e.g. to substitute concrete for masonry) will be automatically rejected.

The aim of the engineer involved in this type of work should be to provide the necessary structural integrity with the least alteration to the original fabric. There may be some conflict with economic considerations but a careful study of the history and construction of the structure will often result in a scheme which meets the requirements of both the conservationists and the maintenance engineer.

It must also be appreciated that many structures are of importance as examples of industrial archaeology as well as for their architectural interest. Original fittings such as lamp posts, balustrades and plaques should therefore be retained; new functional services such as pipes, cables and lighting should be carefully designed and sited so as to avoid harming the appearance of the original construction.

The duty of care which is placed upon the owner of a historic structure, whether scheduled or listed, is described above; some assistance in discharging that duty is available. The local authority and the Secretary of State, acting through English Heritage, may contribute to the cost of the work in the form of grant-in-aid. The proportion of the total cost which may be given in grant depends upon a number of factors, including the historic importance of the structure and the ability of the owner to pay for the work without financial assistance.

In addition, the owner can expect technical and/or professional advice. The normal practice is that, on receipt of an application from the owner for listed building consent or scheduled monument consent, professional officers from the local planning department and/or English Heritage will discuss the proposals and, where appropriate, will suggest modifications or alternative solutions in order to reduce disturbance or preserve the character and fabric of the structure.

The guidance provided can vary from the choice of suitable mortars and pointing techniques to alternative structural or architectural solutions in the form of sketch schemes. Questions may be asked as to whether a correct diagnosis has been made and whether the remedial works are actually necessary. In cases of dispute the owner has the right of appeal, ultimately in the form of a public inquiry by an independent inspector; he reports his findings and recommendations directly to the Secretary of State, who has the responsibility for making the final decision.

25.6 References and further reading

Ashurst, J. and N. (1988) *Practical Building Conservation*: Vol. 1, *Stone Masonry*; Vol. 2, *Brick, Terracotta and Earth*. English Heritage Technical Handbooks (in 5 volumes), Gower Technical Press.

Department of the Environment (1972) *Legislation affecting Ancient Monuments*, AMS Memorandum No. 102, Ref. AA.5201/12.

Department of the Environment (1987) *Historic Buildings and Conservation Areas: Policy and Procedures*, Circular 8/87.

Department of the Environment (1987) *What Listing Means: a Guide for Owners and Occupiers*.

Edwards, J. (1982) *The Ancient Monuments and Archaeological Areas Act 1979. The Practitioner's Companion*, Royal Institution of Chartered Surveyors, London.

Gautier, H. (1714) *Traité des Ponts*, Duchesne, Paris.

Hume, I. (1986) *The Monitoring of Structures and Landscapes*, English Heritage, Historic Buildings and Monuments Commission for England.

Mills, R. L. (1982) Structural remedial works for listed buildings and ancient monuments. *Proc. Conf. on Repair and Renewal of Buildings, Institute of Civil Engineers, London, November*, Thomas Telford.

Ruddock, E. C. (1974) Hollow spandrels in arch bridges: a historical study. *J. Inst. Struct. Eng.*, 52, 281–93.

Ruddock, E. C. (1979) *Arch Bridges and their Builders, 1735–1835*, Cambridge University Press, Cambridge.

Society for the Protection of Ancient Buildings Technical Pamphlets, particularly

No. 1, Macgregor, J. *Outward Leaning Walls*.

No. 4, Ashurst, J. *Cleaning of Stone and Brick*.

No. 5, Williams, G. *Pointing of Stone and Brick Walling*.

Suddards, R. W. (1988) *Listed Buildings: the Law and Practice of Historic Buildings, Ancient Monuments and Conservation Areas*, Sweet and Maxwell, London.

Chapter twenty-six

Coda

A.M. Sowden

Many of the photographs in this book demonstrate the extent to which masonry structures may distort or deteriorate before becoming unserviceable, while some of the examples show how slowly defects may develop. This should inspire a degree of confidence in those responsible for the maintenance of similar structures, but it must never induce complacency. To keep control over the maintenance situation it is essential to be constantly vigilant and aware of the evolving condition of every asset.

The resources at the disposal of Maintenance Engineers are too scarce to be spent prematurely or profligately. Skill in their craft therefore relies critically on their knowing when to intervene and how to do so cost effectively; it demands of its practitioners the patience and persistence to diagnose the root cause of defects, the experience and engineering judgement to assess the urgency for treatment, the imagination and ingenuity to devise appropriate countermeasures and the aptitude and acumen to implement viable and durable remedies. The descriptions, in the foregoing pages, of proven practices may afford a basis for developing that skill.

The demolition of masonry structures may be unavoidable for reasons of redundancy or renewal to meet changing circumstances; otherwise it is only in the last resort that the extreme measures illustrated in Figs 26.1–26.4 should become necessary. In fact, only one of the cases shown was demolition occasioned by structural deterioration.

FIGURE 26.1 Demolition by breaker.

FIGURE 26.2 Demolition by breaker.

FIGURE 26.3 Demolition by explosives.

FIGURE 26.4 Demolition by explosives.

Index

Page numbers in italics refer to tables or illustrations.